CALIFORNIA NATURAL HISTORY GUIDES

FIELD GUIDE
TO MARINE MAMMALS
OF THE PACIFIC COAST

California Natural History Guides

Phyllis M. Faber and Bruce M. Pavlik, General Editors

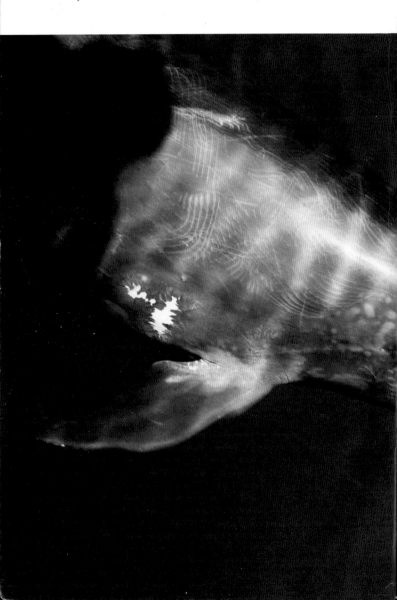

Field Guide to

MARINE MAMMALS

of the Pacific Coast

Baja, California, Oregon,
Washington, British Columbia

Sarah G. Allen
Joe Mortenson

Illustrated by Sophie Webb

UNIVERSITY OF CALIFORNIA PRESS

Berkeley   Los Angeles   London

To the pioneers of marine mammal science and conservation, and especially to Carl Hubbs, Starker Leopold, Karl Kenyon, James Mead, Kenneth S. Norris, Margaret Owings, Robert Orr, Victor Scheffer, Jacqueline Schoenwald, and John Twiss, among countless others.

University of California Press, one of the most distinguished university presses in the United States, enriches lives around the world by advancing scholarship in the humanities, social sciences, and natural sciences. Its activities are supported by the UC Press Foundation and by philanthropic contributions from individuals and institutions. For more information, visit www.ucpress.edu.

California Natural History Guides Series No. 100

University of California Press
Berkeley and Los Angeles, California

University of California Press, Ltd.
London, England

© 2011 by the Regents of the University of California

Library of Congress Cataloging-in-Publication Data

Allen, Sarah G.
    Field guide to marine mammals of the Pacific Coast : Baja, California, Oregon, Washington, British Columbia / Sarah G. Allen and Joe Mortenson ; illustrated by Sophie Webb.
        p.   cm.
    Includes bibliographical references and index.
    ISBN 978-0-520-26544-8 (cloth : alk. paper)
    ISBN 978-0-520-26545-5 (pbk. : alk. paper)
    1. Marine mammals—Pacific Coast (North America)   2. Marine mammals—Pacific Coast (North America)—Identification I. Mortenson, Joseph.   II. Webb, Sophie.   III. Title.

QL715.A45   2011
599.50979—dc22   2010027947

Manufactured in China

16   15   14   13   12   11
10   9   8   7   6   5   4   3   2   1

The paper used in this publication meets the minimum requirements of ANSI/NISO Z39.48-1992 (R 1997)(*Permanence of Paper*).

Cover Illustration: Pacific White-sided Dolphins. Image by Sophie Webb.

CONTENTS

ACKNOWLEDGMENTS

We are grateful to many generous people who have given us their time, images, stories, and encouragement. Frances Gulland, R. L. Pitman, and an anonymous reviewer thoroughly reviewed the entire guide and provided wonderful insights and improvements. We especially enjoyed discussions on the finer points of species identification from R. L. Pitman who spends more time at sea than on land. Many colleagues reviewed and improved portions of the guide, but especially David Ainley, Bea Brunn, Douglas DeMaster, Joseph Gaydos, Wm. Gilmartin, Denise Greig, Jason Hassrick, David Hyrenbach, Norma Jellison, Joanne Mohr, Emma Moore, Brandy Mow, Jim Scarff, Craig Strong and Ramona Zeno. Claire Peaslee and Hazel Flett were exceptionally helpful editing the entire guide, and generating discussions. U.C. Press has been wonderfully kind and supportive, including Kate Hoffman, Lisa Tauber and Lynn Meinhardt; Phyllis Faber provided initial guidance and encouragement throughout; and Jenny Wapner was immensely helpful and extremely patient, shepherding the process to its conclusion. Gary Knoblock stimulated us to complete this guide that had languished for many years.

Numerous people generously loaned us images both as reference for drawings and to include in this guide: Scot Anderson; Francine and Davis Allen; Robin W. Baird, Annie B. Douglas, and John Calambokidis of Cascadia Research Collective; Lisa Ballance, Jim Cotton, Tim Gerrodette, Alan Jackson, Cordelia Oedekoven, and Barbara Taylor from NOAA Southwest Fisheries Science Center; Benjamin Halpern, Nat'l. Center for Ecological Analysis and Synthesis; Lisa Borok of Monterey Bay Research Institute; Judy Bourke; A. Egbert of Phoebe Hearst Museum of

Anthropology of University of California Berkeley; David Ellifrit of the Center for Whale Research; M. Engelbrecht of Bodega Marine Laboratory; David Fierstein; Karin Forney; J. A. Goldbogen of University of British Columbia; Frances Gulland and Denise Greig of The Marine Mammal Center; James T. Harvey of Moss Landing Marine Laboratory; Hans Herrmann and Jeffrey Stroub of the Commission for Environmental Cooperation of North America; Dan Howard and Jennifer Stock of the NOAA Cordell Bank National Marine Sanctuary; Thomas Jefferson; Chris Johnson of EarthOCEAN Media; R. E. Jones; Scott Krafft of Northwestern University Library; Derek Lee of PRBO Conservation Science; Dan McSweeney of Wild Whales Foundation; Sharon Melin of NOAA Alaska Fisheries Science Center; Galen Leeds; Forrest Lyman of Digitalus Media; Lance Morgan of Marine Conservation Biology Institute; Patricia Morris of University of California Reserve System; Phil Myers, Animal Diversity Web; Danial Pauley; Robert T. Pitman; Jan Rolletto of the Gulf of the Farallones National Marine Sanctuary; Susan Snyder of Bancroft Library of University of California Berkeley; Amanda Tomlin, Sarah Codde, Heather Jensen, and Carolla DeRooy of Point Reyes National Seashore; Sue Van Der Wahl; Marc A. Webber; and Jim Wester. Numerous others generously shared their fine images, and the final list and credits are in the appendix.

Researchers, mentors and friends too numerous to name have shared their knowledge, theories and stories with us. Sophie Webb thanks especially Lisa Ballance, Jaime Jahncke and Debi Shearwater who offered opportunities at sea to acquire the experience with marine mammals needed to illustrate them, and Sarah Allen thanks David Ainley for similar opportunities on land and at sea. We also are especially grateful to J. Adams, Scot Anderson, P. Allen, R. Bandar, J. Barlow, T. Baty, R. Brownell Jr., J. Calambokidis, J. Churchman, J. Cordaro, D. Crocker, L. Culp, R. DeLong, D. DeMaster J. Evens, L. Fancher, M. Flannery, M. Follis, K. Forney, K. Frost, Wm. Gilmartin, D. Green, E. Grigg, F. Gulland, J. Hall, L. Hanson, J. Harvey, B. Hatfield, S. Hayes, R. Helm, B. Heneman, E. Hines, H. R. Huber, M. Hester, R. Hoffman, R. Jameson, S. Jeffries, D. Jessup, L. Thomsen Jones, R. E. Jones, S. Kaza, M. E. King, C. Keiper, P. Klimley, D. Kopec, D. Lee, B. J. Le Boeuf, A. Starker Leopold, M. Lowry, H. Markowitz, G. McChesney P. Morris, H. Nevins, D. Nothelfer, G. Oliver,

R. T. Orr W. Perryman, J. Pettee, L. Ptak, P. Pyle L. Querin, T. Ragen, D. Renouf, C. A. Ribic, R. Riseborough, J. Scarff, D. Siniff, L. Spear, R. Stallcup, B. Stewart, Wm. Sydeman, P. Thompson, J. Twiss, E. Twohy, E. Ueber, S. and A. Waber Walker, L. Wheeler, S. Van Der Wahl, M. Webber, R. Wilson, and P. Yochem.

We wish to acknowledge the extraordinary contribution that researchers, particularly from NOAA Fisheries, marine laboratories, and non-profit research and marine mammal care organizations have made towards current knowledge. Their dedication and high standards have ensured that a scientific foundation supports research findings, which sometimes must withstand intense challenges. Many researchers ply the seas for months on this quest of discovery, and bring back reams of data as well as extraordinary tales. Last but by no means least, the crew and officers of the research vessels, MacArthur II, David Starr Jordan and John Martin, have contributed immensely to this vast collection of data. If we have missed facts or muddled interpretations, we ask for comment so that we might correct our errors.

Friendships and family sustain long efforts. Friendships with D. Adams, B. Becker, N. Brown, J. Dell'Osso, G. Fellers, H. C. Foster, N. Gates, L. Grella, J. Hall, W. Holter, B. Huning, J. Mohr, C. Morton, A. Nelson, D. and P. Neubacher, C. Peaslee, D. Press, the Ribars, D. Roberts, L. Rowe, A. Tisei, S. Van Der Wahl and G. White, and family members from sisters and brothers to sons, cousins and nieces have bolstered us when we were flagging. Most of all, we are eternally thankful to our partners in life, Hazel Flett and Dudley Miller.

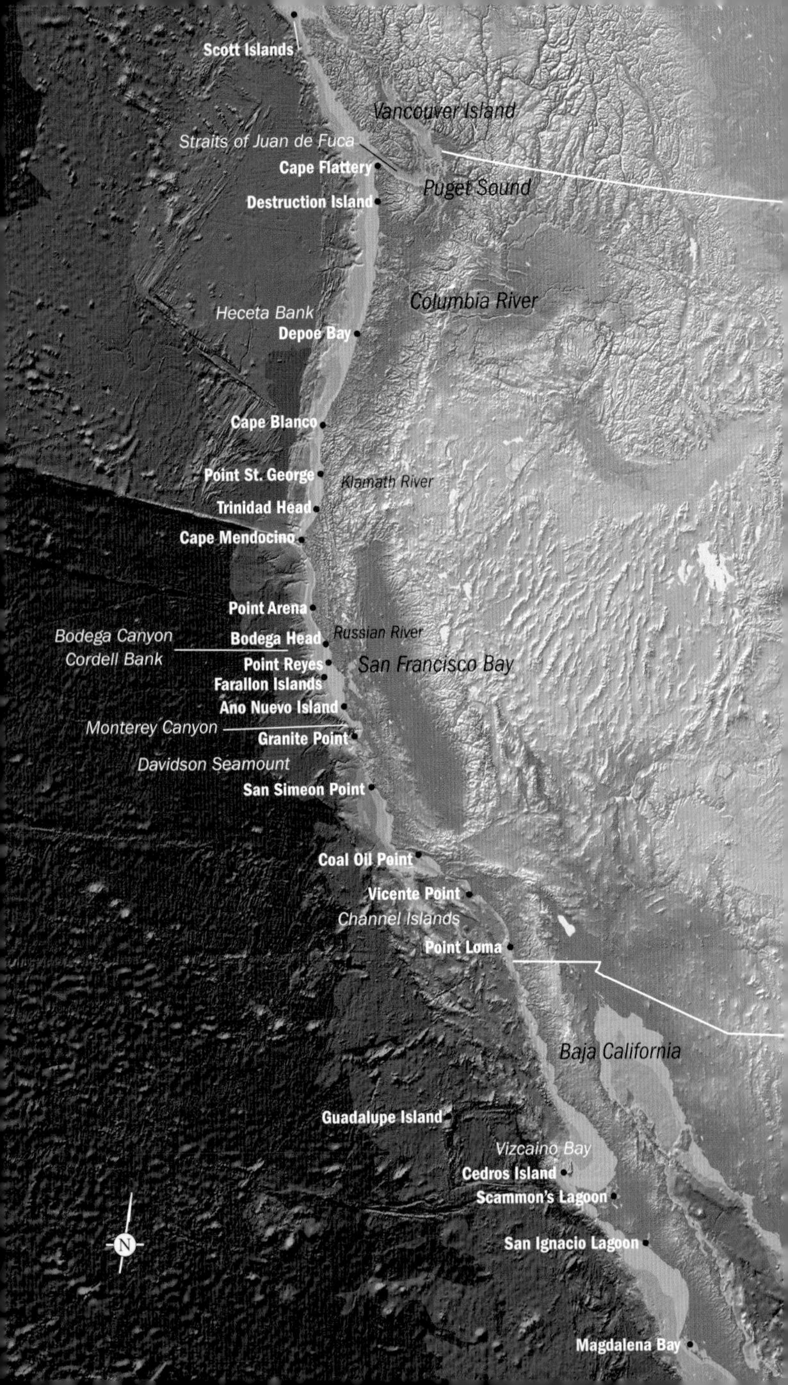

Scott Islands

Vancouver Island

Straits of Juan de Fuca

Cape Flattery

Puget Sound

Destruction Island

Columbia River

Heceta Bank

Depoe Bay

Cape Blanco

Point St. George

Klamath River

Trinidad Head

Cape Mendocino

Point Arena

Bodega Canyon

Bodega Head

Russian River

Cordell Bank

Point Reyes

San Francisco Bay

Farallon Islands

Año Nuevo Island

Monterey Canyon

Granite Point

Davidson Seamount

San Simeon Point

Coal Oil Point

Vicente Point

Channel Islands

Point Loma

Baja California

Guadalupe Island

Vizcaino Bay

Cedros Island

Scammon's Lagoon

San Ignacio Lagoon

Magdalena Bay

N

MARINE HABITAT PLATES

Temperate Nearshore Species

Temperate Offshore Species

Tropical Offshore Species

Temperate nearshore species Gray Whale, Humpback Whale, Common Bottle-nose Dolphin, Harbor Porpoise, Harbor Seal, Sea Otter, California Sea Lion, Leopard Shark, Black Sea Bass, Parasitic Jaeger, Bonaparte's Gull.

Temperate offshore species Blue Whale, Humpback Whale, Fin Whale, Killer Whale, Pacific White-sided Dolphin, Northern Right Whale Dolphin, Dall's Porpoise, Northern Elephant Seal, Northern Fur Seal, herring school, Ocean Sunfish, Western Gull.

Tropical offshore species Sperm Whale, Bryde's Whale, Cuvier's Beaked Whale, Pygmy Beaked Whale, Short-beaked Common Dolphin, Eastern Spinner Dolphin, Spotted Dolphin, Guadalupe Fur Seal, Humboldt Squid, Magnificent Frigatebird.

"*For all at last return to the sea—to Oceanus, the ocean river, like the ever-flowing stream of time, the beginning and the end.*"

—RACHAEL CARSON, *THE SEA AROUND US* (1950)

ALONG THE WESTERN COAST of the North American continent flows an ocean current that contains some of the richest waters in the world when it comes to the study and observation of marine mammals (fig. 1). Worldwide, some 130 different species of marine mammals exist, and more than a third of these are residents or migrants to this rich region. From Killer Whales that dwell around Vancouver Island, Canada, and Gray Whales that give birth in the lagoons of Baja California to Sea Otters that nestle in nearshore kelp beds and elephant seals that plumb the depths of the Pacific Ocean, this region's diversity and abundance of sea mammals surpass those of most places on Earth.

This field guide is meant to stimulate interest in as well as to describe the sea mammals of the Pacific Ocean that inhabit the waters from the tip of Baja California to the northern point of Vancouver Island, British Columbia—the California Current. We hope these fellow mammals will leap from the pages of this volume into your imagination (fig. 2). Think what it is to propel your body to the blackest places on Earth, illuminated only by the bodies of fellow deep-sea denizens, through viscous fluids that press against you with great weight. Then imagine rising to

Figure 1. NOAA Research Vessel the *David Starr Jordan* and Risso's Dolphins.

Figure 2. Synchronized swimming of Short-beaked Common Dolphins.

the surface for a breath, leaping free of the bonds of the ocean, spinning above wave crests in exuberance, communicating the experience to others miles away with a tremendous splash as you fall back into the sea.

Evolution of Sea Mammals

Mammals are not the first forms of life to inhabit the oceans, and their saga begins with the mass extinction, 65 million years ago (mya), of the great reptiles that ruled both land and sea. Life globally underwent a radical change after an asteroid collided with Earth, heating the atmosphere, darkening the sky, and cooling temperatures. Experts debate the exact details of the great extinction, but we do know that the reptiles of the sea (mosasaurs, for example) all perished.

This set the stage for rapid evolution of new species. The lowly mammal life forms that survived the mass extinction grew in number, size, and variety, taking over the empty niches of the vanished reptiles. After 15 million years, some terrestrial mammals, distant relatives of today's Hippopotamus, began to enter

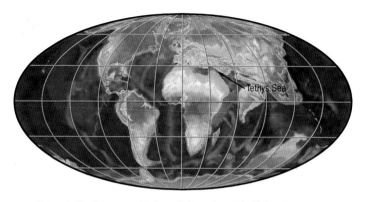

Figure 3. The Eocene world when whales entered the Tethys Sea between Europe and Asia. The continents have separated and are drifting toward the positions they occupy today. (Image redrawn from Ronald Blakey, professor of geology, Northern Arizona University.)

the oceans along the ancient Tethys Sea, part of which lingers today as the Mediterranean (fig. 3). Over the next 10 million years, archaic cetaceans evolved from land ancestors. At first, the sea mammals regularly returned to the land, but as time passed, their feet became flippers and fins, and they ultimately abandoned the continents (fig. 4; table 1).

TABLE 1. Earliest Known Sea Mammals from North Pacific Fossils

	Age (Millions of Years Ago)	Early Fossil Range
Desmostylians (Sirenian-like)	33–10	Oregon, California
Enaliarchids (sea bears)	27–17	Oregon, California
Kolponomos (beach bears)	26–10	Oregon, Washington, Alaska
Desmophocids (link seals)	22–9	California
Odobentids (walruses)	19–present	California, Baja California
Otaroids (sea lions)	12–present	California, Baja California
Hydromalis (sea cows)	8–recent	West Coast of North America
Enhydra (sea otters)	2–present	Oregon

Figure 4. *Ambulocetus*, an early walking and swimming whale of the Tethys Sea.

When much of the Earth and its oceans were warm, two clans of archaic sea mammals evolved: the cetacea—whales and dolphins; and the sirenia—manatees and the Dugong. Over the millennia, continental drift caused sea passages to open and close, alternately creating corridors for ocean currents and migrating mammals, or blocking flows and isolating species. Beginning about 33 mya, global climate again cooled, creating conditions for the rise of toothed and baleen whales (fig. 5). These colder oceans, especially nutrient-rich regions like the California Current, also supported an array of other sea mammals that retained a tie to the land—seals and otters. Most were the descendants of the bear/dog clan of carnivores, and many first appeared in the cold North Pacific.

A field guide to the extinct sea mammals of the California Current would include some amazing creatures. *Desmostylus* was a Hippopotamus-sized mammal whose scientific name means "bonded pillars," referring to its strange, peg-like teeth, likely used for eating plants. Its contemporary, *Kolponomos*, the Beach Bear, swam close to shore and crushed shellfish with

Figure 5. *Aetiocetus*, ancestors to modern-day baleen whales, from 34 mya. A fossil whale was found in Oregon that had both teeth and baleen.

specialized molars. More recently, the California Current was home to fourteen species of walruses over time, including three in Baja alone. Over the last 33 million years, lineages of cetaceans and seal/sea-lion-like animals diversified, replaced older groups, and eventually led to the marine mammals familiar to us today, marine megafauna on the scale of ichthyosaurs of the distant past.

Marine Mammal Ecology

To live in a water world, mammals must possess complex adaptations to surmount great challenges. Simply breathing forces a mammal to rise to the surface, and those that dive deep must also deal with darkness and intense water pressure. Life in a fluid space requires special appendages for movement but also gives weightlessness to large bodies. The needs to regulate salt and water are resolved with impervious skin, highly efficient kidneys, and modified tear ducts (Berta et al. 2006, Kjeld 2006). Sustaining a core body temperature depends upon thick blubber or fur, as well as special internal adaptations. To find, seize, and consume prey, marine mammals have evolved unusual teeth and, in some groups, novel organs that generate biosonar. Their extraordinary abilities represent some of the current known extremes for mammals living on Earth: the largest mammal, Blue Whale; the deepest divers, toothed whales; the longest migrants, Humpback Whales and Gray Whales; and the longest-lived mammals, Bowhead Whales. But little is known about many of their habits, for most lead an entirely aquatic existence far from land. Yet in the past half-century, scientists have discovered many clues about their daily lives and abilities. Here we examine some of the most remarkable adaptations.

While watching marine mammals, you will quickly see how form follows function. When first glimpsed at the surface of the ocean, seals and whales appear sleek and spindle-shaped, a form that helps reduce drag as they swim and dive (fig. 6). Cetaceans are the most hydrodynamic example, with telescoped heads, reshuffled facial features (fig. 7), and reduced or absent

Figure 6. Leaping Eastern Spinner Dolphins with spindle, or fusiform, shape.

Figure 7. Head of a Fin Whale showing the telescoped shape and the evolution of air passages to the top of the head.

pelvic bones and hind limbs. The diversity of head shapes also reflects varied diets; the modified skull of baleen whales is the most extreme example. Indeed, no other animal, marine or terrestrial, has a mouth like that of the baleen whale. Baleen are not teeth but consist of rows of keratin plates with fine ends that entangle prey, allowing the whale to filter-feed large quantities of invertebrates or fish in a single gulp. With the expansion of the gape allowed by pleats or grooves in the throat, a whale's gulp can be voluminous—in the case of the fin whale, as large as a school bus (fig. 8). Another extreme is seen in the asymmetrical skulls of toothed whales with an enlarged right side, related to echolocation.

Body size, blubber, fur, and a specialized structure for recycling blood flow are the main means by which mammals regulate their body temperature, and in this regard, marine mammals are no different from land mammals—just more extreme in their adaptations. Large animals retain heat better than small ones because the ratio of surface area to volume decreases with size, and relative to land mammals, marine mammals are great in size (there is no marine mammal equivalent of a mouse). Large size accommodates a thick blubber layer, which further reduces heat loss and increases buoyancy. Large whales have the thickest

Figure 8. Feeding Humpback Whale with an enormous gulp of fish and water, with pleats fully distended after lunging toward the surface.

blubber [up to 50 cm (20 in.)], accounting for 50 percent of their body mass. Blubber is also elastic and, like the swimsuits of Olympic swimmers, enhances hydrodynamic shape (Reynolds and Rommel 1999). In some deep-diving whales, blubber also contains very firm lipids called "wax esters" that may counter water pressure at depth. A circulatory system for regulating body temperature—the counter-current heat exchange system (CCHE)—is highly evolved in marine mammals. Flippers, dorsal fins, and tails are common locations where heat exchange occurs, as is the tongue, in some whales, to counteract heat loss from swallowing tons of cold water along with prey (Berta et al. 2006).

Diving marine mammals have long enthralled whalers and scientists alike. Among pinnipeds, the record-holders are elephant seals [>1,500 m (5,000 ft)]; among cetaceans, they are beaked whales [1,888 m (6,194 ft)] and Sperm Whales [>2,100 m (6,890 ft)] (Hindell 2002, Tyack et al. 2006, Watkins et al. 1993 and 2002). To dive deep and long, these animals must have sufficient oxygen to fuel muscles, as well as methods for ridding their muscles of lactic acid waste—all while simultaneously keeping the brain alert and the heart pumping. Compared to land mammals, their heart is enlarged to move more blood faster, and their heart rate is lower. The heart rate of a diving elephant seal, for example, can slow to just four beats per minute (Andrews et al. 2000).

If the lungs of land mammals collapse, their lives are in jeopardy, but the lungs of most marine mammals collapse and then re-inflate with every dive below 100 m (330 ft). The "runny nose" of elephant seals is actually fluid expelled from the lungs, which facilitates their inflation. Deep-diving whales, however, can get the bends (decompression sickness), causing bone erosion from nitrogen bubbling into bone tissue (Moore and Early 2004). When whales return to the surface, you can often observe their breath—called the "spout"—and exhaling is their first action upon surfacing, thus ensuring that their first inhalation is not water. Many species have distinctly shaped spouts, which aid in identification.

Overall, the brains of most marine mammals are large compared to their body size, and those of the Sperm Whales and Killer Whales are the largest of all mammals, at 8–10 kg (22–27 lbs); this compares to the average human brain size of 1.1 kg (3 lbs; Tinker 1988). The large brain is wired for sensory organs that are honed for navigation and communication in the marine environment, including echolocation. Sound transmission is five times faster in water than air, so low-frequency sounds emitted by whales can travel great distances—perhaps up to hundreds of kilometers, enhancing communication between members of a population (Tyack 1999).

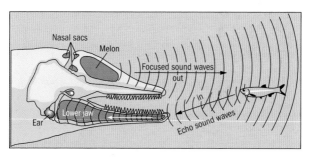

Figure 9. Biosonar structures include a highly modified and asymmetrical skull and a specialized lower jaw with pan bones. The skulls of toothed whales have a scooped-out area that holds a melon-shaped organ containing sound-conducting lipids. Specialized air sacs in the blow hole are called "phonic lips" (redrawn from Berger 2009).

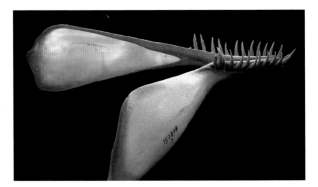

Figure 10. Lower jaw with the pan bones of a toothed whale, *Kogia* genus of Sperm Whale.

Marine mammals are one of several groups of animals (including bats) that use echolocation (biosonar) to navigate and forage. Echolocation depends upon the abilities to project sounds, such as clicks, and to receive and interpret the echoes of the sounds reflected back (fig. 9; Tyack 1999). The biosonar organ is nestled in the toothed whale's asymmetrical skull and the "pan bones" of the lower jaw receive the echoes reflected back (fig. 10). By doing so, an animal can locate, identify, and estimate the distance of an object. Bottlenose dolphins can detect small objects up to 110 m (361 ft) away (Au 2002), and Sperm Whales perhaps up to 16 km (10 mi). Some scientists propose that toothed whales, as a byproduct of their focusing sonar beams, transitioned from prey detection to prey debilitation (Norris and Mohl 1983). Their ability to use sonar to stun fish remains unproven, but the theory is compelling—bottlenose dolphins, for instance, emit sound levels that exceed the lethal threshold for fish.

Marine mammals have further adapted to ocean living with evolved behaviors, and the capacity to pass these learned behaviors on through generations is often associated with a long life span (table 2). One of the longest-lived mammals on record was a bowhead whale that was aged 177 to 245 years (George et al. 1999), and other marine mammals also live long and are highly intelligent, such as killer whales, the Einsteins of marine mammals (fig. 11).

TABLE 2. Length of Gestation and Parental Care between Marine Mammals

	Age of sexual maturity (years)	Gestation* (months)	Lactation (weeks)	Birthing interval (years)	Parental care
Humpback Whale	5–7	11	40–44	2	1–2 years
Gray Whale	9	11–13	28–32	1–2	6–8 months
Sperm Whale	9	15–17	100	3	5–7 years
Bottlenose dolphin	12	12	76	3	7–11 years
Harbor Porpoise	3–6	9–11	32	1	8–16 months
Steller Sea Lion	4–5	11	47	1	1–2 years
Northern Fur Seal	3–7	12	18	1	4 months
Harbor Seal	2–5	10	4	1	4 weeks
Elephant seal	4	11	4	1	4 weeks
Sea Otter	4	6–7	20–30	1	1 year

Adapted from Berta and Sumich 1999.

*Includes delayed implantation.

Figure 11. Male Killer Whale, the Einstein of the sea.

Figure 12. Bow-riding Common Bottlenose Dolphin, leaping before the large vessel *Bow Peace*. Ship strikes that occur when fast-moving ships strike resting or surfacing whales are a frequent cause of mortality to large whales.

Play, too, is a window into intelligence. Toothed whales and sea lions surf waves alongside humans or ride the pressure waves generated by boats, a behavior that likely began with dolphins riding pressure waves around the heads of fast-swimming whales (fig. 12; Wursig 2002). Breaching involves leaping out of the water, but whether its purpose includes play is still a subject of debate. It takes a lot of energy to breach: for a sperm whale to break the surface, it must swim as fast as it can, at least 7 m/sec (17 mph; Whitehead 2002c). Possible purposes of the .breach may be herding or stunning prey, shedding ectoparasites, and communicating. In any case, breaching offers human observers a great opportunity to identify a species (fig. 13).

Predators and parasites influence the survival strategies of marine mammals just as much as their needs to eat and reproduce do. Great White Sharks (*Carcharodon carcharias*) and Seven-gill Sharks (*Notorhynchus cepedianus*) are the most common fish predators in the California Current, but Killer Whales are the most formidable hunters, attacking enormous Blue Whales and even Great White Sharks (figs. 14, 15). Along the California Current, fearsome terrestrial predators such as the Grizzly Bear (*Ursus arctos*) have been replaced by less formidable ones such

Figure 13. Breaching Humpback Whale, the ballerina of whales.

as the Coyote (*Canis latrans*) (fig. 16). Lowly parasites can be just as deadly as predators, and krill and fish are often the intermediate host for parasites, much as fleas carry the internal parasites of dogs.

Figure 14. Great White Sharks are predators and scavenge on many marine mammals, but Killer Whales also eat Great White Sharks.

Figure 15. Killer Whale preying on Harbor Porpoise.

Figure 16. Coyotes, bears, and birds are predators of shore-based pinnipeds. Coyote stalking a Harbor Seal with pup in Bolinas Lagoon, California.

Figure 17. Scientist examining a dead beached Gray Whale in San Francisco Bay.

Dead or live marine mammals that are cast upon the shore have been the source of much knowledge, including the discovery of new species (fig. 17). There are a myriad of reasons why marine mammals strand (abandoned young, sick or injured animals), but the causes of mass strandings are more complicated. Episodes have been attributed to disease, toxins, and human-related causes such as fishery entanglement and military active sonar.

Ocean Ecosystems

The marine waters from Baja California to British Columbia are home to some of the world's most diverse and abundant marine life, including hundreds of thousands of marine mammals. Oblivious to political borders, marine mammals are governed by oceanographic ones, and along the west coast of North America flows a massive "river" of ocean water called the California Current System, providing the foundation for this biological wealth.

When standing upon the edge of a continent or prow of a ship, we see shimmering surface waters and only a glimpse of the seascape below. Beneath the surface, the ocean world is dynamic and complex, constantly transformed by large-scale forces such as the

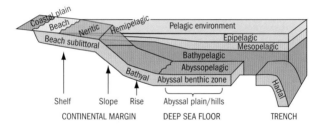

Figure 18. Oceanic environments related to depth and distance from land (redrawn from Berger 2009).

Earth's rotation, winds, and the shape of the seafloor. Oceans are complex habitats because of their chemical, physical, and biological properties (Castro and Huber 2007). Colder, saltier water, being denser, sinks, whereas warmer, less salty, fresher water floats on the surface. These differences in chemistry result in the formation of layers (thermoclines) and boundaries (fronts) between water masses. Water moves along the fronts and around objects, creating eddies and other shapes that become the basic, semipermanent features of ocean ecosystems. The contours of the ocean floor (bathymetry) and the continental shoreline (topography) further steer and force interactions. Visualize a straight line extending from the edge of the continent toward the middle of the Pacific Ocean, and you will see a profile that starts with a submerged ledge called the continent margin (shelf) and reaches out to the deep sea floor (fig. 18). Many marine species, sea mammals among them, are bound to distinct ocean properties and depth ranges, and ocean features provide predictable locations of greater prey availability ("hotspots" of productivity).

Celestial influences force cyclic patterns on the ocean as surely as the Sun rises, the Moon waxes, and Earth spins. The Moon and the Sun generate daily tides that follow monthly and seasonal rhythms, and great gyres of air and water span oceans. In the North Pacific Ocean, the North Pacific Gyre is the dominant ocean circulation pattern, a great circle of currents. This gyre spins clockwise, driven by the strong Westerly winds and the Earth's rotation, and constrained by continents that flank the Pacific Ocean basin. The northerly, east-flowing arm of this gyre collides with the North American continent near the northern tip of Vancouver Island, where it bends southward to form the California Current System (fig. 19).

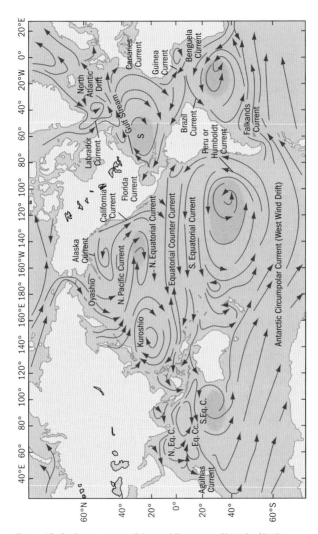

Figure 19. Surface currents of the world's oceans. Note the North Pacific Current and the California Current (redrawn from Berger 2009).

California Current System

The California Current System (CCS) is actually the interplay of three major currents: the California Current, the California Undercurrent, and the Davidson Current (Berger 2009, Hickey 1979). The California Current, flowing along the continental boundary from northern Vancouver Island to the lower tip of Baja, is the dominant current of the three, transporting cold, low-salinity, high-oxygen, subarctic waters along the coast. It is wide [1,000 km (600 mi)] but shallow [from the surface down to 300 m (1,000 ft)], and at Point Conception, where the continental shoreline veers abruptly southeastward, the current weakens and flows offshore. The California Current is one of only five coastal upwelling ecosystems in the world, and the only one in the North Pacific. Although such ecosystems collectively cover only 1% of the area of the world's oceans, they account for nearly one-third of the world's fisheries and are the most productive ocean habitats in the world.

The California Undercurrent is below the surface at a depth of 75 m (250 ft) and flows in the opposite direction—from the Equator toward the pole—with waters that are warmer, saltier, and low in nutrients. When the California Undercurrent rises to the surface in winter, it becomes the Davidson Current and transports warm, equatorial waters north to Vancouver Island. Imagine marine mammals as they migrate through the oceans, getting a lift from these currents flowing in different directions.

The CCS changes in strength and position during three predictable seasons (fig. 20). The upwelling season, March to July, begins when strong winds blow along the coast from the north and northwest, causing waters to upwell to the surface. Episodes of upwelling draw nutrients from the shallow continental shelf to the surface, and are followed by windless days of relaxation. Upwelling is the pump that drives the coastal biological ecosystem, because it breaks down the normally strong layering of the thermocline (a strong vertical change in sea temperature) and brings nutrient-rich waters toward the sunlit surface, where phytoplankton (plant plankton) bloom. At the edges of upwelling plumes, concentrations of plankton and weak-swimming organisms such as krill attract larger

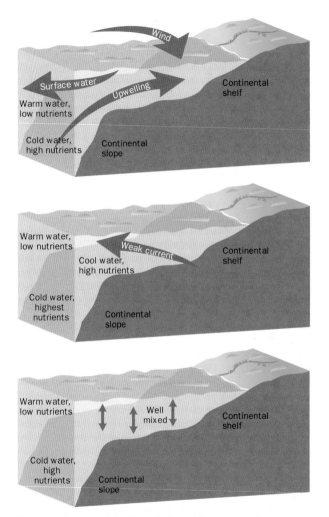

Figure 20. Seasons of the California Current System when there are major shifts in ocean conditions within a year. Redrawn from NOAA.
Upper: Upwelling season, March to August.
Middle: Oceanic season, August to October.
Lower: Winter season, November to March.

Figure 21. Upwelling along the coast of California, as seen from a satellite showing chlorophyll concentrations resulting from algae. Upwelling regions are hot spots of biological diversity. Several distinct small upwelling centers, such as Point Arena and Point Conception, are biological hotspots, seen as hot colors in the satellite image.

consumers, including marine mammals, from higher up the food chain (fig. 21).

During the oceanic season, August to October, the California Current becomes layered when upwelling relaxes. Surface waters are warmed by sunlight but, being below a strong thermocline, are low in nutrients. During this season, whales and sea lions are attracted to warm surface waters where there might be large schools of fish such as sardines. In the winter season, November–February, storms dominate coastal waters with strong winds from the south that push ocean waters onshore and cause warm subtropical waters to "downwell" along the coast. The Davidson Current becomes the dominant surface current along most of the coast to British Columbia. Marine mammals that prefer warm waters (Common Bottlenose Dolphins, for example) shift north, from southern to central California.

Climate Oscillations

Oceanic seasons periodically are altered by large changes in climate. The best known is the El Niño–Southern Oscillation (ENSO), which usually occurs every four to 10 years (Trenberth 1997). Over the past 200 years, there have been some 50 ENSOs; the 1997–1998 one was the strongest on record. ENSO events occur when the trade winds that blow west across the Pacific cease, releasing an oceanic backwash of warm water that surges against the American continents near the Equator. The backwash raises the sea level along the Americas and infuses coastal waters with warmth. With weakened upwelling, marine plants do not bloom, creating a "biological drought" throughout the food web, and marine mammals respond by migrating elsewhere, skipping reproduction, or starving. Converse conditions, called La Niña, occur when prevailing northwest winds are stronger than usual and without pause, resulting in colder-than-average temperatures. La Niñas are not necessarily biologically productive, though, because nutrients and plankton are blown offshore. Optimal conditions for marine organisms occur in years having mild and pulsed upwelling.

Another climate anomaly is the Pacific Decadal Oscillation (PDO), a 20- to 30-year climate pattern that cycles between warm and cold phases. Looking back 100 years, scientists identified several such periods, and noted a positive relationship between cool phases and increases in salmon, anchovy, and zooplankton in the CCS, contrasting with higher sardine biomass during warm phases. Long-term trends in marine mammal populations likely also track the PDO phases.

Ocean Communities and Ecoregions

Creatures that swim through, crawl beneath, or burrow under the oceans are generally categorized into depth zones (see fig. 18). The epipelagic zone, where blooms of phytoplankton underpin

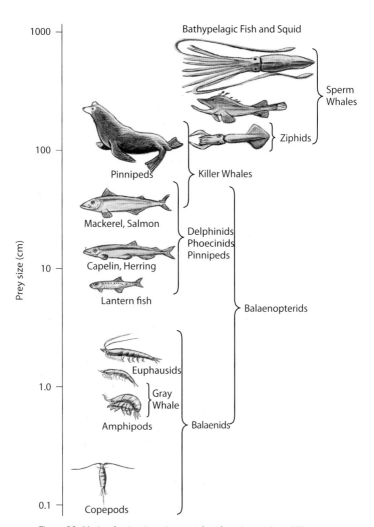

Figure 22. Marine food web and range of preferred prey sizes: Killer Whales, the apex predators, feed on other mammals; seals feed on fish; some baleen whales feed on the "fleas" of the sea, copepods (redrawn from Berta et al. 2006).

the marine food web, holds a staggering biomass of creatures harvested by fish, marine mammals, and humans (fig. 22). Most plants in the oceans are single-celled diatoms, called the "grass of the sea," or dinoflagellates. The zooplankton, minute marine animals, are composed of small crustaceans, fish larvae, and eggs. Marine copepods (crustaceans), the size of a grain of rice and among the most abundant organisms on Earth, are called the "insects of the sea." Some 600 species occur in the northeastern Pacific Ocean; several of them are food for some of the largest whales. A step up the food web are krill (Norwegian for "whale food"), shrimplike crustaceans that provide mouthful meals for whales; two important genera in the eastern North Pacific are *Euphausia* and *Thysanoessa*. Larger invertebrates and fish in the California Current are extraordinarily diverse, but only a few are superabundant and the target of foraging marine mammals; examples are anchovy, hake, sardine, salmon, herring, Market Squid, and rockfish. Though abundances and distributions fluctuate, these animals are reliable annual food for marine mammals.

Dense swarms of krill, squid, and fish feeding upon them form a distinct oceanographic feature called the "deep scattering layer" (named for the layer of echoes that oceanographers detected from the swim bladders of fish). The deep scattering layer rises and falls each 24 hours, up to 460 m (1,500 ft) vertically, as the creatures migrate toward and away from the ocean surface in response to changes in daylight. In pursuit of these prey, many marine mammals similarly migrate up and down within a day.

Kelp forests in the nearshore waters provide many marine organisms with habitat for food, nurseries, and protection from predators. Two important species in the California Current, Giant Kelp (*Macrocystis pyrifera*) and Bull Kelp (*Nereocystis lutkeana*), support complex but different assemblages of species. Kelp beds are generally an indication of rich nearshore habitat for many marine mammals (see page 2).

Oceanographers divide oceans into ecoregions based on latitude, currents, physical features, and biological productivity (Longhurst 1998). From north to south, the California Current System contains three ecoregions (fig. 23; Longhurst 1998, Morgan et al. 2005). The Columbian Pacific ecoregion of the California Current System is distinguished by the Columbia River, the Straits of Juan

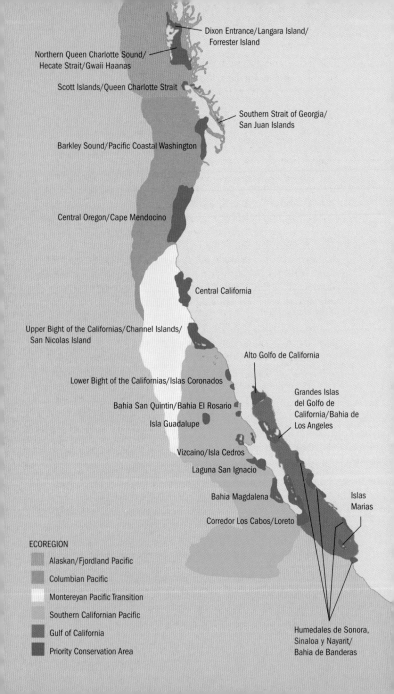

Dixon Entrance/Langara Island/
Forrester Island

Northern Queen Charlotte Sound/
Hecate Strait/Gwaii Haanas

Scott Islands/Queen Charlotte Strait

Southern Strait of Georgia/
San Juan Islands

Barkley Sound/Pacific Coastal Washington

Central Oregon/Cape Mendocino

Central California

Upper Bight of the Californias/Channel Islands/
San Nicolas Island

Alto Golfo de California

Lower Bight of the Californias/Islas Coronados

Grandes Islas
del Golfo de
California/Bahia de
Los Angeles

Bahia San Quintin/Bahia El Rosario

Isla Guadalupe

Vizcaino/Isla Cedros

Laguna San Ignacio

Bahia Magdalena

Islas
Marias

Corredor Los Cabos/Loreto

ECOREGION

Alaskan/Fjordland Pacific

Columbian Pacific

Montereyan Pacific Transition

Southern Californian Pacific

Gulf of California

Priority Conservation Area

Humedales de Sonora,
Sinaloa y Nayarit/
Bahia de Banderas

de Fuca, Vancouver Island, Puget Sound, and the San Juan Eddy. Massive loads of nutrients are discharged from prominent rivers, contributing to enormous algal blooms. The vast inland sea created by Puget Sound embraces islands, bays, and connecting channels that can befuddle a mariner but attract marine animals. The region's features nourish and attract a rich diversity of marine life, but the ecoregion also demarks a break in the ranges of some species.

The Montereyan Pacific Transition ecoregion (see fig. 23 and page 4) has many large capes with upwelling plumes, a vast outflow of freshwater from San Francisco Bay, and several deep marine canyons. The Farallon Islands, thrust up along the edge of the continental shelf, are the only offshore islands. Off San Francisco, the continental shelf is wide, forming the Gulf of the Farallones. This ecoregion is an area of overlap between species, some more aligned with tropical waters, and others with temperate waters. The boundaries are not fixed, though, because warmer and colder waters shift back and forth seasonally, as well as with climate oscillations. The result is a great diversity of species: cooccurring here are warm-water Common Bottlenose Dolphins and cold-water Harbor Porpoises (*Phocoena phocoena*).

The Southern California Pacific ecoregion is dominated by subtropical waters that flow northward from the equator, forming the Southern California Eddy. The only truly oceanic island of the California Current, Guadalupe Island, lies 260 km (160 mi) offshore of Baja, but islands are numerous inshore, where they form chains of complex habitats. The diverse habitats provide marine mammals with rich foraging opportunities and safe havens for reproducing. Some species are limited to this warmer ecoregion, and porpoises are mostly absent (see fig. 23 and page 6).

As scientific tools become increasingly sophisticated, oceanographers continue to refine the description of ocean habitats, especially the realm of the deep ocean, with the aid of marine mammals (figs. 24, 25). The boundaries between ecoregions are "leaky," and major disruptions can cause semistable ones to dissolve or shift. Biological hotspots of diversity may weaken and

Figure 23. Marine ecoregions of the California Current and areas of biodiversity. (Redrawn from *Marine Priority Conservation Areas: Baja California to the Bering Sea*, Commission for Environmental Cooperation, 2005).

Figure 24. Researchers from Monterey Bay Area Research Institute use a wide range of techniques, from satellites to buoys, for collecting ocean data. (David Fierstein, copyright © 2001, MBARI.)

perhaps emerge in new locations—as "hopping hotspots." As global climate alters the oceans, the ability of marine mammals to adapt will be the subject of intense scrutiny. Studies of oceans to date, however, provide a good context for understanding where and how sea mammals of the California Current inhabit their varied marine environment today.

Marine Mammals and Humans

The story of sea mammals is intertwined through all cultures and histories of coastal humans. From before thoughts were put on paper, these animals have loomed large as sources of

Figure 25. A female Northern Elephant Seal, with pup and with a tag attached to her head. Scientists call these animals "autonomous pinniped environmental samplers" of oceanographic data.

food, innovation, and inspiration for humans. Most maritime cultures, however, have altered the ecology of their coasts, and humans as a species are now filling the ultimate keystone role in nature. To acknowledge this role, some scientists have proposed a new epoch: the Anthropocene Age. Paralleling humans' unprecedented ecological engineering has been a conservation voice that is growing in volume and clarity.

The Past

Modern humans emigrated from Africa around 50 thousand years ago (kya), and as they dispersed across continents, the great beasts, the mastodons, mammoths, and predators of the Pleistocene became extinct. Controversy persists over why these megafauna perished, but wherever humans colonized land masses, large animals subsequently vanished. Although seals were hunted in Africa perhaps 100 kya, they and the rest of the megafauna of the sea initially escaped the extinctions that swept the land. Along the west coast of North America, native marine-mammal-hunting cultures included sophisticated societies such as the Chumash in southern California and the Haida of the Pacific Northwest (Arnold 1992). Middens (refuse heaps) from central California north to the Olympic Peninsula trace the harvest of shore-based Northern Fur Seals by Native Americans for thousands of years. In the Channel Islands, where the oldest midden dates from 9 kya, there are also signs of specialized dolphin captures; these are the most ancient evidence for human whaling on the Pacific coast (Porcasi and Fujita 2000). Two northwest coast native cultures specialized in whaling from dugout canoes: the Waskashan of the outer coast of Vancouver Island and the Makah of northwestern Washington State.

There is no evidence that entire sea mammal species perished at the hands of Native Americans, although over thousands of years Native Americans likely significantly suppressed some populations of seals and sea otters. About 1,000 years ago, Northern Fur Seal bones became rare in middens, and the remaining fur seals likely found sanctuary on the remote Farallon Islands, off San Francisco.

Figure 26. Spirit mask of a sea mammal and the shaman who possessed the spirit. Note the toothy grin, perhaps representing that of a killer whale. (Courtesy of Phoebe Hearst Museum of Anthropology, University of California, Berkeley.)

Little comes down to us from these native maritime cultures except for stone artifacts and middens, but the stories and beliefs of a few were collected by folklorists and pioneering anthropologists at the turn of the 19th century. From these tales we know that some native cultures believed that sea animals were their guides, helpers, and life sustainers (fig. 26). Rituals preceded the hunt, honoring the prey in art and song and, in so doing, reflecting an early conservation ethic (fig. 27).

Russian fur traders from Alaska began to arrive along the coast of North America in the 18th century, seeking sea otter and fur seal skins for the lucrative Chinese market. They killed an estimated 2.5 million fur seals from 1786 to 1867. By 1788, British sailing ships had rounded Cape Horn in search of blubber,

Figure 27. "Before caring to practice his dangerous art, the whaler subjects himself to a long and rigorous course of ceremonial purification in order to render himself pleasing to the spirit whale." (From Curtis 1924, courtesy of Charles Deering McCormick Library of Special Collections, Northwestern University Library, Curtis 1924.)

Figure 28. Six-man whaleboat with Greener harpoon gun on bow hunting whales, perhaps near the Farallon Islands on a rare calm day in the late 1800s. (Courtesy of Bancroft Library, University of California, Berkeley.)

Figure 29. Factory whaling station at Trinidad Head, circa 1919–1930. The group includes the owner, F. M. Dietrich; the foreman, Charley W. Majory; whalers; and flensers. (From the collection of R. E. Jones, Napa, California.)

ambergris, and whalebone (baleen). The first shore-based whaling station in California was built in Monterey Bay in 1854, and in 1858, 27 whales were captured at that single station, yielding 26,145 gallons of oil (fig. 28). The trade was more lucrative than gold mining, with an average whale yielding 1,000 gallons of oil at a value of about 45 cents per gallon in 1880 (Bertao 2006). By 1905, advanced steam-powered catcher ships and harpoon cannons made possible the exploitation of whales that had been too fast, and that had ranged too far offshore, to be hunted using earlier methods (fig. 29). Whalers used spotter planes and newly invented wireless radio communication to locate their quarry efficiently.

The number of mammals killed in the Pacific Ocean within a brief 200 years was more than 1 million large whales, and with the inclusion of hundreds of thousands of porpoises and fur seals and thousands of seals, the numbers exceeded several million (fig. 30). Species after species of Pacific sea mammals was brought to commercial extinction. Eventually some, like the California Sea Otter and the Northern Elephant Seal, were thought to be totally extinct, whereas other valuable species became so rare as to be not worth pursuing. The first species to vanish, though, was not a whale but the lone descendant of the ancient family of Pacific sirenians. The Steller's Sea Cow, the largest of the sirenians, had evolved in the northern Pacific. The Russian Captain Vitus Bering discovered a relict population of

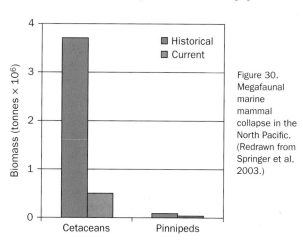

Figure 30. Megafaunal marine mammal collapse in the North Pacific. (Redrawn from Springer et al. 2003.)

perhaps more than 1,500 sea cows, and after only some 27 years, the last one was killed in 1768.

As this tragedy of the commons unfolded, pleas for regulation grew louder. Captain Scammon was one of the first conservationists. He became an advocate for the Gray Whale in the late 19th century, after participating in their devastation in the calving lagoons of Baja. His efforts to protect them were in vain, however; by the 20th century, the Gray Whale was nearly extinct. Other advocates, such as David Starr Jordan, also noted the decline of whales; reviewing the whaling practices of the early

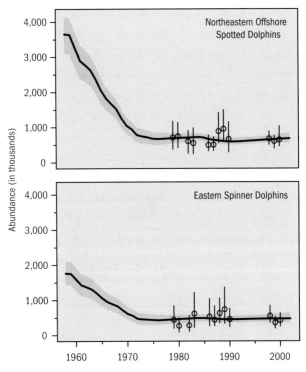

Figure 31. Dramatic population decline of Spinner Dolphins and Pantropical Spotted Dolphins because of the tuna purse seine fishery. The toll amounted to some 350,000 drowned per year before the advent of the Marine Mammal Protection Act in 1972. (Redrawn from Timothy Gerrodette, NOAA Fisheries.)

shore-based whalers, Jordan provided early scientific assessments of the impacts of whaling.

A culture shift in North America began in the 1960s, grounded in ecological understanding and an emotional connection with marine mammals. John Lilly and other unorthodox thinkers wrote about the minds in the waters, and people were disturbed to see dolphins ensnared and drowned in nets set for tuna (fig. 31). Tuna–porpoise fisheries, hunting of polar bears, and uncontrolled commercial whaling fueled momentum in the United States to enact into law the Marine Mammal Protection Act (MMPA) in 1972 and the Endangered Species Act (ESA) in 1973. Responding to public discontent, the MMPA provided broad protections for the conservation of marine mammals, with specific provisions to reduce the mortality of porpoises in the tuna industry and sweeping provisions placing a moratorium on the taking of all marine mammals in U.S. waters. The term "take" is defined as to "harass, hunt, capture, or kill, or *attempt* to harass, hunt, capture or kill any marine mammal." Nations had agreed to a moratorium in hunting right whales in 1937, but not until 1986 did they agree, under the International Whaling Commission (IWC), to a global moratorium on commercial harvesting, allowing only aboriginal hunting and collection for scientific research. Exceptions, however, led to some countries' "collecting" thousands of Common Minke Whales and other species for scientific purposes. Nevertheless, under these acts and agreements, some species of marine mammals have recovered: some, perhaps, to more than their prehistoric numbers; others, never to their former numbers.

The saga of ocean exploitation through the 20th century has begun to pale, though, in light of the cataclysmic changes unfolding in the nascent 21st century.

The Present and Future

> "Now, I truly believe that we in this generation must come to terms with nature, and I think we're challenged as mankind has never been challenged before to prove our maturity and our mastery, not of nature, but of ourselves."
>
> —RACHAEL CARSON, SILENT SPRING (1962)

Geologists demark some major periods in Earth's past by mass extinctions, and the pace of species extinctions today is quickening. An armageddon is predicted for the oceans of the world, precipitated by an "Anthropocene mass extinction" now in progress (Jackson 2008). The oceans as we know them at the beginning of the 21st century are predicted, within the next 100 years, to become barren from overfishing, warmed and acidic from global climate change, polluted from massive runoff with widespread anoxic dead zones, and invaded by exotic species. The ultimate consequences could be lower and simpler food webs, loss of species diversity, and the disappearance of whole ecosystems, including the California Current ecosystem.

The most imminent challenges to marine mammals and their ecosystems are imbalances from global human effects (Halpern et al. 2008). The byproducts of industrial technology have begun to alter the chemistry of the air and water, interrupting life functions on land and in the sea, and changing global temperatures. The U.S. Census Bureau predicts that the human population will expand to 9.4 billion by 2050, with commensurate expansion in the use of natural resources and the amount of garbage tossed upon the land and sea (fig. 32). Fishing down the food web, which begins with removal of top predators first and then of pelagic schooling

Figure 32. California Sea Lion entangled in gill net around its neck. Such animals usually die unless scientists can capture them and remove the net.

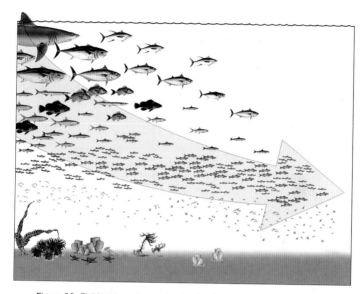

Figure 33. Fishing down the marine food web means that fisheries remove first larger fishes at the top of the food chain, and then remove species situated lower and lower down the food chain until they are targeting small fishes and jellyfish. (Courtesy of Daniel Pauly and artist Rachel "Aque" Atanacio, from Pauly and Maclean 2003.)

fish, causes a shift from biologically diverse food webs to simple, "slimy" ones (fig. 33; Coll et al. 2008, Pauly et al. 1998). Jellyfish, the preferred food of sea turtles, thrive in such conditions—but few marine mammals consume jellyfish. With the depletion of resources, hotspots of high species diversity shift geographically, and marine mammals likely will follow these shifts.

Ultimately, the effects of climate change will trump other threats to the lives of marine mammals (Glanz 2003, Simmonds and Isaac 2007). Predicted changes in oceanic conditions include increased sea temperature, sea level rise, alterations of oceanic processes such as upwelling, and higher ocean acidity. Among the consequences for marine mammals may be alterations in their movements and migrations, as well as in what they eat; changes in abundance, reproductive success, and timing of breeding; susceptibility to disease; and local extirpation or extinction. Whatever the combination and intensity of effects, by the end of the 21st century, the California Current and the animals associated with it will be different from what we know today.

With increased sea temperature, some species may range farther north and others may contract their range (Ragen et al. 2008, Harington 2008). A look at the effects of ENSO events on marine mammals provides a glimpse into the possible results of globally warmer sea temperatures. During past ENSOs, pinniped productivity and survival declined significantly (Trillmich and Ono 1991). Increases in harmful algal blooms, which often cause marine mammal mortality, also intensified, resulting in spikes of sick and dead California sea lions (Van Dolah 2005, Gulland et al. 2002).

Delays in the onset of seasonal cycles disrupt species' lives as much as an ENSO event. The life cycles of most marine species are tightly aligned to ocean seasons, and delayed transitions cause mismatches in normally stable food webs, cascading through the food web to whales (Palacios et al. 2004). Scientists are not entirely sure how oceanic processes will change, but most agree that the

Figure 34. Northern Elephant Seal colony awash in rising waters during an El Niño–driven storm event.

timing and strength of the winds that mark the upwelling season in the California Current will alter over the next century.

Sea level rise will reduce pinniped haul-out areas by submerging low-lying coastlines but may also create new habitat through erosion. The young of most pinnipeds cannot swim at birth, and with elevated seas, places where seals give birth may be eliminated (fig. 34). Other changes predicted in ocean conditions—such as changes in ocean acidity—are ones whose consequences are yet unknown. Ultimately, environmental pressures may lead to extinction for some species, such as polar bears, that are unable to evolutionarily adapt.

The outlook for marine mammals in the next generation and beyond is bleak. How bleak—and how much of today's biodiversity can persist—depends on what we do now. There are actions that can stall or reverse today's negative trends, beginning with the creation of a global network of marine protected areas (MPA). As we begin the 21st century, less than 0.01 percent of the world ocean's area is designated as marine reserves, and only a few congressionally designated marine wilderness areas exist (fig. 35).

Figure 35. Harbor Seals resting on a tidal sandbar in Drakes Estero, part of the only congressionally designated marine wilderness south of Alaska and within the California Current System.

Figure 36. A friendly Gray Whale off Baja, California.

A network of MPAs from Baja to British Columbia is our best hope for ensuring the resilience and perpetuation of the biodiversity of the California Current in the face of a radically changing world ocean.

Human culture that previously devastated sea mammals is also changing, and a new conservation ethic is emerging, spun from the tales of humans interacting with sea mammals. Friendly whales of Baja, marine mammal hospitals, and animals in oceanariums have stirred renewed reverence for sea mammals (fig. 36). Whether this attitude can translate into urgently needed, worldwide changes in policy, economies, and individual lifestyles remains to be seen. As marine mammalogist Douglas DeMaster noted, "Loving marine mammals is not enough; as a species, we have to reduce greenhouse gases significantly and immediately. Nothing else is as important now" (D. DeMaster, NMFS-AFSC, pers. com., July 2009). With action, there is hope that future generations of humans will study, observe, and take inspiration from future generations of marine mammals.

Taxonomy

Marine mammals are not easily grouped because of their evolutionary roots, and are so diverse that they fall into several taxonomic orders and families. Current classification separates the entire group into three of the 25 orders under the class Mammalia (Rice 1998): carnivores, cetaceans, and sirenians. There are 130 marine mammal species currently described (2 to 3 percent of all mammal species), but with genetic analyses, new species seem to be named almost annually. Recently, a species of beaked whale new to science *(Mesoplodon perrini)* was discovered from analyzing the tissue of a dead beach-cast specimen (Mead and Brownell 2005, Perrin 2009). A specimen of the mythological mermaid, however, is yet to be collected.

ORDER CARNIVORA

SUBORDER MUSTELIDAE

SUBORDER PINNIPEDIA

Family Phocidae

Family Otariidae

ORDER CETACEA

SUBORDER MYSTICETI

Family Balaenidae

Family Balaenopteridae

Family Eschrichtiidae

SUBORDER ODONTOCETI

Family Physeteridae

Family Kogiidae

Family Ziphiidae

Family Delphinidae

Family Phocoenidae

In this guide, we limit our accounts of living species to those that occur within the California Current (CC), and so will present information on just two of the three orders of marine mammals—carnivores and cetaceans [although fossils attest to the presence of the extinct species of sirenian, Steller's Sea Cow *(Hydrodamalis gigas),* as far south as Baja]. Here we provide general information comparing and contrasting orders and families. The species accounts that follow provide greater detail for each of 45 species that reside in, migrate through, or forage within the CC.

Using This Guide

The scientific naming of species used here conforms to the taxonomy of the Society of Marine Mammalogy by Rice (1998) and the Committee on Taxonomy (2009), the World Cetacea Database by Perrin (2009), and *Mammals of the World,* with a section on Cetacea, by Mead and Brownell (2005). Some modifications have been made per recent discoveries of new species (beaked whales) or genetic fine-tuning of existing known species (right whales).

Each account provides a description as well as a comparison to other similar species; a section on natural history with information on behavior, reproduction, diet, and mortality; another on distribution with information on overall range and migratory patterns, where applicable, as well as viewing locations; and a range map. The final section includes issues specific to conservation, describing past and present threats and providing a current estimate of the population.

Color Plates

Color plates illustrate in great detail the 45 species covered in this guide. Identification is difficult in the field, though. Marine mammals spend much of their time submerged and are usually far away, swimming in unsettled seas, whereas observers are viewing from a heaving boat. To aid in identification in the field, some plates also provide a comparison of head, tail, and blow shapes. Accompanying each species account except those for the few very rare beaked whales is a photograph of what you might see at the surface of the ocean or, in the case of pinnipeds, on land.

Range Maps

Range maps are regionally limited to the California Current from British Columbia to Baja, though many species included in this guide are found throughout the Pacific Ocean and even beyond. Range maps are inherently general and lack the detail of a species' preferred habitats, so each species account provides particular information on where a species can be seen, and with which habitats it is most associated, if known. Some species are rare, and their numbers few worldwide, whereas others are fairly abundant, and easily seen nearshore. Range maps include breeding locations for some species of pinniped that gather onshore, and breeding ranges for some species that are known. In a few instances, range maps identify the distribution of sub-species or ecotypes.

Conservation Status

The conservation status for each species is provided in four categories: the U.S. Endangered Species Act (ESA), the U.S. Marine Mammal Protection Act (MMPA), the Committee on the Status of Endangered Wildlife in Canada (COSEWIC), and the International Union for the Conservation of Nature (IUCN). In the United States, the ESA applies to endangered and threatened species that might perish without special protection, and the MMPA gives designation to marine mammal species whose populations are "depleted" but are not covered under ESA; depleted is used to describe a species or a population that is below its optimum sustainable population level. The COSEWIC is an independent advisory organization of the Canadian government, with experts that designate which wildlife species are in danger of extirpation from Canada, much as the ESA does. The listing also includes categories for candidate species to be listed. The IUCN is a venerated international environmental organization founded in 1948, providing guidance on the protection of species worldwide with designations that range from "endangered" to "least concern." Limited information is a serious challenge to conservation strategies—hence the occasionally encountered status "data deficient."

Marine-Mammal Watching

You can see pinnipeds and otters from land; even some species of whales, dolphins, and porpoises occur nearshore. Whales are trickier to detect from land, but cast your eyes to the horizon and look for a bushy spout of water from a blow, or a tail being lifted above the surface (fluke up) (fig. 37). For the hardy observer, though, a voyage by boat or kayak can deliver you into the habitats of many more marine mammals.

Distinctive natural markings can be used for identification when an animal surfaces to breathe. These include size; head

Figure 37. Humpback Whale fluking.

shape and type of teeth; presence, shape, and location of a dorsal fin; presence and shape of flukes; shape and length of flippers; and coloration, skin pigmentation, and scarring. Field identification also includes aerial or surface behaviors, such as resting at the surface in a head-up "bottling" or horizontal "logging" posture, and, for whales, the shape and height of the blow (fig. 38).

Figure 38. Group of Sperm Whales logging at the surface.

Figure 39. Wake-riding Short-beaked Common Dolphin.

Friendly animals do approach boats to "people-watch," but most sea mammals are wary of people and avoid them, or even flee if approached too closely. Moreover, some species can be dangerous, and may attack people who try to get close. To avoid disturbing marine mammals, the National Marine Fisheries Service (NMFS) recommends approaching no closer than 300 ft when observing whales and seals by land or boat, and no closer than 1,000 ft by air. Bow-riding dolphins are an exception (fig. 39).

SPECIES PLATES

PLATE 1 Large Whales I

Four very distinctive whales in four different genera.

SPERM WHALE *(Physeter macrocephalus)*, PAGE 185
Adult: 11–16 m (36–53 ft)

GRAY WHALE *(Eschrichtius robustus)*, PAGE 172
Adult: 12.8–16.8 m (42–55 ft)

HUMPBACK WHALE *(Megaptera novaeangliae)*, PAGE 161
Adult: 14–17 m (39–42 ft)

NORTH PACIFIC RIGHT WHALE *(Eubalaena japonica)*, PAGE 105
Adult: 13.7–14 m (45–46 ft)

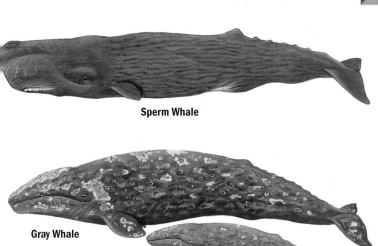

Sperm Whale

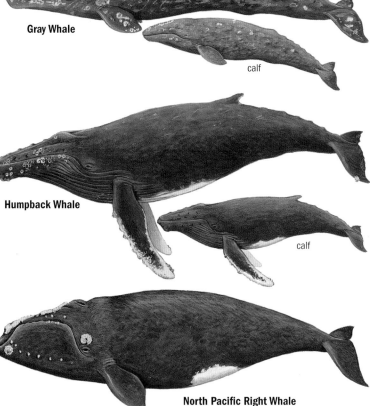

Gray Whale

calf

Humpback Whale

calf

North Pacific Right Whale

PLATE 2   Large Whales I at Sea

SPERM WHALE *(Physeter macrocephalus)*

Unmistakable if seen well. Often seen at surface "logging," blow cants off to left, huge head (one-third of body), dorsal hump rather than fin, wrinkled skin behind large head. Humped shape when diving. Frequently raises flukes when dives.

GRAY WHALE *(Eschrichtius robustus)*

Unmistakable if seen well. No dorsal fin but a bit of a hump and knuckles extending back from hump to fluke, low puffy blow, gray to blue gray mottled skin. Can show flukes when diving.

HUMPBACK WHALE *(Megaptera novaeangliae)*

Unmistakable if seen well. Long white pectoral flippers, fairly tall blow, humped back when diving with oddly shaped dorsal fin, frequently flukes when diving. Underside of flukes varies: entirely black, blotchy, or all-white. The most acrobatic species of whale in the region.

NORTH PACIFIC RIGHT WHALE *(Eubalaena japonica)*

Rare. Unmistakable if seen well. Broad, smooth, dark back lacking any dorsal fin, V-shaped blow, callosities down rostrum, high-arching jawline.

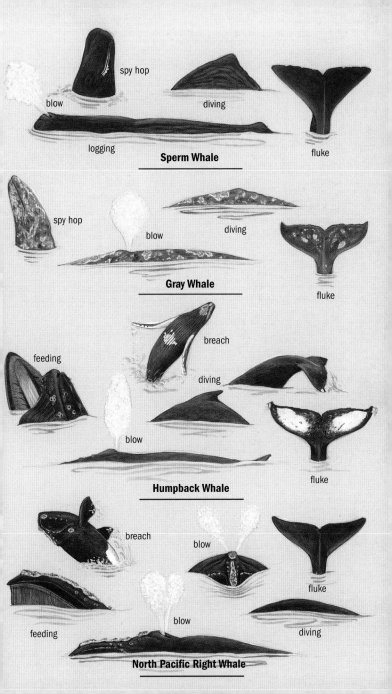

Sperm Whale

spy hop

blow

logging

diving

fluke

Gray Whale

spy hop

blow

diving

fluke

Humpback Whale

breach

feeding

diving

blow

fluke

North Pacific Right Whale

breach

blow

fluke

feeding

blow

diving

PLATE 3   Large Whales II

Five closely related baleen whales in the genus Balaenoptera.

COMMON MINKE WHALE *(Balaenoptera acutorostrata)*, PAGE 152
Adult: 7.8–8.8 m (25.5–29 ft)

BRYDE'S WHALE *(Balaenoptera edeni)*, PAGE 145
Adult: 12–14 m (39–46 ft)

SEI WHALE *(Balaenoptera borealis)*, PAGE 136
Adult: 13.1–13.7 m (43–46 ft)

FIN WHALE *(Balaenoptera physalus)*, PAGE 127
Adult: 18–19 m (59–61 ft)

BLUE WHALE *(Balaenoptera musculus)*, PAGE 117
Adult: 23–24 m (75–80 ft)

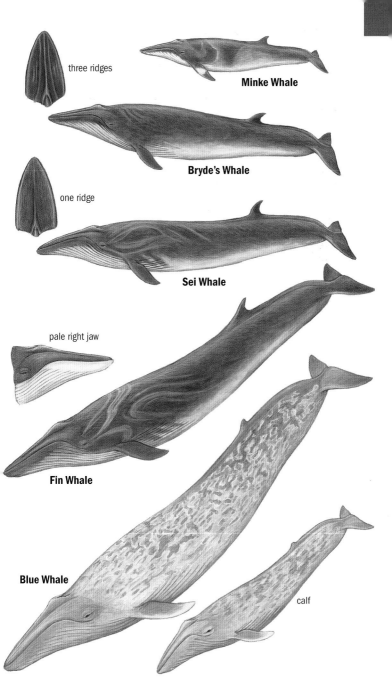

three ridges

Minke Whale

Bryde's Whale

one ridge

Sei Whale

pale right jaw

Fin Whale

Blue Whale

calf

PLATE 4   Large Whales II at Sea

COMMON MINKE WHALE *(Balaenoptera acutorostrata)*

Relatively inconspicuous small whale. Very falcate dorsal fin and pointed rostrum. Pectoral flippers have a white band across them. Due to its small size, the dorsal fin is usually seen simultaneously with the back when it surfaces. Usually does not show flukes.

BRYDE'S WHALE *(Balaenoptera edeni)*

Medium-sized whale that is difficult to distinguish from the Sei Whale. Note the shape of the smoothly curved dorsal fin and the three ridges on rostrum. Blow is often indistinct. Usually does not show flukes.

SEI WHALE *(Balaenoptera borealis)*

Medium-sized whale that is difficult to distinguish from Bryde's Whale. Note the very upright dorsal fin, often with a kinked or bent appearance. There is a single ridge on the rostrum. Rostrum can have a bent curve. Blow can be quite distinct and fairly tall. Usually does not show flukes.

FIN WHALE *(Balaenoptera physalus)*

Large whale that can be difficult to distinguish from the Sei Whale and Bryde's Whale. The right jaw is pale to whitish. A pale chevron is behind head. Note that the dorsal fin often has a long, swept-back appearance. Blow is strong, columnar, and tall. Usually does not show flukes.

BLUE WHALE *(Balaenoptera musculus)*

Largest whale. Overall mottled gray to blue-gray. Tiny dorsal fin located far back on the back, so often not seen. Blow is very strong, columnar, and tall. Frequently shows flukes when dives. Blue whales can look turquoise under water.

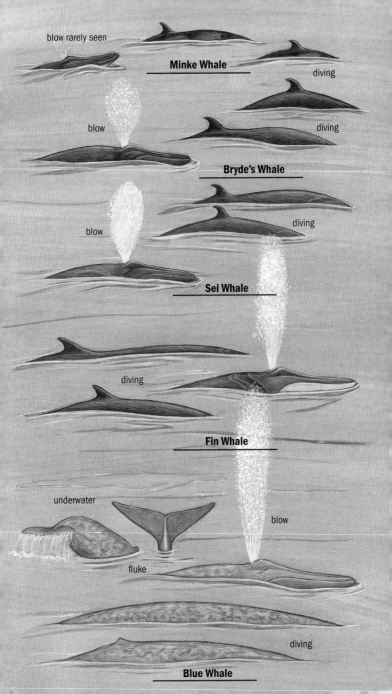

blow rarely seen

Minke Whale

diving

blow

diving

Bryde's Whale

blow

diving

Sei Whale

diving

Fin Whale

underwater

fluke

blow

Blue Whale

diving

PLATE 5 Large Beaked Whales

All beaked whales are difficult to see and identify due to long dives and a deep ocean habitat. Cuvier's Beaked Whale and Baird's Beaked Whale are the two most easily seen in the California Current.

CUVIER'S BEAKED WHALE *(Ziphius cavirostris)*, PAGE 215
Adult: 5–7 m (16–23 ft)
Medium-sized beaked whale seen as individuals or in small groups of two to four. Falcate dorsal fin. Males have a white head. All have a sloping forehead curving into a short beak. Color varies from ochre gray to gray. Has an inconspicuous, puffy blow.

LONGMAN'S BEAKED WHALE *(Indopacetus pacificus)*, PAGE 251
Adult: 7–9 m (23–30 ft)
Rare. A medium-sized beaked whale usually seen in groups of six to 12. All individuals have a large bulbous forehead and a bi-colored beak and are often covered with white spots from cookie cutter shark bites. Small, distinct, puffy blows. When moving away rapidly, sends up rooster-tail splashes. May associate with pilot whales.

BAIRD'S BEAKED WHALE *(Berardius bairdii)*, PAGE 208
Adult: 11.9–12.8 m (39–42 ft)
Large beaked whale usually seen in groups of from eight to 30 or more. All individuals have a steep, somewhat bulbous forehead and a long, heavy beak. Coloration varies from ochre gray to deep reddish gray. Adults are covered with numerous scars; immatures lack these scars. Fairly falcate dorsal fin can vary. Distinct, puffy blows. Can be acrobatic: breaches.

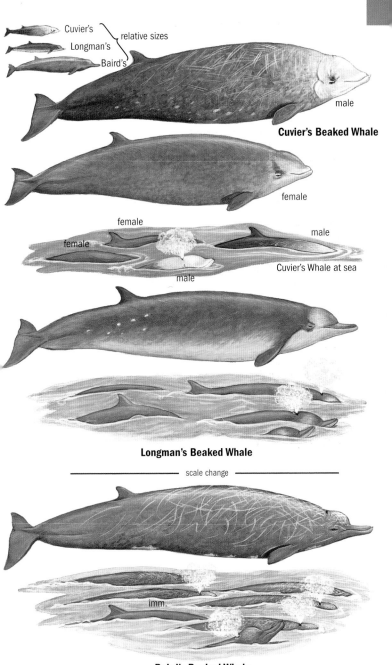

Cuvier's
Longman's — relative sizes
Baird's

Cuvier's Beaked Whale

male

female

female

female

male

male

Cuvier's Whale at sea

Longman's Beaked Whale

——— scale change ———

imm.

Baird's Beaked Whale

PLATE 6 Mesoplodon Beaked Whales I

Beaked whales in the genus *Mesoplodon* are extremely difficult to identify due to a number of factors, including their general scarcity, small size, deep ocean habitat, and, perhaps, shyness of ships, making most views brief and poor. Some of the species depicted have yet to be identified at sea, so the illustrations are here to aid in giving clues to identity but may not have all definitive field marks. Note, in particular, the placement of erupted teeth on males, as well as the jawline, if you are lucky enough to see one of these enigmatic whales well enough.

GINKGO-TOOTHED BEAKED WHALE *(Mesoplodon ginkgodens)*, PAGE 229
Adult: 4.5–4.9 m (14.8–16 ft)

STEJNEGER'S BEAKED WHALE *(Mesoplodon stejnegeri)*, PAGE 246
Adult: 5–7 m (16.5–23 ft)

HUBBS' BEAKED WHALE *(Mesoplodon carlhubbsi)*, PAGE 237
Adult: 4.8–5.3 m (15.7–17.4 ft)

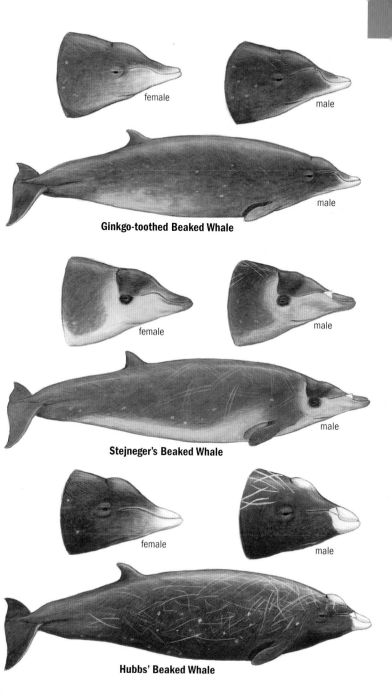

Ginkgo-toothed Beaked Whale

female

male

male

Stejneger's Beaked Whale

female

male

male

Hubbs' Beaked Whale

female

male

PLATE 7   Mesoplodon Beaked Whales II

Beaked whales in the genus *Mesoplodon* are extremely difficult to identify due to a number of factors, including their general scarcity, small size, deep ocean habitat, and, perhaps, shyness of ships, making most views brief and poor. Some of the species depicted have yet to be identified at sea, so the illustrations here are to aid in giving clues to identity but may not have all definitive field marks. Note, in particular, the placement of erupted teeth on males, as well as the jawline, if you are lucky enough to see one of these enigmatic whales well enough.

PERRIN'S BEAKED WHALE *(Mesoplodon perrini)*, PAGE 233
Adult: 3.9–4.4 m (13–14.8 ft)

PYGMY BEAKED WHALE *(Mesoplodon peruvianus)*, PAGE 241
Adult: 3.8–4.1 m (12.5–13.5 ft)

BLAINVILLE'S BEAKED WHALE *(Mesoplodon densirostris)*, PAGE 223
Adult: 4.5–6 m (15–20 ft)

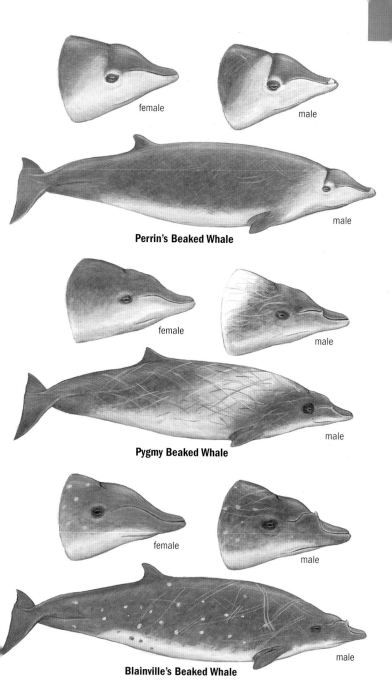

Perrin's Beaked Whale

female

male

male

Pygmy Beaked Whale

female

male

male

Blainville's Beaked Whale

female

male

male

PLATE 8   Mesoplodon Beaked Whales at Sea

Little is known about what *Mesoplodon* beaked whales look like at sea. These illustrations show what one might see, and what one should look for, in combination with the previous two plates. Usually Mesoplodon beaked whales are seen individually or in groups of two to six.

PERRIN'S BEAKED WHALE (*Mesoplodon perrini*)
Known only from beach strandings. Note erupted teeth at the tip of the male's lower jaw, and the male's faint scarring. Both sexes have blunt, somewhat curved dorsal fins, and perhaps a dark eyespot with paler lower jaw.

PYGMY BEAKED WHALE (*Mesoplodon peruvianus*)
One of the easier beaked whales to identify. Male has large pale bands with extensive scarring. Note males' teeth erupting from raised area in middle lower jaw. All age/sexes have a hint of an eye patch. Distinctive triangular dorsal fin.

BLAINVILLE'S BEAKED WHALE (*Mesoplodon densirostris*)
Male's erupted teeth are located on an oddly exaggerated raised area in the middle of the lower jaw. Male is often scarred. Both sexes can be spotted with numerous cookie cutter shark bites and have a hint of a darker eye patch. Female's jawline is a less exaggerated version of the male's.

GINKGO-TOOTHED BEAKED WHALE (*Mesoplodon ginkgodens*)
Male's teeth erupt from center of jaw. Male has no obvious scarring. Male and female have contrastingly pale beaks. Both sexes are overall dark gray to blackish.

STEJNEGER'S BEAKED WHALE (*Mesoplodon stejnegeri*)
Male's triangular teeth erupt from the center of the lower jaw. Both sexes may have a dark cap, pale partial collar, dark eye patch, and paler lower jaw. Male has fairly distinct scarring.

HUBBS' BEAKED WHALE (*Mesoplodon carlhubbsi*)
Most distinctive of the mesoplodon beaked whales in the region. Male has a white "beanie" on top of head, a thick white beak, large teeth erupting from the center of the jaw, and distinct white scarring. Both male and female are dark gray to blackish. Female also has a paler beak and can have some spotting from cookie cutter shark bites.

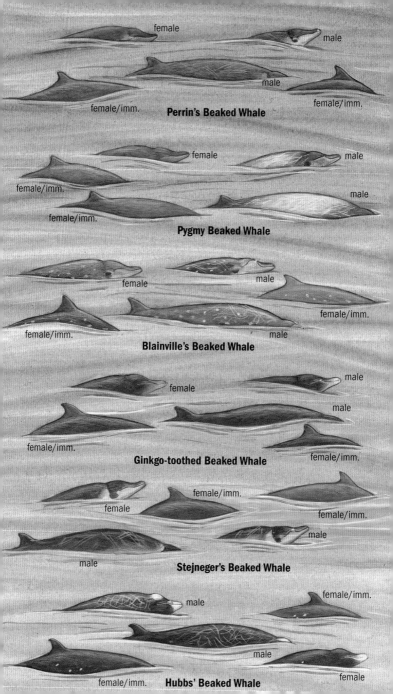

Perrin's Beaked Whale

Pygmy Beaked Whale

Blainville's Beaked Whale

Ginkgo-toothed Beaked Whale

Stejneger's Beaked Whale

Hubbs' Beaked Whale

PLATE 9   Killer Whales

Killer whales are medium-sized, distinctive black-and-white whales. There are three ecologically and somewhat morphologically distinct groups found in the California Current. Differences between these groups are subtle, and the research on them is ongoing. Additional things to note when identifying killer whale ecotypes are location and group size.

RESIDENT KILLER WHALE *(Orcinus orca)*, PAGE 255
Male: 9.8 m (32 ft). Female: 8.5 m (28 ft)
Male's dorsal fin has a faint S-curve to it. Female's is falcate, narrow at the base with a slightly blunt tip. Saddle patch is usually open, with black bleeding into it. Eye patch is straight, running parallel to the back but fairly thick.

TRANSIENT KILLER WHALE *(Orcinus orca)*, PAGE 255
Male: 9.8 m (32 ft). Female: 8.5 m (28 ft)
Male's dorsal fin usually very straight-sided, broad-based, and upright. Female's is falcate, broad-based, and pointed. Saddle patch is usually closed, with no black bleeding into it. Eye patch is perhaps angled slightly downward.

OFFSHORE KILLER WHALE *(Orcinus orca)*, PAGE 255
Male: 9.8 m (32 ft). Female: 8.5 m (28 ft)
Male's dorsal fin is usually fairly straight-sided, with a distinctly blunt tip. Female's is falcate, with an obvious blunt tip, and frequently can have a ragged trailing edge. Eye patch runs parallel to the back but is often quite narrow.

lob tail

calf

male

Resident Killer Whale

breach

female

Resident Killer Whale

female

female

male

female

Transient Killer Whale

spy hop

male

female

Offshore Killer Whale

spy hop

PLATE 10 **Blackfish**

A group of attractive all-dark, blunt-headed, beakless whales. Risso's Dolphin is often included in this group.

MELON-HEADED WHALE *(Peponocephala electra)*, PAGE 281
Adult: 2.5–2.8 m (8–9 ft)
Most easily confused with the False Killer Whale. Note the dorsal fin shape, white lips, and position and shape of saddle. When moving quickly, schools often form a chorus line.

FALSE KILLER WHALE *(Pseudorca crassidens)*, PAGE 274
Adult: 4.5–5.3 m (15–17 ft)
Most easily confused with Melon-headed Whale. Note shape of dorsal fin and dark (almost black) color, with long, narrow, indistinct saddle and more rounded head.

SHORT-FINNED PILOT WHALE *(Globicephala macrorhynchus)*, PAGE 266
Male: 5.3–7.2 m (17–24 ft). Female: 4.1–5.1 m (13–17 ft)
Unmistakable if seen well. Note the dorsal fin shape, which in the male is huge, broad-based, and curved. Head is very round and bulbous. The low puffy blows can be distinct. Pale saddle is variable.

Melon-headed
Whale

False Killer
Whale

Short-finned Pilot Whale

relative sizes

chorus line

dorsal fins

Melon-headed Whale

scale change

at sea

breach

False Killer Whale

male

male

female

imms.

Pilot Whales at sea

female

Short-finned Pilot Whale

male

PLATE 11 Rough-toothed Dolphin, Common Bottlenose Dolphin, and Risso's Dolphin

Three quite different species that when seen at sea can cause some identification problems.

ROUGH-TOOTHED DOLPHIN *(Steno bredanensis)*, PAGE 288

Adult: 2.8 m (9 ft)

Most easily confused with bottle-nose dolphin. Note the sloping forehead and long slender beak, the white "lips" of the adult, and the slightly broad-based dorsal fin and overall pattern. Generally does not approach ships.

COMMON BOTTLENOSE DOLPHIN *(Tursiops truncatus)*, PAGE 308

Adult 2–4 m (6–12 ft)

Most easily confused with Rough-toothed Dolphin. Note the distinct crease of the melon above the thick stubby beak, mostly uniform gray. Often very acrobatic; often bow-rides.

RISSO'S DOLPHIN *(Grampus griseus)*, PAGE 301

Adult: 3 m (10 ft)

Lacks a beak. Adults are heavily scarred. There is huge variation in dorsal fin size. Unscarred immatures, if not seen well, can be confused with the Common Bottlenose Dolphin. Risso's Dolphins are acrobatic; when breaching they show a pale narrow anchor pattern from chest to flukes.

at sea

imm.

Rough-toothed Dolphin

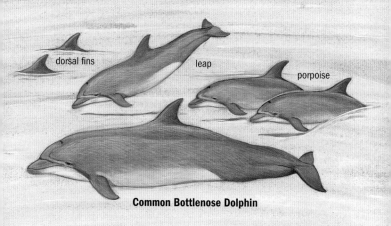

dorsal fins

leap

porpoise

Common Bottlenose Dolphin

imm.

breach

Risso's Dolphin

PLATE 12   **Tropical Dolphins**

Four species of attractive dolphins found mostly in tropical water.

SPINNER DOLPHIN *(Stenella longirostris occidentalis)*, PAGE 325
Adult: 1.65–1.8 m (5.5–6 ft)

PANTROPICAL SPOTTED DOLPHIN *(Stenella attenuata)*, PAGE 318
Adult: 1.7–2.4 m (5.6–7.9 ft)

STRIPED DOLPHIN *(Stenella coeruleoalba)*, PAGE 332
Adult: 2.2–2.4 m (7.2–8 ft)

FRASER'S DOLPHIN *(Lagenodelphis hosei)*, PAGE 358
Adult: 2.7 m (8.9 ft)

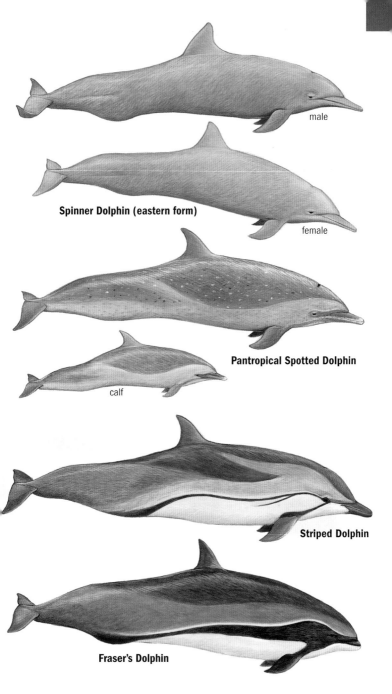

Spinner Dolphin (eastern form)

male

female

Pantropical Spotted Dolphin

calf

Striped Dolphin

Fraser's Dolphin

PLATE 13   Tropical Dolphins at Sea

SPINNER DOLPHIN *(Stenella longirostris occidentalis)*

The eastern form of Spinner Dolphin is a uniform gray, slender, long-beaked dolphin. The dorsal fin shape is variable, and is often canted forward in adult males. Adult males also have an enlarged anal keel area, giving them an odd profile. Spinner dolphins can be very acrobatic, coming vertically out of the water and spinning. Often schools with Pantropical Spotted Dolphins.

PANTROPICAL SPOTTED DOLPHIN *(Stenella attenuata)*

Variably patterned dolphin. Near-shore populations tend to be more heavily spotted than offshore. Immatures can lack spots. Dorsal fin is slender and falcate, with white tip to beak distinct when seen from above. Spotted dolphins can be very acrobatic, porpoising and leaping into frozen arches. Often schools with Spinner Dolphins.

STRIPED DOLPHIN *(Stenella coeruleoalba)*

A stocky, fast, and beautiful dolphin that is unmistakable but sometimes difficult to see well. Immatures are less distinctly marked than adults. Striped dolphins are extremely acrobatic, somersaulting and leaping high into the air. Tends to be shy; avoids ships. Small groups of Striped Dolphins frequently associate with large schools of Short-beaked Common Dolphins.

FRASER'S DOLPHIN *(Lagenodelphis hosei)*

A stocky, fast, and beautiful dolphin that is unmistakable but difficult to see well. Has proportionally small pectoral flippers and dorsal fin. Immatures are less distinctly patterned than adults and pale gray. Forms a froth of splashes when moving quickly, but does not show much above the water surface. Tends to be shy; avoids ships.

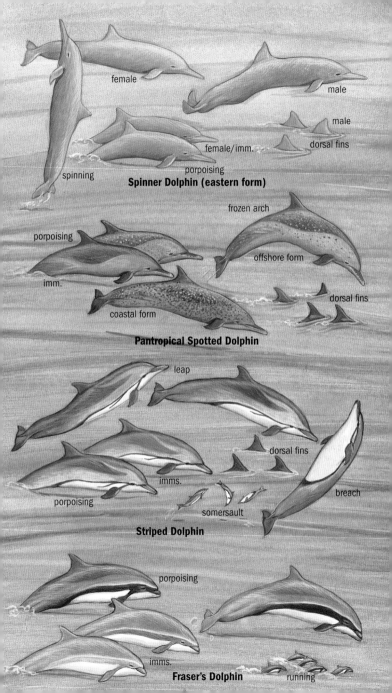

Spinner Dolphin (eastern form)

female

male

female/imm.

male

dorsal fins

porpoising

spinning

Pantropical Spotted Dolphin

frozen arch

porpoising

offshore form

imm.

dorsal fins

coastal form

Striped Dolphin

leap

dorsal fins

imms.

porpoising

somersault

breach

Fraser's Dolphin

porpoising

imms.

running

PLATE 14 Temperate Dolphins

These four dolphins are the most commonly encountered species in the California Current. Long-beaked and Short-beaked Common Dolphins can be difficult to distinguish at sea.

PACIFIC WHITE-SIDED DOLPHIN *(Lagenorhynchus obliquidens)*, PAGE 294
Adult: 1.7–2.5 m (5.6–8.2 ft)

LONG-BEAKED COMMON DOLPHIN *(Delphinus capensis)*, PAGE 349
Adult: 1.9–2.5 m (6–8 ft)

SHORT-BEAKED COMMON DOLPHIN *(Delphinus delphis)*, PAGE 338
Adult: 1.6–2.0 m (5.2–6.5 ft)

NORTHERN RIGHT WHALE DOLPHIN *(Lissodelphis borealis)*, PAGE 363
Adult: 2.3–2.8 m (7.5–9.2 ft)

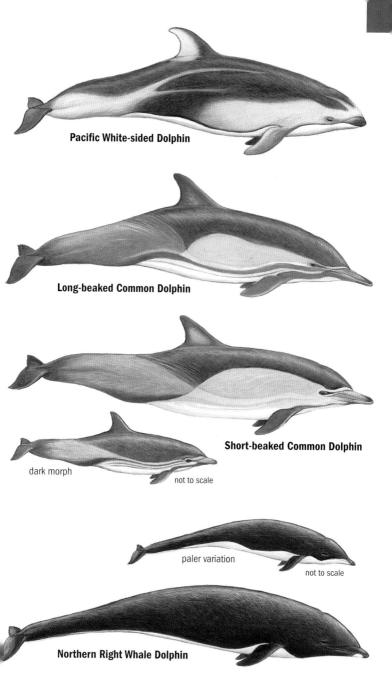

Pacific White-sided Dolphin

Long-beaked Common Dolphin

Short-beaked Common Dolphin

dark morph

not to scale

paler variation

not to scale

Northern Right Whale Dolphin

PLATE 15   Temperate Dolphins at Sea

SHORT-BEAKED COMMON DOLPHIN (*Delphinus delphis*)

An attractive dolphin most easily confused with the Long-beaked Common Dolphin. Overall markings tend to be sharp, although there is great variation. Thoracic patch deep cream to ochre, beak stout, eye noticeably surrounded by pale, thoracic line from jaw back (to vent) tends to be pale and not continuous, dorsal fin frequently has an extensive pale mottled center. Immatures can be very washed out. Has a fairly uncommon dark color variation. Forms large schools of hundreds to thousands. Friendly; readily bow-rides. Schools can have a few Striped Dolphins associated with them.

LONG-BEAKED COMMON DOLPHIN (*Delphinus capensis*)

An attractive dolphin most easily confused with the Short-beaked Common Dolphin. Overall markings can be washed out or look slightly smudgy, although there is great variation. Thoracic patch pale yellow, beak long, eye does not stand out, face can be very dark. Thoracic line from jaw back is continuous and can be pale to dark and distinct. Immatures can be very washed out. Dorsal fin infrequently has pale mottling in center. Can be in large schools of hundreds to thousands. Friendly; readily bow-rides.

PACIFIC WHITE-SIDED DOLPHIN (*Lagenorhynchus obliquidens*)

A beautiful, distinctly patterned dolphin with a very short beak. The bicolored dorsal fin varies in shape from fairly falcate to broad-based and rounded in adult males. Acrobatic and friendly; readily bow-rides. Frequently schools with Northern Right Whale and Risso's Dolphin.

NORTHERN RIGHT WHALE DOLPHIN (*Lissodelphis borealis*)

An unmistakable, beautiful, streamlined dolphin. Lacking a dorsal fin and having tiny flippers and flukes, it is most easily confused with distant porpoising sea lions. Acrobatic and friendly; readily bow-rides. When breaching they show a white narrow hourglass pattern from chest to flukes. Frequently schools with Pacific White-sided Dolphin and Risso's Dolphin.

CALIFORNIA SEA LION—SEE PLATES 17 AND 19.

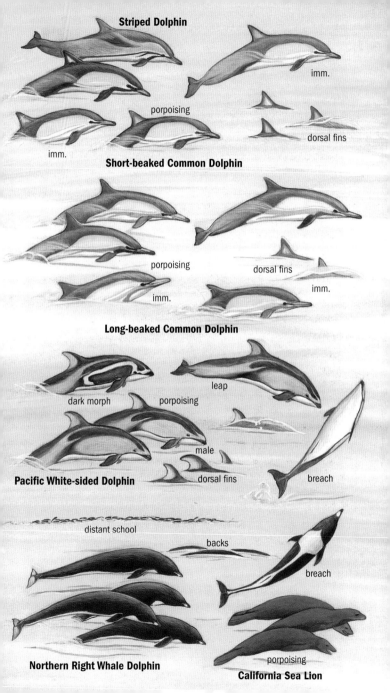

Striped Dolphin

imm.

porpoising

dorsal fins

imm.

Short-beaked Common Dolphin

porpoising

dorsal fins

imm.

imm.

Long-beaked Common Dolphin

dark morph

porpoising

leap

male

dorsal fins

breach

Pacific White-sided Dolphin

distant school

backs

breach

porpoising

Northern Right Whale Dolphin

California Sea Lion

PLATE 16   Porpoises and Sperm Whales

The two species of porpoises that occur in the California Current look very different and are easily told apart. To identify Dwarf Sperm Whales and Pygmy Sperm Whales, though, calm flat conditions are needed. In the distance, both small sperm whale species look like logs with a bit of branch sticking up at the back. The back of the animal behind the dorsal fin hangs down in the water column. The Harbor Porpoise and Dwarf and Pygmy Sperm Whales are shy. When approached by a ship, the small sperm whale species sink rather than arching and diving.

HARBOR PORPOISE (Phocoena phocoena), PAGE 377
Adult: 1.4–1.6 m (4.6–5.2 ft)
A small porpoise that is difficult to see well; all that is usually seen is a bit of gray-brown back topped with a small triangular dorsal fin as the porpoise rolls at the surface. Shy. Usually seen as individuals or small groups of three or four. Generally, the Harbor Porpoise is found closer to shore than the Dall's Porpoise.

DALL'S PORPOISE (Phocoenoides dalli), PAGE 370
Adult: 1.8–2.3 m (6–7.5 ft)
A stocky, powerful, strikingly patterned porpoise. Rooster tails as it moves rapidly through the water. Dorsal fin has a white trailing edge and is canted forward, particularly in adult males. Usually seen in spread-out groups of three to fifteen. Friendly; readily bow-rides.

DWARF SPERM WHALE (Kogia sima), PAGE 202
Adult: 2.7 m (8.9 ft)
Most easily confused with the Pygmy Sperm Whale. Note the falcate dorsal fin and perhaps proportionally smaller head with blowhole dip closer to the front of the head. Usually seen singly, or in small groups of up to 10, logging at the surface.

PYGMY SPERM WHALE (Kogia breviceps), PAGE 196
Adult: 3.78 m (12.5 ft)
Most easily confused with the Dwarf Sperm Whale. Note the rounded dorsal fin and perhaps proportionally larger head with blowhole dip further from the front of the head. Usually seen singly, or in small groups of up to 10, logging at the surface.

porpoising

rolling

Harbor Porpoise

rooster tail

rolling

male

truei form

female/imm.

Dall's Porpoise

Dwarf Sperm Whale

logging

logging

Pygmy Sperm Whale

PLATE 17   Fur Seals and Sea Lions

Medium-to-large otariid, eared seals, that when on land sit upright on foreflippers and walk on both foreflippers and hind flippers.

GUADALUPE FUR SEAL *(Arctocephalus townsendi)*, PAGE 423
Adult male: 1.8–2.4 m (6–8 ft). Adult female: 1.2–1.5 m (4–5 ft)
Note the long pointed muzzle with sloping forehead. Males have grizzled fur on neck. Fur extends onto front flippers. Can be confused with California Sea Lion female and immature, usually darker, muzzle narrower and more pointed. Male lacks sagittal crest of male California Sea Lion. See also plate 19.

NORTHERN FUR SEAL *(Callorhinus ursinus)*, PAGE 412
Adult male: 2.7 m (8.9 ft). Adult female: 1.3–1.5 m (4.5–5 ft)
With a good view, unmistakable head shape. No forehead, short muzzle, long prominent whiskers. Fur does not extend onto front flippers. All ages and sexes mostly dark brown. See also plate 19.

CALIFORNIA SEA LION *(Zalophus californianus)*, PAGE 403
Adult male: 2.1–2.4 m (7–8 ft). Adult female: 1.8–2 m (6–6.5 ft)
Male has a distinct forehead to bump/sagittal crest. Has a doglike face. Some color variation: males are dark chocolate brown but some are pale brown. Females and immatures pale reddish brown to gray. See also plates 15 and 19.

STELLER SEA LION *(Eumetopias jubatus)*, PAGE 392
Adult male: 2.7–3.7 m (9–12 ft). Adult female: 2–2.7 m (6–9 ft)
Largest of the group. Male has a very heavy muzzle with a pushed-in look to the face and rounded forehead. Very thick neck with heavier fur (mane). All sexes pale brown to reddish brown to blond. Female/immature easily confused with California Sea Lion female/immature but has heavier, more blunt muzzle and usually a paler color. See also plate 19.

Guadalupe Fur Seal
female
male
pup

Northern Fur Seal
female
male
pup

California Sea Lion
female
male
pup

Steller Sea Lion
female
male
pup

PLATE 18   Otters and Seals

Otters are voracious predators related to weasels and badgers and most live on land, only feeding in the water. Sea Otters are the exception, living almost entirely at sea. Two species of phocids, the earless seals or true seals, move on land by humping along on their bellies.

RIVER OTTER *(Lontra canadensis)*, PAGE 468
Adult: 0.9–1.2 m (3–4 ft)
Long, slender, weasellike body; usually seen foraging not far from shore in bays and river mouths; agile on land and in the water. Much smaller than the Sea Otter, it has a narrower two-toned face and more prominent ears.

SEA OTTER *(Enhydra lutris)*, PAGE 454
Adult: 1.2–1.8 (4–6 ft)
Usually this species is seen floating on its back in kelp, resting, feeding, or grooming. The often-pale prominent head and the large prominent feet give it an odd profile. The fuzzy pup is secured with a piece of kelp or rests on its mother's belly.

HARBOR SEAL *(Phoca vitulina)*, PAGE 431
Adult: 1.5–1.8 m (5–6 ft)
Sweet, doglike face with dark eyes. Usually pale with dark spots, but some individuals are mostly dark with pale spots. Harbor Seals are shy and easily disturbed when hauled out, but curious when in the water. See also plate 19.

NORTHERN ELEPHANT SEAL *(Mirounga angustirostris)*, PAGE 442
Adult male: 3.7–5.5 m (12–18 ft). Adult female: 2.7–3.1 m (9–10 ft)
Largest of all pinnipeds, the unmistakable male is dark gray- or brown with a huge, wrinkled nose and scarred and checkered neck shield. Female is smaller, a uniform gray or -brown with a smooth forehead and slightly pointed Roman nose. See also plate 19.

River Otter

Sea Otter

feeding

pup

Harbor Seal

pup

Harbor Seal

scale change

relative sizes

Northern Elephant Seal

male

female

pup

Northern Elephant Seal

PLATE 19   Seals, Fur Seals, and Sea Lions at Sea

NORTHERN ELEPHANT SEAL *(Mirounga angustirostris)*

Usually floats upright with only the head breaking surface and the nose pointing skyward. It looks like an odd buoy in the distance, particularly in adult males. Note the female's and immature's muzzle shape: a Roman nose with a pointy tip. When approached, sinks and swims away. See also plate 18.

HARBOR SEAL *(Phoca vitulina)*

Often floats upright with nose pointing skyward, but will also swim with head held horizontally. Has a cute, doglike face. Generally shy. See also plate 18.

NORTHERN FUR SEAL *(Callorhinus ursinus)*

With a good view, unmistakable head shape: no forehead and a short muzzle. The long pale whiskers are prominent, and contrast with the dark brown face and body. Note the very long floppy toes of hind flippers. The Northern Fur Seal commonly "jug handles," and is curious. See also plate 17.

GUADALUPE FUR SEAL *(Arctocephalus townsendi)*

Most easily confused with California Sea Lion female and immature, but is darker with narrow and more pointed muzzle. Jug handles like the Northern Fur Seal, but flippers are not so long and floppy. See also plate 17.

CALIFORNIA SEA LION *(Zalophus californianus)*

Adult male has a distinct forehead to bump/sagittal crest. Females and immatures have doglike muzzle. Some color variation: males are a dark chocolate brown or light gray. Females and immatures are reddish brown to gray-brown. Can form large feeding groups of several hundred. Often raises fins at surface. Curious. See also plates 15 and 17.

STELLER SEA LION *(Eumetopias jubatus)*

Largest of the group. Heavily muzzled adult male has broad head and pushed-in look to face. Females and immatures have blunter muzzle than the California Sea Lion does. All sexes are pale brown to reddish brown to blond. Occasionally fins. See also plate 17.

imm. male

imm.

male

Northern Elephant Seal

Harbor Seal

male

jughandling

male

female/imm.

male

Northern Fur Seal

male

male

jughandling

female/imm.

male

Guadalupe Fur Seal

male

male

female/imms.

finning

flip

California Sea Lion

female/imms.

male

Steller Sea Lion

"CETACEAN" IS DERIVED both from the Greek *ketos,* for "sea monster" or "huge fish," and the Latin *cetus,* for "large sea animal." The order includes hairless, fishlike mammals that live entirely in water (fig. 40). They shed their fur and evolved an impermeable skin and thick blubber. They also shed their land-based hind limbs to assume a fishlike shape with flukes as tails. The tail is the primary means of propulsion through the water.

Although they are not carnivores per se, all cetaceans eat other animals, ingesting prey varying from crustaceans to other marine mammals. A primary difference between kinds of whales is their mouth, which can feature a variety of specialized adaptations for feeding in nearly all the ocean habitats of the world, from the deepest trenches to within meters of land. Species often segregate spatially, based on ocean depth, temperature, and oceanographic features.

Cetaceans are classified into two suborders (see table 3): baleen whales *(Mysticeti)* and toothed whales *(Odontoceti).* Around 86 species are classified into 13 to 14 families, but identification to species continues to be refined with genetic analyses (Perrin 2009).

Baleen Whales (Suborder Mysticeti)

The word "mysticeti" has a confused derivation, but the most apt may be the "mustached" whale, referring to the appearance of a fringed upper lip caused by the baleen. Instead of teeth, this type of whale has baleen plates. To accommodate the baleen, whales have very bowed jaw bones with a loose joint between the two sides of the jaw, allowing it to open wide for feeding. In most

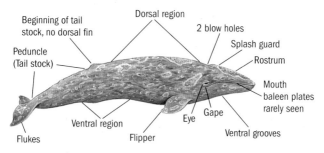

Figure 40. The Gray Whale, a baleen whale of the Mysticeti suborder.

TABLE 3. Comparison of Differences between Baleen and Toothed Whales

	Toothed Whale	Baleen Whale
Skull symmetry	Asymmetrical	Symmetrical
Teeth/baleen	Teeth	Baleen
Echolocate	Yes	No
Sounds produced	Wide range; ultrasonic	Sonic and subsonic
Lower jaw	Mandible fused	Mandible not fused
Hairs on head	Usually absent	Always present
Occlusion of jaw	Complete	Incomplete
External blow hole	One blow hole	Two blow holes
Ear plug	Absent	Present
Cervical plug	Present	Absent
Larger sex	Male	Female
Stomach divisions	3–13	3
Food	Squid, fish, larger animals	Plankton, small fish

Modified from Tinker 1988.

species, ventral pleats extend from the tip of the lower jaw and expand like an accordion when feeding, allowing the whale to capture large masses of prey in one gulp (fig. 41). Mysticeti are the largest of all the large whales except for Sperm Whales, and females tend to be larger than males. And there are other differences from toothed whales: baleen whales have two blow holes, instead of the one seen in toothed whales (fig. 42), and do not appear to echolocate as do many toothed whales (although they do produce low-frequency sounds that travel long distances).

Their diversity is best seen in the different head shapes, throat pleats/grooves, and baleen size and shape—all reflecting differences in feeding styles and diet. Long baleen with fine bristles are associated with invertebrate prey, and short, coarse baleen more with fish.

Baleen whales are divided into four families including about 14 species, a number that continues to change. Three of the four families worldwide occur in the California Current (CC): Balaenidae, Balaenopteridae and Ecschrichtiidae.

Balaenidae whales have the longest baleen plates (with fine bristles), a robust body shape, and no dorsal fin. They have a

Figure 41. Fringes of baleen in the upper jaw are exposed as a Humpback Whale opens its mouth while feeding.

large head to accommodate the long baleen, but few throat grooves and no pleats. Around the face they have distinctive callosities (pale, hardened skin patches) not found on any other whale. One species, the North Pacific Right Whale, occurs in the

Figure 42. Cetacean blow hole with two openings (found in mysticeti whales such as this Humpback Whale).

Figure 43. Member of the Balaenoptera family, Blue Whale.

CC, and three to four other species (the number to be determined by genetic studies) occur in other oceans of the world. These are among the most endangered of marine mammals, and the North Pacific Right Whale may become extinct within this century.

The members of the Balaenopteridae family are also known as "rorquals," from the Norwegian for Furrow Whale, referring to their numerous throat pleats. Most species have a slim, long body shape, evolved for fast movement, and all have a dorsal fin and short to midsized baleen (fig. 43). The top of the head, called the rostrum, is V-shaped and about one-quarter the length of the body. Rorquals include the distinctive Humpback Whale in a separate subfamily. The CC is home to six of the eight identified species worldwide.

Gray Whales of the Ecschrichtiidae family have short baleen plates, an arched head, and a robust body shape, and are the only species within the family. They have few throat grooves and feed on the bottom, sucking and filtering sediment for invertebrate prey. They lack a dorsal fin and are slow swimmers.

Toothed Whales (Suborder Odontoceti)

Toothed whales are among the most highly evolved and widely distributed of all mammals (fig. 44). They have teeth, though not all species have many. Most species have a dorsal fin and are built for fast swimming, and a few are the deepest divers of

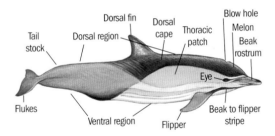

Figure 44. Short-beaked Common Dolphin, a toothed whale of the Odontoceti suborder.

any mammal. They have a one-holed blowhole (except for the Sperm Whale, which has two, one of which is internal). A number of them use biosonar to detect and perhaps capture prey, and their skulls are asymmetrical to accommodate their biosonar structures. There are some 10 families of toothed whales, five of which occur in the CC: Physeteridae, Kogiidae, Ziphiidae, Delphinidae, and Phocoenidae.

The Sperm Whale is a box-headed whale and belongs to its own family, Physeteridae. The Dwarf Sperm Whale and Pygmy Sperm Whale, two lesser-known relatives, are much smaller, and are assigned to the Kogiidae family. Genetic analyses, however, confirm that the three are close cousins, and they are sometimes joined in the same superfamily. Features that they have in common are an extremely asymmetrical skull, to accommodate an enlarged biosonar organ; a blunt-shaped head; and numerous but simply shaped teeth in the lower jaw. They are all specialized deep divers that feed in the open ocean, and their diet consists mostly of soft-bodied invertebrates.

Beaked whales of the Ziphiidae family include the most remote and the least known of cetaceans. They all possess a distinct "beak," hence the name, and have a spindle shape, a slightly curved dorsal fin, and a large melon head housing the biosonar organ (fig. 45). Teeth are few and usually only identifiable in males as a pair of tusks, used for social interactions rather than seizing prey (fig. 46). They suck prey into their mouths and are called teuthophages, for "squid eaters." Males of different species vary in size, shape of beak and forehead, position and number of teeth, and markings. Some 21 species are recognized, nine of them

Figure 45. Female beaked whale of the genus Mesoplodon breaching. Females lack distinguishing features such as tusks and are not identifiable at sea, even by experts.

associated with the CC. Though they constitute about a quarter of all recognized cetacean species worldwide, they are the rarest of whales, and new species continue to be discovered. The family presently consists of several genera, four of which are associated with the CC: Ziphius, Berardius, Indopacetus, and Mesoplodon. Most species of beaked whales fall into the genus Mesoplodon, distinguished by a compressed body and smaller size [4 to 6 m (13 to 20 ft) long]. Separation to species is extremely difficult, and often left to genetic analyses, although the scarring caused by male tusks can aid in identification. Baird's Beaked Whale (Berardius) is the largest species of the family [11 to 12 m (36 to 39 ft) long], and Cuvier's Beaked Whale (Ziphius), with a shorter beak and a slight "smile," is one of the deepest divers.

Of all the marine mammal families, Delphinidae are the most diverse, occurring in all the oceans of the world in around 36 species, nearly half of which (15) are associated with the CC. Identification to genera is somewhat muddled because of recent genetic analyses, and there will be taxonomic revisions in the future. Based on morphology, though, the family is split into two groups in the Northern Hemisphere, the small beaked dolphins (fig. 47) and the "blackfish" whales (fig. 48). The 10 species of small beaked

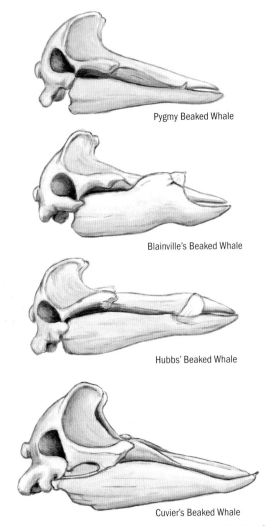

Figure 46. Comparison of beaked whale skulls, showing diversity of shapes and shape of teeth of males. Not to scale.

Figure 47. Small beaked member of the Delphinidae family, Short-beaked Common Dolphin.

dolphins possess an array of beak and dorsal fin shapes and coloration; the Northern Right Whale Dolphin is the only species that lacks a dorsal fin. Small beaked dolphins can be further grouped by habitat and habits. Pacific White-sided Dolphins are strongly associated with cold, temperate ocean waters, whereas Long-

Figure 48. Blackfish member of the Delphinidae family, Short-finned Pilot Whale.

Figure 49. Member of the Phocoenidae family, Dall's Porpoise, a fast swimmer, with rooster tail.

Figure 50. A school of Short-beaked Common Dolphins.

beaked Common Dolphins occur in both warm tropical and temperate waters. Five species are grouped under blackfish, including the well-known Killer Whale. In addition to their blunt-shaped heads, blackfish are generally larger than short-beaked dolphins and are mostly dark-colored. Risso's Dolphins sometimes are grouped with short-beaked dolphins, but their blunt head shape and genetics align them more with blackfish.

There are only six species of porpoises worldwide that constitute the family Phocoenidae, or "seallike ones." Porpoises are small, stout cetaceans that evolved in the North Pacific (fig. 49). Their blunt jaws rest beneath their melons, giving them a characteristic round-headed look, which distinguishes them from dolphins (fig. 50). Their flattened, spadelike teeth are unlike the conical teeth of dolphins, and they all have a short triangular dorsal fin. The two species in the CC share the chunky porpoise form but seem complete opposites in personality. The inshore Harbor Porpoise has somber markings and habitually evades boats, whereas the mostly offshore Dall's Porpoise has distinctive marks and exuberantly bow-rides boats and whales, providing excellent opportunities for close observation.

NORTH PACIFIC RIGHT WHALE *Eubalaena japonica*
Pls. 1, 2; Fig. 51.

> Adult male average length is 13.7 m (45 ft). Adult female average length is 14 m (46 ft). Average weight for both is 50,000 kg (55 tons). Calf average length is 4 to 6 m (13 to 20 ft).

SCIENTIFIC NAMES: The North Pacific Right Whale (fig. 51) is one of several members of the genus *Eubalaena,* of the Balaenidae family. Naturalist John E. Gray provided the genus name for all right whales—*Eubalaena,* Latin for "true whale"—but taxonomy for the species has been unclear for several decades. The North Pacific population was first given a separate species status in 1818, by French naturalist Bernard Germain de Lacepede, who coined the name *Eubalaena japonica.* North Pacific Right Whale populations of the Northern Hemisphere, though, were usually lumped into one species under the name *Eubalaena gracialis,* a name provided by the naturalist J. Muller in 1776.

Figure 51. North Pacific Right Whale.

Recent molecular analyses have confirmed that the populations represent two distinct species, and that they separated around 1 to 2 mya. *Balaena sieboldii* is a recent synonym for North Pacific Right Whale that was withdrawn in favor of the earlier taxonomic name.

COMMON NAMES: Common names coined by various whalers and biologists include True Whale, Right Whale, Black Whale, Black Right Whale, and, recently, North Pacific Right Whale. In early accounts, the name Right Whale was loosely applied to both right whales and Bowhead Whales.

DESCRIPTION: Whales in the right whale group are possessed of robust size and enormous mouth unrivaled by any other whale. The head is huge—up to one-third the body length. The back is very broad and lacks a dorsal fin. Right whales can reach remarkable lengths, and females are larger than males. The largest female recorded was nearly 18 m (60 ft), and the heaviest was 106,500 kg (117 tons). Their girth can reach 60 to 70 percent of the length of the body, with blubber up to 25 cm (9 in.) thick.

The head is large to accommodate a mouth that is radically different from that of all other baleen whales except for Bowhead Whales. The mouth is formed by an upper jaw that is extremely arched and narrow, in contrast to the exceptionally wide, bowed lower jaw. The whale lacks throat pleats to expand the gape; instead, the bowing of the lower jaw accommodates the numerous (about 225 to 270) long baleen plates suspended from the highly arched upper jaw. The length of the dark gray or black baleen is unique among Balaenidae whales: right whales' baleen reaches up to 3 m (9 ft), averaging 2 m (7 ft); only the Bowhead Whale's is longer. The baleen has extremely long and fine bristles to ensnare the smallest of prey. Their very large lower lips are unique among baleen whales, too, and extend far above the lower jaw. The upper border of the lower lip at the peak of the arch often has serrations rather than smooth edging.

Their uniform dark black color is sometimes disrupted by white patches on the lower flanks and belly and, around the face, by many whitish "callosities." Callosities are unique to right whales and consist of hardened skin that is colonized by cyamids, commonly called "whale lice" (although they are small crustaceans, not true lice). The cyamids add prominent shades of pink, orange, and yellow against the black. The horny skin is like a callus, and patches are distributed around the two nostrils, along the edge of the chin and snout, and as islands between the blowhole guard and the tip of the snout. Callosities on the tip of the rostrum are called the "bonnet," a name first applied by whalers. Facial hairs are often but not always associated with the callosities; males tend to have more callosities than females. The hardened skin on newborn calves is gray but is soon invaded by whale lice, which burrow into the skin, increasing the roughness of its surface. Within a few months, the patterning on the rostrum is distinctive enough to identify not only the species but also individuals.

The flippers are large and blunt, about 10 percent of the body length (1.7 m; 5.6 ft) and well adapted for turning quickly, like a paddle for a canoe. The flukes are up to 40 percent of the whale's body length and deeply notched, with a smooth trailing edge. When the tail is lifted, the tail stock can also be identified as wide and thick. The two blow holes are separated, so the blow is often distinctly divided, forming a wide, broad V up to 5 m (16 ft) tall.

FIELD MARKS AND SIMILAR SPECIES: North Pacific Right Whales are so unusual in their appearance that there is little chance of confusion with other large whales in the California Current. Key features include a large, broad, uniformly black body; the presence of callosities around a large head; and the lack of a dorsal fin or throat grooves. North Pacific Right Whales may be confused with Gray Whales, which also lack a dorsal fin and have thick infestations of whale lice; however, Gray Whales are a mottled gray, have a smaller head with no callosities, and have pronounced bumps along the peduncle. No other whales have callosities. When the North Pacific Right Whale's large, notched flukes are raised, they are usually all-black. The tail stock is thick and wide from all angles, whereas for the rorquals, the tail stock is narrow when observed straight on. The humpback is the only other whale with a V-shaped blow, but the North Pacific Right Whale's blow is much wider [4 to 8 m (13 to 26 ft)] and is directed slightly forward.

NATURAL HISTORY: The North Pacific Right Whale is the most endangered of all the large whales, with less than 50 believed to be in existence in the eastern North Pacific. Little is known about their habits in the Pacific, and much of our understanding of them is based on studies of the closely related Northern Atlantic Right Whale and Southern Right Whale.

A review of the North Pacific Right Whale's morphology gives some insight into the lifestyle of this strange-looking whale. Its large size and thick blubber enable it to remain in high subpolar waters for long periods, but those features also prevent it from swimming fast. North Pacific Right Whales are slow-swimming whales, reaching top speeds of only 8 kph (5 mph), in sharp contrast with the greyhounds of the sea, Fin Whales and Sei Whales. They also spend a lot of time at or near the surface, although they are able to dive up to 305 m (1,000 ft) deep and remain submerged for 20 minutes. During these dives, they are pursuing zooplankton in patches dense enough to sustain their large bodies, a size contrast likened to humans feeding on bacteria. To ensnare the enormous amounts of small-sized prey, right whales and Bowhead Whales have the longest and broadest net of baleen of any whales. Rather than gulping or lunging for fish prey, they merely open their mouth and swim straight through dense patches of crustaceans. The slow-swimming prey become caught in the baleen as water flows through the whale's mouth. The whale then swabs the krill from its baleen with its large tongue, and swallows.

This simple description belies the complicated evolution of such a strange strategy for feeding. Copepods are small, so right whales need to filter enormous amounts of water to get a meal. They do so by swimming through dense swarms of copepods with large amounts of water flowing continuously through their open mouths. As the whale swims in a straight line, more water flows along the outside edge of the baleen than into the center of the mouth, in part because the water is channeled against the outer baleen by the massive lips that fan out to the side. The relatively faster flow of water on the outer edge of the baleen creates suction at the rear of the mouth, which draws the water all the way through the mouth—and through the entangling baleen.

How the whales find crustacean patches sufficiently large to sustain them is still a mystery. One theory suggests that they return to feed to a general area first, then to a specific area where they were successful in previous years, and that they then seek out prey patches that are large and dense, with the highest energetic value. One study of the Atlantic right whales examined the baleen's food-filtering efficiency by sampling the zooplankton in the path of feeding whales; the researchers discovered that 95 percent of the zooplankton met a minimum size limit. Indeed, researchers in the Atlantic can predict where right whales will forage for a season based on the density of zooplankton early in the year. But although humans can detect prey density with satellite images, how do right whales do so? They seem to quickly find prey and dive to the depths where concentrations and caloric content are highest. Some researchers suggest that to aid in detecting differences in patch density, the vibrissae around the mouth may be a sensory organ enabling the whale to sense changes in water current speed or temperature, which in turn limit where their prey occur.

Whatever the method, North Pacific Right Whales may have many competitors, from Sei Whales to fin fish, for this superabundant food. Although Sei Whales are also rare, fin fish may be abundant and may outcompete right whales for copepods. Such competition may explain in part why right whales are not recovering after the moratorium on whaling. There is no direct evidence, though, of interference competition by Sei Whales or fin fish.

Another reason for their nonrecovery could be the comparatively late onset of sexual maturity of females when compared to that of other baleen whales. From analyzing hunted whales, biologists determined that females were sexually mature from ages 5 to 15 (the age of sexual maturity of males is unknown). Whales are considered physically mature at 25 to 30 years of age, and live for up to 60 years; a maximum age of 70 years has been documented in one Atlantic whale.

No matter their age, groups are usually small, consisting of one to three animals, and the mother–calf pair bond is long-term. Groups of up to 30 have been observed in the Atlantic species, but the largest group recently seen of North Pacific Right Whales was 17, on feeding grounds in Alaska.

For such a rotund species, North Pacific Right Whales are remarkably nimble. They can lift their bodies almost completely out of the water in breaches, turn in a tight radius, and vigorously slap their flukes and flippers on the surface. Tail "sailing" has been described in the Southern Hemisphere species: the broad tail is lifted out of the water and functions like a sail, with wind pushing the whale through the water; however, this surface behavior has not been seen in the North Pacific species.

North Pacific Right Whales make a variety of sounds, but they do not sing. Rather, they produce low-frequency groans and belches, with the tone of their calls often sweeping upward. The calls have not been heard when whales are feeding (as if used for echolocation) but are likely used for communication. This theory is reinforced by data collected using remote acoustic buoys (sonobuoys), which have detected whale calls occurring together within short periods. Not only can researchers use the sonobuoys to detect the presence and communication of whales by recording whale sounds, they can also determine the direction and distance of the whale call and, from this, calculate which direction a whale is moving.

Recent tagging studies have provided some understanding of North Pacific Right Whale behavior. In a study of the Atlantic species, researchers attached radio tags with suction cups to 28 whales. The suction cups remained attached for only 30 minutes, but the tags measured the length and depth of the whale dives, as well as the temperature of the water by depth. The average dive time was 12 minutes, but the longest dives lasted up to 16 minutes. The average dive depth was 121 m (397 ft), but whales also swam down to 174 m (571 ft), at sluggish speeds of 1 to 2 km/hr (0.6 to 1.2 mi/hr). A surprising discovery was that pregnant females and females with calves spent much more time on the surface to recover from their dives; this disproportionate time exposed might suggest the females' greater vulnerability to being hit by passing ships.

REPRODUCTION: Whalers and scientists alike have speculated about the location of this species' calving grounds. In the Atlantic species of right whale, calving occurs in coastal waters in December through April, but historic records of right whale calving do not exist in the eastern Pacific. The prevailing theory is that the calving grounds were never coastal, because there is almost no evidence of right whale calves along the coast and there are extremely few records of adult right whales in Native American middens of the eastern Pacific (Brownell et al. 2001). Instead, most researchers believe that they calve offshore, but where remains a mystery.

Gestation is longer than 1 year (357 to 396 days), which causes females to calve every 3 years on average; the interval between calving can last up to 7 years. Such a long time between births accommodates 1 year of gestation, 1 year of nursing, and 1 year of recovery in weight to sustain the next pregnancy. The exact weaning period is not known, but ranges from 8 to 17 months and may be up to 2 years, the range noted in the Atlantic species.

Calving has never been documented in the North Pacific Right Whale (and rarely in the other species of right whales). Recent sightings of cow–calf pairs in the North Pacific were in 2006, on summer feeding grounds. In other species of right whale, after calving females lead their young to subarctic latitudes to feed, calves return to the same feeding areas in the following years. Presumably this lesson learned from their mothers about prime foraging areas is important to the survival of young whales.

In the Bering Sea, the species has been observed courting but has not been confirmed mating. In the Atlantic species, courtship can occur at any time of the year. Right whales' testicles are the largest of any mammal [they weigh around 500 kg (1,100 lbs) each]; as is also true for Gray Whales. The "gladiators" are males that produce more sperm than their competitors. A female ready to mate moans to alert males from miles around, and males then converge, pushing and shoving each other, to maneuver next to her for a brief encounter. Some direct aggression between males does happen, when males rub up against rivals with their callosities. The jagged edges of the callosities are likely painful to the sensitive-skinned cetacean, and the behavior causes immediate and strong avoidance by rivals. Although females mate with many males, the male with the most sperm has a competitive advantage.

FOOD: One of the largest whales of the world feeds exclusively on one of the smallest organisms, calanoid copepods. Even the largest copepods, *Neocalanus cristatus,* are only the size of a grain of rice—7.5 mm (0.3 in.) in length with a body weight of 1 mg or more—and the right whale's main prey, *Calanus marshallae,* is even smaller. In the North Pacific, right whales have also been seen feeding on krill *Euphausia pacifica* and *Thysanoessa raschii,* and likely also on pteropods, planktonic snails. In the Atlantic species, right whales have been noted feeding on large concentrations of salps, a gelatinous invertebrate.

The whales skim the weak-swimming prey, both at the surface and at depth, during both day and night. In an Atlantic right whale study, the whale dive pattern followed the diurnal movements of the copepods, but the same study discovered that the whales also targeted large concentrations of copepods in diapause. Diapause is a suspended state between stages of life (the fifth copepodite and adult form), used by copepods when their food is scarce. To remain in this suspended state, copepods have a large oil sac, which also makes them relatively high in calories and more appealing to hungry whales. Consequently, whales feeding at night will bypass concentrations of copepods not in diapause at the surface and instead will dive down to 150 m (492 ft) to feed on masses of plumper copepods in diapause.

Because their prey is very small, right whales require food in large concentrations to achieve their energy needs. By watching whales and counting crustaceans, the researchers estimated that a whale fed on a density of some 3,600 copepods per m^3, and that whales may ignore prey concentrations lower than 1,300/m^3. Within a day, a whale might feed on a patch many kilometers across. An individual North Pacific Right Whale's daily average consumption has been estimated at 4.07×10^5 Kcal/day.

MORTALITY: Right whales of the Pacific are so rare that currently there are no known natural sources of mortality for them. Killer Whales are an obvious predator of such a slow-moving whale, and in other oceans, right whales respond to Killer Whales by forming a circle with tails pointed out in a rosette pattern. In this formation, the whales expose less of their vulnerable parts, and tail slapping can inflict injury on the predators.

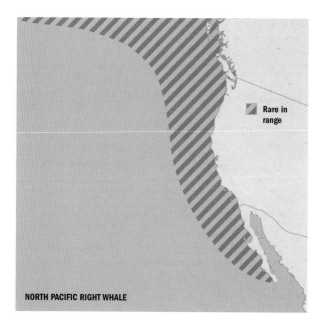

Rare in range

NORTH PACIFIC RIGHT WHALE

DISTRIBUTION: Before whaling decimated right whale populations, the species was documented to range broadly throughout the North Pacific, north into the Bering Sea and south to at least latitude 35° N around Monterey Bay. Within this range, North Pacific Right Whales migrate annually between feeding and calving grounds. Unlike other species of right whales, the North Pacific species does not appear to migrate to nearshore calving grounds. Early whaling records and sightings show a seasonal presence of right whales in the Gulf of Alaska and the southeastern Bering Sea from March through September, and absence from the region from October through February. This winter period of the whales' life remains a mystery.

Although there were a few sightings of North Pacific Right Whales along the coast of California during the 1980s and 1990s, there have been no sightings south of Canada in the last decade. One satellite-tagged North Pacific Right Whale exemplifies the species' offshore but restricted movements; its own were confined just to the southeast Bering Sea for over 40 days. From spring to fall, these whales reside in temperate and subarctic waters in the southeastern Bering Sea and in the Gulf of Alaska south of Kodiak Island, feeding on zooplankton. Between 1996 and 2009, researchers annually monitored whales in these areas using aerial surveys, as well as acoustic recorders left in the water to detect the whales' presence. The acoustic sonobuoys captured most North Pacific Right Whale calls from May to November, and a few as late as December.

As light levels decrease in fall, the availability of food in the subarctic waters decreases, and the whales migrate south to destinations unknown. A handful of confirmed sightings have been recorded over the past several decades in the nearshore waters of the eastern Pacific, including off San Clemente Island, California; off the Farallon Islands and Half Moon Bay of Central California; and in Washington State. One sighting was as far south as latitude 20° N, just 23 km (15 mi) south of Baja, and another as far south as Hawaii, in the open ocean.

HABITAT: Although right whales may range widely throughout the North Pacific, they prefer cool temperate waters with depths of 100 to 225 m (330 to 660 ft). In Alaska, they feed over the middle shelf, shelf break, and thermal fronts. Presumably the shallow shelf waters create places where prey are concentrated in swarms dense enough to attract the whales. Deep-water oceanic gyres might also provide attractive sources for dense patches of prey. For example, copepods are most abundant along the continental shelf and slope where currents concentrate them, but copepods also are passively aggregated in major mixing zones offshore and around fronts and eddies that form near bottom features such as seamounts.

VIEWING: Despite these fairly predictable habitat associations, to encounter a North Pacific Right Whale is an exceedingly rare event within the California Current. Nevertheless, almost all of the sightings during the last 30 years south of Canada have been made very close to shore, often by shore-based observers or recreational whale-watchers. They are often associated with Humpback Whales, so try to identify every individual whale in a group of whales. If you see a North Pacific Right Whale, immediately contact the nearest NMFS office.

CONSERVATION: Right whales were first hunted in the eastern Atlantic Ocean during the 11th century, by Basque whalers;

were hunted in North America during the 17th to 19th centuries by early European settlers; and were hunted in the Pacific by Japanese whalers during the 16th century and European whalers during the early 1800s.

Agreement	Conservation Status
U.S. ESA	Endangered with critical habitat
U.S. MMPA	Depleted
COSEWIC	Endangered
IUCN	Endangered

Early whalers targeted the "right" whales because they were slow-moving and, when killed, usually floated because of their large blubber layer, in contrast to other faster-swimming species that generally sank when killed. The purpose of the hunt varied over time, initially being to obtain the whales' meat and later, the oil derived from the blubber, and sometimes the baleen. The Aleuts hunted small numbers of right whales along the Aleutians; other aboriginal tribes along both coasts of the Pacific hunted right whales—though rarely, even when they hunted other species of whales. Not until commercial whalers arrived in the 19th and 20th centuries were right whales hunted intensively. In a single decade, between 1840 and 1850, as many as 30,000 North Pacific Right Whales, including 19,000 in the eastern Pacific, were killed. The number of whales hunted was underreported for years, because some two to three whales were lost to each one eventually landed by whaling ships. During 12 whaling expeditions from 1838 to 1860, 327 North Pacific Right Whales were injured or killed, but less than half (133) were recovered.

The first North Pacific Right Whale was slaughtered in 1835 in the Gulf of Alaska, by a French whaler, and by the 1870s, the species was considered rare. By 1937, there was a ban on whaling all species of right whales, but Japan and Russia were not parties to the agreement. By 1951, both Russia and Japan joined the IWC but legally hunted whales under scientific permits for more than a decade afterward. These scientific collections included five females, a couple of which were pregnant. Russian whalers also illegally hunted right whales from the 1950s through the early 1970s, killing some 508 whales (Clapham et al. 2004).

The combined effects of the scientific collections and the illegal hunting were devastating to the North Pacific Right Whale species, and particularly to the eastern Pacific population.

The North Pacific Right Whale was first listed as endangered under the Endangered Species Conservation Act of 1970; in 2008, after identification as a distinct species, the North Pacific Right Whale was listed separately as an endangered species. Critical habitat also was designated in a small area of the Gulf of Alaska, near Kodiak Island, and in a larger area in the southeast Bering Sea, based on consistent sightings of whales feeding there. Right whales are one of only a few marine mammal species with critical habitat identified under the Endangered Species Act.

Despite these protections, continued direct and indirect contact with humans is the most likely reason for right whale species' failing to recover in the Northern Hemisphere; by way of contrast, the Southern Hemisphere species are recovering. Atlantic right whales are often hit by ships, because their feeding grounds, migratory routes, and wintering grounds are all situated in or near major commercial shipping channels. Additionally, the whales swim slowly, near the surface, and females are disproportionately struck by boats because they spend longer times on the surface while resting. In summer, whales can also become entangled in fishing gear; 57 percent of the living right whales in the Atlantic have scars from entanglement in fishing gear. In the Pacific, noise pollution because of seismic exploration for energy may be a much larger deterrent to North Pacific Right Whale recovery.

Right whales may also be especially vulnerable to extreme oscillations in weather or climate that may affect the distribution and concentration of their preferred food. A long-lived species might normally survive such shifts in prey, but the right whale's population is dangerously small, and copepods, the whale's main prey, are particularly sensitive to changes in sea water temperature, salinity, and currents. Invertebrates such as these are also sensitive to ocean acidity, as their exoskeletons dissolve in acidic sea water.

POPULATION: The North Pacific species is currently classified into two stocks—the northeastern Pacific stock and the Sea of Okhotsk stock off Russia. In the 20th century, only 1,956 right whales were seen in the North Pacific, and of these, only 690 came from the northeastern Pacific, associated with the California Current.

In the 21st century, though, the total population estimate for the North Pacific is fewer than 500, with only tens of whales found in the northeastern Pacific. Recently in Alaska, 17 individual whales were observed together in 2004, including three cow–calf pairs, and 24 whales in 2001. In 2007, for the first time, NMFS saw no right whales during their annual surveys. The potential for recovery of the stock in the eastern Pacific is bleak, and most researchers believe that this population will become extinct in the 21st century, when the final ancient members die off.

COMMENTS: When species become extremely rare, unpredictable factors can have catastrophic effects, leading to their extinction. Genetic anomalies may arise in a population because of a genetic bottleneck, and inbreeding may lead to harmful recessive genes' reducing the survival of individuals. Known as the 50/500 rule, a population below 500 individuals is at risk for a loss of genetic variability due to genetic drift, and a population below 50 individuals is at risk of inbreeding. North Pacific Right Whales are on the brink of becoming extinct, per the predictions of the 50/500 rule, despite efforts to protect them from hunting, and despite the protection of critical habitat.

REFERENCES: Baumgartner and Mate 2003, Baumgartner et al. 2006, Brownell et al. 2001, Clapham et al. 2004, Gendron et al. 1999, Kenny 2004, Kenny et al. 2001, Knowlton et al. 1994, Lui et al. 2005, Mayo et al. 2001, Scarff 1986, Scarff 2001, Shelden et al. 2005, Wade et al. 2006.

BLUE WHALE *Balaenoptera musculus*
Pls. 3, 4; Figs. 43, 52.

> Male and female average length is 21 to 24 m (75 to 80 ft); average weight is 90,000 to 136,000 kg (100 to 150 tons). Newborn calf average length is 7 m (23 ft); average weight is 5,400 kg (6 tons).

SCIENTIFIC AND COMMON NAMES: The generic name *Balaenoptera* comes from the Latin word *balaena* for "whale" and the Greek *pteron* for "wing." The specific name *musculus* can mean "muscular" in Latin, but another meaning is "little mouse."

Figure 52. Blue Whale cow and calf.

Linnaeus called the common house mouse *Mus musculus* (*Mus* signifying "mouse," and *musculus* "little mouse"), and probably intended the Blue Whale's species name to have a wry double meaning. Today most biologists frown on puns in scientific nomenclature, but a few slip in, such as the famous mosquito genus *Aiyiyi*. Worldwide four Blue Whale subspecies are recognized, but only one occurs in the North Pacific: *Balaenoptera musculus musculus* (fig. 52). The English common name is Blue Whale, but older names include Great Blue Whale, Great Northern Rorqual, Sibbald's Rorqual, and Sulfur-bottomed Whale. Sir Robert Sibbald (1641 to 1722) was the Scottish physician and antiquary who first described the Blue Whale. The Sulfur-bottomed Whale was the 19th-century sailor's term (the whale can have an orange or yellow cast to its belly when a film of diatoms clings to it). French common names include *rorqual bleu* (Blue Rorqual) and *baleine bleu* (Blue Whale). Spanish terms include *ballena azul* (Blue Whale) and *rorcual azul* (Blue Rorqual).

DESCRIPTION: The Blue Whale is the largest of all mammals—indeed, the largest living animal on Earth—but its elegant, streamlined body belies its great mass. Long, slender, and somewhat tapered, the Blue Whale's body slips smoothly through the water. The whale's head makes up about one-fourth of its body length. The top of the head is wide and flat—almost

U-shaped—and a prominent ridge leads from near the tip of the jaw to a very prominent splash guard just before the blowholes. Some 50 to 90 pleatlike grooves run down the outer wall of the mouth and throat from the tip of the jaw to the navel, allowing the throat to distend outward. The pointed flippers are 3 to 4 m (13 ft) long, and the dorsal fin is far aft, and short for its elongated body—less than 33 cm (1 ft) tall. The dorsal fin is variable in size and shape, ranging from hooked and prominent to a mere hump. The deeply keeled triangular flukes are broad and notched.

The whale is a shade of blue-gray, typically dappled with light or dark spots, and the paler undersides of the head may be yellowish from algal staining. The flippers may be mottled gray lined with white above, and can be mottled gray or white below. The flukes are dark below.

Although its body is built for speed and ease in cruising, the Blue Whale's shape transforms when it feeds. The mouth gapes open as the whale presses forward, and the tongue falls back and down. The pleats in the wall of the mouth and throat widen, and the floor of the mouth drops. Water—as much as 70 tons of it—rushes into this chamber as the whale lunges forward, and the graceful whale now looks more like a "bloated tadpole." The drag and weight of the water-filled mouth slows the whale as it presses ahead. Then the powerful mouth closes, and krill are captured on the approximately 300 paired rows of baleen that hang from the upper jaw. Each plate is about a meter deep at its far edges, and 3 m (10 ft) wide. As this most gigantic of mouths begins to close, the tongue forces the engulfed water out through the black baleen plates, and the pleats close. The 2.7 ton tongue weighs as much as an elephant. The krill are caught on the baleen and swept down the throat. The gulp of the Blue Whale has been called "the largest biomechanical action in the animal kingdom." The sheer volume and weight of the water taken in this gulp outclasses the oral capacity of any other vertebrate.

The bones of the head are specially modified to engulf prey. There are four depressions in the front of the skull that permit the attachment of powerful jaw muscles. The lower jaws themselves are the heaviest of any animal's bones, and can be more than 6 m (20 ft) long. The jaws can rotate outward when opening, since they meet each other in front at a flexible joint. All this permits the whale to eat 1 ton of krill in a feeding bout, and up to 4 tons in a day—more than 40 million individual krill.

According to whalers' logs, the longest Blue Whale measured 33 m (110 feet). The maximum measurement recorded in modern times was 29 m (95 ft), for a Southern Hemisphere female. Whales in the Northern Hemisphere are shorter than those in the Southern Hemisphere. Additionally, females tend to be longer and weigh more than males, and can add 26 percent to their weight when pregnant. The Blue Whale's skeleton alone may weigh 20 tons, a third of that is oil impregnated in the bones. A Blue Whale would smother on land from its own weight, and in water sinks if it does not swim, because its body is heavier than the water it displaces.

FIELD MARKS AND SIMILAR SPECIES: A surfacing Blue Whale might be confused with a Fin Whale or Sei Whale. Fin Whales average only slightly smaller than Blue Whales in the northeastern Pacific but are brownish gray above and lighter below, not darker below like a Blue Whale, and do not have mottling of spots. Fin Whales have light coloration on the undersides of the flukes (apparent rarely when they dive), whereas Blue Whales have darkly colored undersides. The coloration of a Fin Whale's head is strikingly asymmetrical—this extends even to the baleen, which is yellowish on the left but gray on the right—whereas the Blue Whale's head is symmetrical, and its baleen is black. The Fin Whale's head is more V-shaped, in contrast to the Blue Whale's U-shape. The striking vertical columnar blow of a Blue Whale can be 12 m (39 ft) tall, but the Fin Whale's is shorter, from 4 to 8 m (13 to 26 ft). During a long low roll at the surface, the Blue Whale's blow is followed by long dorsal and sometimes flukes, but the Fin Whale's blow is followed shortly by its backswept dorsal, its roll arches higher, and it does not show flukes. Sei Whales differ from Blue Whales in size, head shape, color, and blow. Although they are the third largest of the rorquals, they are half the size of the Blue Whale. Blue Whales can look turquoise blue under water.

NATURAL HISTORY: Huge as it is, the Blue Whale is today more often heard than seen. Since whaling days, few whales of this species have been spied from the flying bridges of survey ships transecting some quarters of the northern seas. Where once whalers sighted blows, many scientists now listen with hydrophones to locate the small populations that survived hunting. But hydrophones reveal more than Blue Whale numbers: they record distinctive regional accents in songs, discover seasonal movements, and find whales where no one has seen them for decades. Although Blue Whales still are infrequently seen north of California, they can be heard there on occasion. As electronic instrumentation becomes more miniaturized and sophisticated, we are learning more and more about the life of this leviathan.

The hydrophones hint that the world may have as many as nine stocks of Blue Whales. In the north Pacific are two populations: one in the northwest, off Alaska, and another in the eastern Pacific, along the west coast of North America. These Blue Whales are the best studied in the world, in part because of the concentration of universities and marine research institutions along the Pacific coast. Each population of whales has distinctive songs. Only the higher harmonics of the calls would be audible to our unaided ears if we merely listened on a hydrophone, but when we speed up recorded songs, we can hear the fundamentals that fall below 20 Hz (cycles per second), the lower limit of human hearing. These songs are simpler than those of Humpback Whales, and have often been described as a series of moans.

Eastern Pacific Blue Whales have unique "A" and "B" elements in their songs, sometimes called trills or moans. Males sing in series starting with an A element, or in a series of Bs. Blue Whales can sing briefly, or for days. Because only males are known to sing, the songs may be meant to attract females. On the other hand, males sing year-round, so the songs may serve more than one function. Blue Whales also call in shorter, variable sounds that change in frequency—downsweeps, upsweeps, or reversing low tones. Were these tones not low, they might be reminiscent of toothed whale calls. The downsweeps are made by females as well, and have been linked to feeding bouts. These calls might serve to draw other whales to krill swarms, but their exact significance is uncertain. Blue Whales make A and B calls more frequently at twilight and night, and when their prey is most abundant nearby, and can be heard over great distances—perhaps as far as 20 miles. The result is consistent with the concept of the "range herd"—whales' (and other large animals') staying in acoustic contact over long distances.

Blue Whales have no obvious social structure or schooling behavior. They are seen alone or in small groups of two to three; rarely, larger groups of 19 to 80 are seen in the California Current. Because their lifespan may be 50 to 70 years, social interactions are likely more complicated than can be gleaned from merely listening to their voices.

REPRODUCTION: Blue Whales are sexually mature at 7 to 12 years. Females calve about every 2 to 3 years. Mating likely occurs in winter in warmer waters off Baja and Central America, and after a gestation period lasting 1 year, the largest calf of any mammal is born, weighing up to 6 tons—as much as an adult hippopotamus.

Calves increase in weight by around 90 kg (200 lbs) per day, and after about 7 months of nursing, they are weaned on the feeding grounds. When weaned, at 4 to 10 months, calves weigh 20 tons. While nursing, calves drink 380 L (100 gal) of milk a day, and gain 3.8 kg (8.4 lbs) an hour. When weaned, the calf is a whopping 16 m (53 ft) long, the length of an adult Humpback Whale. The female then migrates south to mate in winter, and the cycle begins again, so that every 2 to 3 years a female bears a new calf.

FOOD: Blue Whales, the largest of mammals, feed almost entirely on one of the smaller invertebrates of the ocean, krill. Four species of krill dominate their diet in the Pacific: *Euphausia pacifica, Thysanoessa inermis, T. lonipes,* and *T. spinifera.* Blue Whales also eat copepods, amphipods, and squid. Along the coast of Baja, pelagic red crabs were also reported as being eaten. Small schooling fish such as anchovies are consumed, but fish are a small part of their diet when compared to the role they play in the diets of Humpback Whales and Fin Whales.

Blue Whales are oceanic wanderers that usually roam and feed singly or in pairs; greater numbers sometimes gather at feeding grounds. One theory is that they feed in small groups because they eat diminutive krill rather than fish, and so require much greater masses of prey to sustain each individual. They migrate in twos and threes, while fasting.

Their feeding dives are short and shallow, a surprise in light of their great size. During dives, a whale glides downward and then lunges vigorously upward perhaps 30 m (100 ft), from once to several times during a single dive. A whale feeding this way accelerates as its flukes drive it forward but slows and stops as its mouth fills with water. Sometimes Blue Whales lunge all the way up to the surface, their jaws breaking water.

From tagged whales, scientists are able to track the foraging dives of Blue Whales, and have noted that they lunge after swarms of krill at a rate of up to six times per dive, but that lunges take so up much energy that dive time is limited, averaging only 4 to

8 minutes. Blue Whales focus on swarms of krill at depths of up 250 to 300 m (820 to 980 ft), but many dives are shallow, at less than 16 m (52 ft). As with other baleen whales, dive depths become shallower at dusk and deeper during the day, following the deep-scattering layer of food. To maintain this large a body, Blue Whales consume up to 2 to 4 tons of food per day.

MORTALITY: Blue Whales are too large for most predators, but Killer Whales have attacked them. One of the two cases reported occurred on the west coast of North America, off the tip of Baja. A frenzy in the calm waters drew the attention of researchers who saw an 18 m (60 ft) Blue Whale trying to escape a seemingly coordinated attack by 30 Killer Whales. The Killer Whales ripped off strips of skin and blubber as the young whale left a trail of blood in the sea. The attack continued for at least 5 hours, after which the Killer Whales suddenly vanished before nightfall, leaving the Blue Whale maimed and possibly mortally wounded.

Blue and other baleen whale populations, according to one theory, evolved or learned to migrate to subtropical waters to calve and escape Killer Whales, since transient Killer Whales are most frequently reported in temperate or subarctic seas. Most attacks by Killer Whales on mysticetes seem to be on young whales. Internal parasites are rare in Blue Whales, likely because they eschew eating fish, which harbor many common parasites found in most fish-eating marine mammals.

DISTRIBUTION: By using hydrophones to listen to the waters along the eastern Pacific Rim, researchers have learned that the eastern Pacific Blue Whales range from the Chukchi and Bering Seas in Alaska in summer down to Panama in Central America in winter. In the Gulf of Alaska, calls of both the northwestern Alaskan and eastern Pacific populations can be heard, which indicates that some Blue Whales of both north Pacific stocks seasonally share these cold temperate waters. Associated with the Costa Rican Dome, a large semistationary bulging eddy off Central America, Blue Whales feed in the cool upwelling waters rich in krill year-round, but they are more abundant there in winter, January through March.

Though it is known that Blue Whales migrate south in winter to subtropical waters and to temperate waters in summer and fall, their migratory routes are not well mapped, because of the nomadic movements of individuals. During the northward

BLUE WHALE

Range

migration, whales apparently travel far offshore, but they are closer to the central California coast when they migrate south. Peak numbers of whale sightings off Baja show a fall migration south and a spring migration north through the region, with peak numbers in April to June. Off California, sightings are made seasonally between June and December in the Southern California Bight and between May and November in northern and central California. From photo-identification records, researchers have identified individual whales that congregate in the warm waters of Baja and then migrate to California, and occasionally to Oregon, to feed.

Blue Whales usually cruise across the vast ocean at about 20 kph (12 mph) but can sprint up to 48 kph (30 mph). When sounding, they will perform 10 to 20 shallow dives followed by a deep dive of 10 to 30 minutes.

HABITAT: Depth ranges of water where Blue Whales are seen are from 80 to 3700 m (270 to 1200 ft), but mostly they congregate in upwelling regions, such as off northern and central California and the southern Channel Islands. Blue Whale reliance on krill ensures that they occur within 370 km (230 mi) of the continental shelf, and especially where concentrations of krill are high downstream from upwelling areas and around islands and seamounts.

VIEWING: Blue Whales gather in areas with intense and predictable upwelling, such as the Costa Rica Dome, the Southern California Bight, Santa Barbara Channel, the edge of Monterey Canyon, the Farallon Islands, and Cordell Bank to the west of San Francisco. Late summer and fall are the best times for krill and their cetacean grazers.

CONSERVATION: In the 19th century, the sleek Blue Whale could outrun whalers in stout wooden sailing ships, and thus escaped the fate of the slower Gray Whales and right whales. But by 1900, steam engines instead of sails powered whaling ships, and steel plates replaced oaken planks. Harpoons were no longer thrown by hand but were fired out of rifled cannons, exploding inside the whale's body. Mother ships could retrieve carcasses of the most massive whales on ramps, and crews could dismember bodies on deck, discarding their brains and guts as waste. Blue Whales were highly prized, because one whale could yield 100 barrels of oil; in later years, they were so valued in comparison to other whales that quotas were measured against the Blue Whale as a unit: 1 Blue Whale = 2 Fin Whales = 6 Sei Whales.

Agreement	Conservation Status
U.S. ESA	Endangered
U.S. MMPA	Depleted
COSEWIC	Endangered
IUCN	Endangered

For 60 years, mechanized whaling ships hunted down the greatest mammal on Earth, and by the 1960s roughly 10,000 of the original 300,000 or so Blue Whales were left worldwide. In the Southern Hemisphere, where most Blue Whales had lived, more than 97 percent of the population perished. Between 1913 and 1937, a reported 1,378 Blue Whales were killed off California and Baja, with a peak take of 239 in 1926. Blue Whale populations may have been the least hunted of all the great whales;

although after their full protection in 1966, there is little evidence for recent increases in their numbers.

Modern-day conservation issues for Blue Whales include boat strikes, low-frequency sounds, and climate change. At least nine Blue Whales have been struck by ships in Southern California since 1975, three of them having been killed in September 2007. Since Blue Whales sink when killed, there have probably been more unrecognized fatal strikes along the west coast of North America.

The oceans have become considerably noisier since the advent of steamship and sea traffic continues to grow, with increasingly longer vessels plying the coast. Most of the mechanical sounds that vessels make are low—under 1000 Hz—and intense. Seismic explorations in the sea and naval sonar have added to the undersea hubbub. Most baleen whales' signals are less than 100 Hz, and the background noise might limit their ability to hear one another.

Global warming may ultimately disrupt ocean currents and upwelling areas, with unforeseeable consequences for Blue Whales that rely on krill. In 2006, a small seabird called the Cassin's Auklet that feeds almost exclusively on krill failed to reproduce in response to the absence of krill in central California; few Blue Whales were seen that year. The presence and productivity of krill, auklets, and Blue Whales are sentinels of the sea, providing warning of changes in ocean conditions.

POPULATION: Surveys have produced an estimate of some 1,400 Blue Whales off Baja, California, Oregon, and Washington. In Canada, the estimated population has fallen to under 250. Having this many Blue Whales along the west coast of North America is heartening, although it is unclear whether this represents an increase in the regional population or immigration from other areas. Worldwide, Blue Whales have never recovered to prewhaling levels.

COMMENTS: In the course of 60 years, the Blue Whale was almost eliminated from Earth, and the ecology of the seas has been altered with the virtual absence of this mighty consumer of krill. In the north Pacific, all the great whales originally may have consumed over 60 percent of the total primary marine biological production, but with their decimation, their current consumption may have dropped to 8 percent. The consequences of this depopulation of great whales are unclear. Killer Whales, which preyed on baleen whales, may have turned to other sea mammals for food; this may in part explain the dramatic decline of Alaskan pinniped and Sea Otter populations (though this theory is controversial). In Antarctica, the 250,000 or so Blue Whales that were taken may have been replaced, in part, by krill-consuming Common Minke Whales, seals, and penguins.

REFERENCES: Bortolotti 2008, Calambokidis et al. 2008, Lagerquist et al. 2000, Oleson et al. 2007, Sears 2002, Sears and Calambokidis 2002.

FIN WHALE   *Balaenoptera physalus*
Pls. 3, 4; Figs. 7, 53.

> Adult male average length is 18 m (59 ft). Adult female average length is 19 m (61 ft). Average weight for both is 45,000 to 64,000 kg (50 to 70 tons). Calf average length is 6 to 7 m (20 to 23 ft); average weight is 1,800 kg (2 tons).

SCIENTIFIC NAMES: The species name *physalus* comes from the Greek root for blow or bellows. For derivation of *Balaenoptera*, see Blue Whale species account. The whales of the northern and southern hemisphere are genetically isolated, forming two separate subspecies, and that of the northern hemisphere is named *Balaenoptera physalus physalus* (fig. 53).

COMMON NAMES: Common English names include Fin Whale, Finner Whale, and Finback Whale, all of which refer to the prominent fin often visible when the whale surfaces. The alternate name razorback is suggested for the distinct tail stock ridge that runs between the dorsal fin and the flukes. The Common Rorqual is another English term, which is equivalent to the French name *rorqual commun*, and to the Spanish *rorcual commun*. Similarly, the English Fin Whale is the translation of the French term *baleine fin* and the Spanish *ballena aleta*.

DESCRIPTION: The Fin Whale is the second largest whale, mammal, and animal on Earth. It reaches a length of 26 m (88 ft) and a weight of 80 tons, and like all baleen whales, it is sexually dimorphic, with females being perhaps 5 to 10 percent

Figure 53. Fin Whale with a pale right jaw.

longer than males. Southern hemisphere whales average longer and larger than northern fins, reaching 27 m (89 ft) in females.

Despite its mass, the Fin Whale has been called the "greyhound of the sea." They are among the fastest whales in the world, sustaining speeds of 37 kph (23 mph), and sometimes reaching 40 kph (25 mph) when alarmed, although no one has ever organized a formal race between cetaceans. The Fin Whale is slender, with a girth that is between 40 and 50 percent of its length. Its head is flat and V-shaped, with a central ridge sweeping back toward a splash guard around the twin blowholes. Its flippers are tapered and relatively short, allowing it to ease its way through the water. Its notched flukes are broad but narrow, which lessens turbulent drag behind it as the whale powers ahead.

The bodies of cetaceans are often more asymmetrical than other mammal's. The striking color asymmetry of Fin Whales represents an extreme example in vertebrates. The Fin Whale's lower jaw, which takes up one-quarter to one-fifth of its length, is black or gray on its left, whereas its right side is creamy. The inside of the mouth and baleen on the left side is also dark, whereas the inside of the mouth is white on the right and the tip of the baleen on the right is yellowish. However, this pattern is reversed on its tongue. The rest of the Fin Whale's coloration is more conventional. The body is gray or gray-brown above and whitish below, as are its flippers and flukes. Delicate chevrons of lighter color run behind the blowholes toward the tail, and then sometimes reverse, returning toward the head. The chevrons may be more prominent on the right side. Blazes or patterns of stripes may originate near the ear and flipper base. Distinctive chevrons, blazes, scars, and fluke shapes can serve to identify individual Fin Whales.

The namesake dorsal fin of this whale is slightly curved, about two-thirds of the way down the body. It joins the back at a low angle and can reach 66 cm (26 in.) high. Between 260 and 475 pairs of baleen plates rest in the mouth. The plates are keratin, like human fingernails, and fray into fine ends. The maximum length of a baleen plate is about 70 cm (28 in.) and the maximum width is 30 cm (1 ft). Running down the underside of the whale from chin to navel are from 50 to 100 throat pleats. These pleats are part of an accordionlike blubber layer, which expands as the Fin Whale feeds.

FIELD MARKS AND SIMILAR SPECIES: On the west coast of North America, Fin Whales are most readily confused with their near relatives, Blue Whales and Sei Whales. The Blue Whale is the largest animal on Earth, but the Fin Whale is a close second, making it hard to tell the two greatest cetaceans apart based on size. The head of the Fin Whale is V-shaped, whereas that of the Blue Whale is U-shaped. The Blue Whale is typically a mottled blue and is dark below, whereas the fin is light below and solidly silver or brownish gray above. The Fin Whale's blow is lower, and its head tends to be tilted upward, whereas the blue blows higher with its head on the level with the sea. The bigger dorsal fin of the Fin Whale is about two-thirds back, and appears earlier than that of the Blue Whale. The Blue Whale may raise its tail (fluke) on sounding, whereas the Fin Whale arches its back before diving and rarely flukes.

The asymmetrical coloration of the Fin Whale's face is the best field mark. If you see black on the left lower jaw and white on the right in a large baleen whale, then you most likely have a fin. Occasionally, some individual Common Minke Whales show a similar pattern, but they are much smaller. Sei Whales have no asymmetrical coloration and are often marked with the oval scars caused by cookie-cutter sharks, while Fin Whales typically show far less disfigurement.

NATURAL HISTORY: The natural histories of Fin and Blue Whales are very closely related, and genetically, the Fin Whale is so closely related to the Blue Whale that hybrids occur in nature. Male hybrid offspring are infertile, but females can be fertile, and occasionally hybrids are seen off British Columbia. The similarity between the two species is as close as humans and chimpanzees, and the two species may have separated as recently as 5 mya. Fin Whales, like Blue Whales, can be long-lived, attaining 85 to 90 years of age. The oldest one documented was 114 years old, based on examination of the inner ear plug.

Fin Whales can be solitary, but groups of three to seven are common, and feeding groups of up to 50 whales have been noted rarely. The only known stable bonds are between mothers and calves, which last at least 6 to 8 months, and some mothers and their offspring return to the same foraging areas in different years. Whaling records indicate segregation of the sexes and age groups, but current observations have not confirmed this. They sometimes form mixed schools with Blue Whales, which would present an opportunity for the infrequent hybridization mentioned earlier.

Such large whales cannot dive deep because of blubber buoyancy, but they can dive up to 500 m (1,640 ft) and remain underwater for 25 minutes, which might allow them to access krill resting at depth during the day. Indeed, Fin Whales are more active divers during the day than at night, although their size likely limits their diving abilities. Fin Whales breach, and they use their mass to create very loud songs, as well as loud splashes.

If you could listen below the surface, you might hear a Fin Whale singing as it cruises along the continental break of the west coast of North America. Over and over, the whale repeats a deep series of downsweeping calls, so low in pitch that they drop below the limit of human hearing. The downsweeping moans are sung at regular intervals, perhaps 17 to 40 seconds apart. The regular downsweeps might continue in a song bout of 10 minutes or so and then be resumed later, with the series of bouts lasting for many days. All Fin Whales in the northern hemisphere sing in this kind of repeated pattern. The songs are seasonal, and the singers are all males. We are not sure why male Fin Whales sing, but the overlap of calling with the mating season suggests that the calls may attract mates. Sometimes there is a hint of a duet between singers. The cadence can vary with the interval between pulses in a series longer or shorter. Sometimes the downsweeps sound doubled, which is characteristic of Fin Whale song in some areas, such as the Gulf of California. These distinctive doublets may be part of the song, or they may be the result of the singer's pulses arriving on more than one path at the recording hydrophone. The first pulse of a doublet may in fact be the echo from the sea floor, and the second may be the waterborne pulse. Indeed, it has been suggested that the booming pulses penetrate the softer sediments of the sea bottom to bounce off the crustal rock of the continents. It is as if the whale were using the deep rock as a sounding board. One wonders if the callers themselves can recognize distinctive resonances off islands, like the Farallones and other undersea "seamarks" along their migratory routes. And you can think of silent Fin Whales listening to the passing singer, sizing up his strength and thinking of mating or challenging the wanderer.

REPRODUCTION: Fin Whales in the northern hemisphere reach sexual maturity at 17.5 m (57 ft) in males and 18.5 m (61 ft) in females, and the age of sexual maturity has dropped from 10 years to 6 to 7 after whaling depleted the population. They do not congregate in specific areas to mate or give birth; instead, females likely do so in the open ocean in southern latitudes. Conception occurs in the winter over a 5-month period peaking from December to February, and a calf is born 1 year later. In some regions, there is evidence that reproduction occurs year-round, but off California, calves are only observed in September. The calf is weaned at perhaps 6 or 8 months, when it attains 11 to 12 m (36 to 39 ft) in length, and the female mates again the following winter. Based on this breeding cycle, a female may calve every 2 to 3 years. Twinning has been recorded for Fin Whales in utero, as with a couple of other species of rorqual.

Little is chronicled about the Fin Whale's mating system. Because male testes are proportionately small to their overall size, males likely physically compete for access to females, but such interactions have not been noted.

FOOD: When feeding, the Fin Whale can lunge forward at a speed of about 11 kph (7 mph), and then literally drop its lower jaw to engulf a swarm of prey. The biomechanics of this work is like "opening a parachute at high speed" (Goldbogen et al. 2007). About 68,000 L (18,000 gal) of water can rush in, approximately the volume of a school bus, bigger even than the body size of the whale. As the water is expelled, about 10 kg (22 lbs) of krill or other prey may be entangled on the baleen and swept down the throat. The 6- to 10-second lunges are intensely energetic and limit the number of dives the whale can make. The Fin Whale may lunge perhaps four times during an 8 to 12 m (26 to 39 ft) dive and then surface for three to eight breaths. Data on krill swarm density suggests that about 3 hours of lunging would suffice for the whale's food needs for a day—an average of 1,300 L (295 gal) per day of krill. One researcher attempted to calculate the total annual prey eaten by all the Fin Whales of the northeastern United States, and came up with a number of 602,370,667 kg (664,000 tons)—clearly the biomass processed in the marine food web by this one species may be substantial.

Fin Whales sometimes feed on the surface, turning on their right side to engulf prey, and they have also been seen to circle schools of fish, driving them into a tight ball before lunging. Some scientists believe that the white lower right jaw of the Fin Whale may serve as a lure or prod in concentrating prey.

The food of Fin Whales is very diverse and likely reflects seasonal and local conditions. Food items consist of schooling fish, small squid, and crustaceans such as krill and copepods. Anchovies, herring, Capelin, mackerel, sardines, hake, and sandlance are fish taken in the northern hemisphere. From looking at the stomachs of harvested whales in the northern Pacific, biologists estimated that more than half of the diet consisted of krill, a quarter was copepods, and less than 5 percent was fish. But a study off California found that krill and anchovies were the main diet.

To find these dense swarms of food, some scientists speculate that the whales cue on the bioluminescence from ctenophores. Others have speculated that Fin Whale vocalizations may alert their cohorts of good feeding, when the researchers observed whales converging from vast distances on large concentrations of prey. However, once converged, there is no apparent cooperation in feeding except to "stay out of each other's way."

MORTALITY: Killer Whales attack Fin Whales, but direct observations are rare. More frequently, bite marks are reported on flanks, fins, and flukes. Flight seems to be their basic defense, and that of other streamlined baleen whales—a defense that would suit fast swimmers. A giant nematode, *Crassicauda boopis,* is the only other documented nonhuman potential cause of death in

Fin Whales. This pathogen inflames the arteries of the kidneys, eventually leading to kidney failure.

DISTRIBUTION: Fin Whales are widely distributed in the world's open oceans between latitudes of 20 to 75° N, but they are not found near the polar ice packs or in a few peripheral areas such as the Red and Black seas. As with many large whales, Fin Whales migrate annually between winter breeding and calving grounds and summer feeding grounds. Migration between high and low latitudes occurs in the open ocean, so routes and timing are not well mapped. Pregnant females tend to arrive first at summer feeding areas, but the location of winter breeding and calving is not known, so knowledge about timing is not available. How the whales navigate in the open ocean is not known for sure, but the routes appear to follow low geomagnetic gradients. Their ability to detect magnetic fields with crystals of magnetite in their brains (called "biomagnetism," also observed in other migratory species, such as birds) was recently discovered and could explain, in part, their migratory prowess in large oceans,

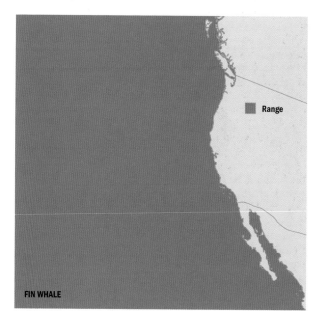

Range

FIN WHALE

where few geographic features serve as signposts for navigating (Walker et al. 1992).

Ship surveyors have noted year-round presence of Fin Whales off central and southern California that peaks in summer and fall, and acoustic surveys off the west coast record Fin Whale calls year-round, with a concentration between September and February. Whether this means that there are more Fin Whales, more Fin Whales calling, or some whales calling more than others is not clear. Whaling records off British Columbia suggested that Fin Whales stayed for longer periods off Vancouver Island and in nearby coastal waters, but females promptly departed in September, when they would have migrated south to give birth.

This fastest of whales, however, could easily travel between destinations in record time throughout the year, which might explain year-round sightings in some regions. The longest distance traveled in a day was 292 km (181 mi), and one study found a female with her calf roaming over 1,500 km (932 mi) of ocean over 43 days, and up to 157 km (97 mi) in one day.

HABITAT: Fin Whales are most often seen in areas with very high productivity in temperate and coastal waters along the continental shelf and slope ranging in depth from 77 to 3,200 m (252 to 10,506 ft). Sea temperatures where Fin Whales occur range widely from 5 to 28 degrees C (41 to 82 degrees F), the highest temperature recorded in the Gulf of California. Peak abundances of whales in the Southern California Bight are reported following periods of maximum upwelling.

VIEWING: Marine surveyors see Fin Whales off southern and central California year-round, and tour boat operators from Baja to British Columbia boast of fin sightings. On any given whale-watching trip, a Fin Whale sighting is possible, and in even rarer instances, glimpses from points of land such as the Farallon Islands and Point Reyes are possible.

CONSERVATION: Perhaps half a million Fin Whales plied the world's oceans before whaling. At first, Fin Whales were only rarely sought, because of their speed and offshore habitat, but eventually when steam replaced sail, they were targeted. The IWC has reckoned that about 750,000 Fin Whales have been taken since 1900, which would make this species the most heavily harvested cetacean of the 20th century. Perhaps 45,000 Fin Whales swam in the northern Pacific before the start of mechanized whaling after 1900, including 2,500 to 2,700 along the eastern Pacific. About the same number were taken over the next six decades. Off California, 600 were killed by the San Francisco whaling stations between 1956 and 1962, but under the IWC, females and calves were protected. Historically, Fin Whales may have thrived in the coastal waters of British Columbia, but whaling for this species peaked in 1958, with 573 caught, and some 7,605 were killed there in all between 1908 and 1967. There has been speculation that British Columbian waters were originally home to a distinctive, local population.

Agreement	Conservation Status
U.S. ESA	Endangered
U.S. MMPA	Depleted
COSEWIC	Pacific population threatened
IUCN	Endangered

Modern-day threats to Fin Whales include fishery entanglements, "scientific whaling," ship strikes, and pollution. Fishery interactions are low with such a large animal, and since improvements in nets and techniques on the U.S. west coast were made after 1997, only a single Fin Whale entanglement has been observed. However, driftnets set for sharks and Swordfishes may still drown Fin Whales off the coast of Baja. The Japanese government began collecting Fin Whales with "scientific" permits under the rules of the IWC in 2005, but it has not yet taken them in the northern Pacific.

Fins are the most frequently struck of the great whales. Moreover, in some areas, a third of all Fin Whale strandings seem to involve ship strikes. Strikes of all great whales were first noted when steamship speeds reached 24 kph (15 mph) in the late 19th century. Ship strikes mounted after 1950, with increasing vessel size, speed, and numbers. Whales are usually struck by the bows of vessels and not by propellers, and thus can suffer fatal fractures of the jaw and skull.

POPULATION: Something on the order of 16,000 Fin Whales survive today in the northern Pacific, and along the west coast of the United States, surveyors estimate a population of 3,500. Sightings along the west coast of North America seem to have increased over the last 30 years, but the trend is not statistically significant. The IWC has assumed that there is a single main management stock in the northern Pacific, but the population structure in the northern Pacific is not well understood. A small, isolated stock of Fin Whales resides in the Gulf of California, and there is another regional stock in the East China Sea.

COMMENTS: The Fin Whale sings in a voice as loud as a jet taking flight. To its thundering song, the rocks and bones reply. The great voice shakes the sides of islands, rattles the edges of continents, and is reflected off the darkened plain below.

REFERENCES: Aguilar 2002, Croll et al. 2002, Goldbogen et al. 2007, Laist et al. 2001, Walker et al. 1992, Watkins et al. 1987.

SEI WHALE *Balaenoptera borealis*
Pls. 3, 4; Fig. 54.

> Adult male average length is 13.1 m (43 ft). Adult female average length is 13.7 m (46 ft). Average weight for both is 20,000 kg (22 tons). Calf average length is 4.5 to 5 m (14 ft).

SCIENTIFIC AND COMMON NAMES: Sei Whales are one of seven species of baleen whales of the genus Balaenoptera, "winged whale"; the species name borealis means "northern" (fig. 54). Sei Whales are classified into two subspecies to distinguish those in the northern and southern hemispheres—*B. b. borealis* and *B. b. schlegellii,* respectively. The primary common name Sei is a variant on the Norwegian whaler name for pollack fish, Seje, upon which the whales feed in the North Atlantic Ocean. Other names include Sardine Whale, Pollack Whale, and Coalfish (another name for pollack) Whale, after the various fish with which they have been associated. A common Pacific name is Japanese Finner, and in the Atlantic, the name Rudolphi's Rorqual is used, after the Swedish-born naturalist, Karl Asmund Rudolphi, who described the species.

DESCRIPTION: The overall impression of Sei Whales is one of large size and slim shape. The sleek profile is accented by a pointed head and short, slender flippers that are only 10 percent of the body length. The largest female recorded was nearly 20 m (65 ft) long, and the heaviest weighed 40,000 kg (44 tons). Blubber

Figure 54. Sei Whale with tall dorsal fin.

accounts for 18 percent of the total body weight, and muscle for 60 percent.

The color of this whale's back is black to gunmetal gray, and there is white countershading on the belly. Back and flanks are pockmarked with scarring, which gives the impression of very light spotting; this is thought to be caused by cookie-cutter sharks, lampreys, or skin disease. Sei Whales commonly have a countershading pattern that extends from the navel forward to the chin and upward to a line from the eye back to the flipper. The chin can be white or light gray. An uncommon pattern has blazes across the back as a series of arched lines that are light gray to white. The lines originate from the eye and radiate across the back and shoulder, one to the insertion of the front flipper and others that curve back toward the flank. From above, the blazes appear like an hourglass. The blazes give the impression of pronounced eye shade and disrupt the silhouette of the whale. In side view, some Sei Whales have a white line along the leading edge of the flipper and two dark lines that extend from the dark flank into the white of the belly just past the flipper. Unfortunately, the marks are too dark to be seen from the surface and end well in front of the dorsal fin when the back is exposed.

The rostrum is narrow, with a slightly arched profile, and has a single distinct, narrow ridge when viewed from above. The

40 to 65 throat pleats extend across the belly and end well before the navel. The baleen is mostly light gray to black, with light gray to white on the tips, and is distinctly different from that of any other baleen whale. As many as 220 to 400 baleen plates have been counted on each side of the mouth. The plates are medium in length [48 cm (19 in.)], ending in very fine, white-tipped bristles that are less than 1 mm thick (0.004 in.). The fineness of the bristles can be used to identify the species.

The dorsal fin is set well back on the whale and is taller in height than that of most baleen whales [25 to 61 cm (10 to 24 in.)]. In profile, the dorsal fin is erect but variable in shape, from falcate to straight, with a slight curve at the tip. The tail is notched, with a fairly straight smooth edge. When the whale surfaces, the dorsal fin and blowhole are often seen at the same time, because the species usually surfaces at a shallow angle compared to others.

FIELD MARKS AND SIMILAR SPECIES: Sei Whales may be confused with Fin Whales and Bryde's Whales. Key marks for the Sei Whale are a single ridge on the rostrum, compared to the three ridges of the Bryde's Whale, and a taller dorsal fin compared to those of Fin Whales and Bryde's Whales. The Sei Whale has a number of throat grooves similar to that of the Bryde's Whale, but they do not extend as far as the navel. Sei Whales are smaller than Fin Whales and larger than Common Minke Whales; they lack the asymmetrical markings on the head of the Fin Whale and also the banding on the flippers that Common Minke Whales have. When surfacing to breathe, the blow is tall [3 m (10 ft)] and, although similar in shape, is shorter than the Fin Whale's [4 to 6 m (13 to 20 ft)].

NATURAL HISTORY: Sei Whales, although abundant elsewhere in the world, are extraordinarily rare in the northeastern Pacific; consequently, little has been learned about their habits and haunts in the Pacific in the past 30 years. Sei Whales make an annual trek, from high-latitude feeding areas in summer to low-latitude breeding areas in winter, but the destinations are not consistent through the years. Whalers and biologists alike have spoken of "Sei Whale years," when the species invades a feeding area for a few years—but then disappear. The whale invasions may coincide with abundant prey such as schooling fish, but the phenomenon is poorly understood. Perhaps a threshold of prey abundance is necessary before Sei Whales will stay to exploit it, or perhaps larger Fin Whales, seeking the same superabundant prey, muscle them out.

One theory regarding the Sei Whale's movements into varied marine locales is that its preferred prey attracts it to offshore or onshore areas. Its baleen tips are very fine for entanglement of small invertebrates (similar to the Right Whale's) but are also short, which is more advantageous for feeding on fish. The combination of shape and size of the overall body and of the baleen means that the Sei Whale can exploit food both on and off the continental shelf. It is adaptable enough to switch eating habits when prey abundance fluctuates and to move onshore periodically.

Regardless of where they journey, Sei Whales, with their moderate size and sleek shape—like that of a smaller and lighter marathon runner—may be the fastest of all rorqual whales. They have been clocked at bursts of 56 kph (35 mph). Until recently, there was one reliable record of a Sei Whale traveling 4,000 km (2,485 mi) in 10 days, or a sustained average of 17 kph (10 mph). For traversing vast areas of the ocean to spots where food is superabundant, these whales' speed and endurance matter.

Sei Whales occur in groups of one to five, but larger groups of 30 have been noted during surveys in the California Current. Although the average group size is two, cow–calf pairs are often associated with a third whale.

Sei Whales are not acrobatic, and when they do rarely breach, they belly flop because of their large size. They also are not particularly deep divers, and when they slip underwater by arching their back, they rarely lift their tails to dive. The pumping of the tail during these shallow dives produces a series of "fluke prints" on the surface that reveal the whale's location under the water. Shallow dives may explain why Sei Whales do not hyperventilate before dives, as do other baleen whales. Most dives follow a single breath and last 1 to 3 minutes, and dives of 6 to 11 minutes are uncommon.

Sei Whales appear to avoid boats more than Fin Whales do, which might contribute to the low number of sightings. Signature sounds made by Sei Whales have been recorded in the Atlantic and Antarctic populations, and this may be a way to distinguish subspecies. The sounds were loud, low-frequency moans, emitted in a series of 10 to 20 sweeping signals, as well as some higher-frequency sounds. Sei Whales also produce loud growls and whooshes. Researchers have speculated that the sounds correspond to breeding "songs" or may have to do with feeding. Low-frequency sounds do not transmit far and likely are used for communicating over short distances of only a few kilometers.

REPRODUCTION: From analyzing hunted whales, biologists determined that females were sexually mature at 5 to 15 years, were physically mature at 25 to 30 years, and lived for up to 60 years.

In the northeastern Pacific, calving occurs in subtropical waters in winter, but it has never been actually observed. Conception occurs in winter, likely during the time when a female–calf pair is attended by a third whale, a male. Gestation lasts 11 to 13 months, and the calving interval is 2 to 3 years. Females nurse for 6 to 9 months, and calves are weaned in the summer feeding areas, when they are about 11 m (36 ft) long. Pregnancy rates of adult females were 30 to 70 percent in the northern Pacific, when measured during the end of the 20th-century whaling period, but current rates are unknown.

FOOD: Despite the Sei Whale's rarity, quite a bit is known about its diet, because biologists analyzed the partially digested animal bits in the stomachs of hunted whales. Their food consists of invertebrates and fish, as is seen in other baleen whales, but the proportions vary significantly. With their fine baleen, Sei Whales appear to feed more on zooplankton, such as copepods, in the open ocean. Biologists sifting through the stomachs of 1,453 whales killed off British Columbia over a 5 year span noted that copepods were the dominant food in three of the years, fish (herring, hake, saury, lanternfish) in one, and krill in one. Seasonal differences in diet showed that they fed more on saury from July through September, and more on copepods in spring and early summer. Overall, large zooplankton accounted for 80 percent of their diet, and the rest consisted of small squid and numerous types of small fish, including mesopelagic and pelagic fish such as herring, sardines, and even salmon. The diversity of diet implies that Sei Whales eat what is available—and that they will switch when necessary.

Sei Whales use two strategies to capture different prey, skimming and lunge-feeding. Skimming targets weak-swimming copepods that reside near the surface and involves the whale either rolling on its side, about 3 m below the surface, or swimming at the surface in a straight line with its mouth slightly open. By skimming the surface, the whale entangles copepods in its fine bristles, and water flows out of its partially open mouth. The speed of this whale enables it to swiftly pursue schooling fish or squid, lunging to the surface with an open mouth and engulfing the school. Some whales use the two methods in alternation, which suggests that they may feed on copepods and then on the fish feeding on copepods.

In rare instances, when researchers have seen this species feeding, they have reported that the whales feed at dawn and early morning, which implies that they may be targeting food that is more available when the deep-scattering layer is present near the surface. The size of the area where they were recorded feeding for an hour or more was around 0.5 km^2 (250 ac).

An individual whale's daily average consumption has been estimated to range from 277 kg/day (740 lbs/day) to 900 kg/day (2,000 lbs/day), although some estimates are suspect, because commercial fishing interests used them to justify the resumption of whaling of Sei Whales as competitors.

MORTALITY: Current natural sources for mortality are not known, because Sei Whales rarely strand onshore. From whaling records in the 1970s, however, biologists estimated natural mortality was 6 percent for adults and 10 percent for juveniles. During whaling days off California, biologists noted a disease in 7 percent of Sei Whales that caused the baleen plates to fall out, but they also noted that the "toothless" whales had fish in their stomachs. Killer Whales do prey on Sei Whales, but the consequence for the population is unknown.

DISTRIBUTION: Sei Whales occur throughout most of the temperate and subtropical oceans of the world but rarely venture above latitude 55° N in the northern Pacific or south of California. Researchers do not agree about how many separate populations occur in the northern Pacific because of inconclusive evidence, which is based on distribution, blood types, and baleen morphology. The species may consist of one stock composed of several populations. Regardless, the northern Pacific stock ranges almost exclusively in pelagic waters and rarely ventures into coastal waters. This rarity limits our understanding of Sei Whales' current distribution and migratory routes.

Researchers who analyzed 20th-century whaling records determined that Sei Whale migration followed a strong seasonal

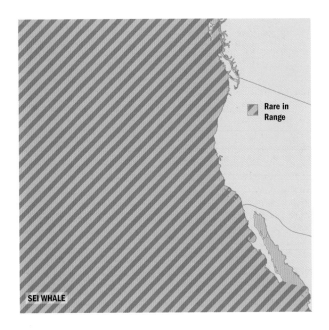

Rare in Range

SEI WHALE

pattern, which varied by age and sex. In winter the whales traveled to lower latitudes of the northern hemisphere, and in summer they migrated north to high-latitude feeding areas. Based on modern-day data collected by satellite-tagged whales in the Atlantic, Sei Whales appear to cover these distances directly and very quickly.

In the past, the number of Sei Whales peaked off California and British Columbia in July. Pregnant females were the first wave of migrants both north and south, followed by males, and lastly by young. The migratory route passed close by the western shore of Vancouver Island in June and July but then veered offshore in August. Sei Whale invasions followed by periods of absence occurred within a decade, and in California, 1926 was a known Sei Whale year, when there was a spike in Sei Whales hunted. Whaling records also revealed a seasonal onshore–offshore movement of Sei Whales in the eastern Pacific, with more whales present closer to shore from March through May and more of them much farther from shore (100 km; 62 mi) from July through September. These patterns of movement, however, may no longer exist.

HABITAT: When Sei Whales occur offshore, they have been associated with strong temperature gradients where there may be higher copepod densities, and their known water temperature range is 8 to 25 degrees C (46 to 77 degrees F). Copepods are more abundant in major mixing zones offshore, along the continental shelf and slope, around fronts and eddies that form near-bottom features such as seamounts, and following eddies that drift offshore such as the Haida in Washington. They are also associated with larger, semipermanent oceanic gyres such as the Alaskan and the northeastern Pacific gyres.

VIEWING: Sei Whales are very rarely seen, and viewing by plane or boat is the only way to see them in pelagic waters between California and Washington.

CONSERVATION: Before whaling, population estimates of Sei Whales ranged between 42,000 and 62,000 in the northern Pacific, but commercial whaling of this species was intense. Sei Whales were not initially targeted by whalers because of their speed and relatively low yield of oil compared to Blue Whales and Humpback Whales. Whalers eventually intensified efforts on Sei Whales after the other rorquals became rare. Whalers killed an estimated 61,500 from 1947 to 1987 in the northern Pacific, mostly between California and Washington. Around 384 were killed offshore of California between 1958 and 1965. In 1926, 25 were killed off California, but between 1919 and 1925, only two in all were killed. The intensity of hunting was sustained up to the end; in Canadian waters, Sei Whales were the last species hunted commercially, up to 1971, with a total of around 4,000 taken off British Columbia.

Agreement	Conservation Status
U.S. ESA	Endangered
U.S. MMPA	Depleted
COSEWIC	Data deficient
IUCN	Endangered

Only one stock of Sei Whale is officially recognized in the northern Pacific, but there is evidence for several stocks, including one between longitude 155° W and the western edge of North America. The current population estimate in the California Current, excluding Canada, is fewer than 50. With limited sightings, understanding population trends is not possible.

Japan issued itself permits to kill 50 Sei Whales per year for scientific purposes. Other potential reasons for the modern population to be depressed in the eastern Pacific include ship strikes and entanglement in drift gillnets. One ship strike off Washington was reported in 2003; given the rarity of the species, this rate of mortality might have a disproportionate effect on the population. There have been no reports of fishery entanglement, but offshore fisheries are not well monitored.

POPULATION: The population of Sei Whales in the northern Pacific in 1977—the most recent reliable estimate—was around 9,110. In the 1970s, near the end of whaling, population estimates ranged from 7,200 to 12,600 in the northern Pacific, but these numbers do not likely represent the current population. In the northeastern Pacific, Sei Whales are extremely rare. During aerial and boat surveys between 1991 and 2005, researchers observed them only five times off California and never off Oregon or Washington; none has been seen in British Columbian waters since whaling ended. In 2001, only one Sei Whale was recorded during an annual NOAA cetacean survey cruise covering more than 9,500 km (5,900 mi).

COMMENTS: One theory explaining the rare distribution of Sei Whales revolves around competitive exclusion by larger Fin Whales or perhaps fin fish. One bit of evidence for competitive exclusion is gleaned from whaling records, which reveal a divergence in the occurrence of the two whale species caught off British Columbia. Sei Whales were killed near shore in spring and offshore in fall, whereas Fin Whales were killed near shore in the fall and offshore more commonly in spring. Some researchers have speculated that pelagic fish also are Sei Whales' competitors, because they feed on similar invertebrate prey. Competition from Fin Whales and fin fish may explain the Sei Whale's continued low numbers and apparent nonrecovery from whaling.

REFERENCES: COSEWIC 2003, Flinn et al. 2002, Gregr et al. 2000, Horwood 1987, Horwood 2002, Jefferson et al. 1993, Locker and Waters 1986, Perry et al. 1999, Shilling et al. 1992.

BRYDE'S WHALE
Balaenoptera edeni

Pls. 3, 4; Figs. 55, 56.

> Adult male average length is 12 to 13 m (39 to 43 ft). Adult female average length is 13 to 14 m (43 to 46 ft). Average weight for both is 13,600 to 15,000 kg (15 to 16.6 tons). Calf average length is 3.5 m (11 ft).

SCIENTIFIC AND COMMON NAMES: The first Bryde's Whale specimen was described by J. Anderson from a specimen found in Burma in 1878 and named after a British commissioner there named Eden. Then, another specimen from South Africa, thought to be a separate species, was described by Olsen in 1913, and he named it after one Johan Bryde, a South African whaler—hence the common name of Bryde's. There continues to be confusion over the species, however, because of morphological and genetic differences, and as many as five different species may exist worldwide. Currently, the *Balaenoptera edeni* is the recognized scientific name for the species in the eastern Pacific, but synonyms include *brydei*. The English common names are Bryde's Whale and Eden Whale, and the Spanish is *ballena de Bryde*.

DESCRIPTION: Of the great whales, Bryde's Whales are among the smallest, but even still, they reach maximum lengths of 15.5 m (51 ft) (fig. 55). Males are, on average, 0.3 to 0.6 m (1 to 2 ft) shorter than females. The head is proportionately broad for a *Balaenoptera* whale, has a pointed tip like that of the Sei Whale, and accounts for 25 percent of the body length. The throat has 54 to 56 pleats to the navel and beyond. If you see the top of the head, three parallel ridges on the rostrum between blowhole and tip of snout are prominent. The middle ridge is more pronounced and aligned with the blowholes, ending in the splash guard around them.

The mouth holds 285 to 350 baleen plates, each of which is wide [50 cm (20 in.)] but short [20 cm (8 in.)]. The slate gray, short, coarse bristles are more designed for catching fish than copepods. The baleen is light-colored in front of the mouth and dark in the rear. Baleen varies with geographic location, likely reflecting the diet differences with variations in length of baleen

Figure 55. Bryde's Whale; note three ridges on rostrum.

Figure 56. Bryde's Whale with scarring.

plates and fineness of bristles. At the tip of the snout, there is a gap between the baleen plates.

Bryde's Whales have the typical countershading pattern of many marine mammals with a dark blue-gray back and light belly. The dark color continues down to some of the throat grooves and the flippers. Oval scars mar the skin of whales that migrate to tropical waters where the cookie-cutter sharks of the species *Isistius* occur (fig. 56). Along the back is a prominent dorsal fin [46 cm (18 in.)] that is very falcate and narrows to the tip. The flukes are moderately wide at 24 percent of the body length and with a fairly straight edge, but the fins are short, narrow, and pointed. If you found a dead whale and could handle the ribs, you would find them thin in width but broad, and you would discover that the first rib had a double head.

FIELD MARKS AND SIMILAR SPECIES: Bryde's Whales are commonly confused with Sei Whales, and occasionally with Fin Whales. Bryde's Whales generally are smaller than Sei Whales and Fin Whales and larger than Common Minke Whales. The key field marks for Bryde's Whales, however, are the three ridges along the rostrum from snout to blowhole. All other *Balaenoptera* whales, including seis and fins, have a single ridge. The dorsal fin is smaller than that of Sei Whales, but it is likewise taller and more falcate than that of the Fin Whale. Throat grooves extend to and beyond the navel in Bryde's Whales but fall short of the navel in Sei Whales. When surfacing to breathe, the blow is 3 to 4 m (10 to 13 ft), forming a column, but it has also been described as bushy. The dorsal fin surfaces after the blow, and rarely does the whale raise flukes before a dive.

NATURAL HISTORY: Bryde's Whales have the classic long, sleek shape of other *Balaenoptera* whales, but ecologically, they are quite different. They are a rare species of the cooler waters of the California Current, but instead, they are strongly tied to warm waters that straddle the equator between latitudes 20° N and 10° S and east of longitude 150° W. No other large baleen whale has such a restricted range, occurring only on the southern edge of the California Current along the coast of Baja. Their relatively small size compared to other baleen whales and their generalist diet may explain, in part, why this species is able to avoid migration to the arctic, where there is superabundant food to nourish and sustain the enormous mass of the larger whales.

The Bryde's Whale, although a large mammal, is one of the smallest great whales, and consequently, it was avoided during periods of intensive and unregulated whaling. Ironically, this is one reason why we know so little about this elusive whale. They reach sexual maturity when 11 to 12 m (36 to 40 ft) at 10 to 13 years, but continue to grow until they are 20 to 25 years old. The oldest recorded individuals were 72 years old.

Wide differences in estimates of size and aging in Bryde's Whales occur because different populations inhabit different areas of the Pacific Ocean. Some populations are associated with nearshore, highly productive upwelling waters, and others with offshore pelagic waters. Different populations may eventually be recognized as different species, but presently, they are described as different forms based on ecological, biological, and genetic characters (Yoshida and Kato 1999).

Bryde's Whales occur in groups of two to three in the California Current, and the largest group seen had only 12 members. The species is more abundant and seen in larger groups of two to 10, and of up to 100 in other parts of the world, but such numbers are unlikely in the California Current.

Although Bryde's Whales are rarely seen, their vocalizations have been recorded and studied off Baja and elsewhere in the world. They produce low-frequency moans that are short but loud. Six separate call types were identified in the eastern Pacific. Most of the sounds included two types emitted simultaneously, which would be like whistling and humming at the same time. A low-frequency downswept tone was superimposed over a continuous upper tone. Calls were repeated every 1 to 3 minutes, and many were produced while the whales were moving. Adult and calf pairs produced three call types, including a pulsed and unpulsed moan in the 90 to 500 Hz range, and one calf produced higher pulses in the 700 to 900 Hz range. Researchers could tell that the whales were calling back and forth, and they guessed that different calls were used for different reasons because the type of call varied with the size of the group (Oleson et al. 2003).

There has been no clear evidence of communication while feeding, however, as occurs with a few other baleen whales. Nevertheless, they have a very complex repertoire of feeding styles from skimming to lunging and even to using bubble nets. Bubble net foraging consists of swimming alone in a circle underwater releasing bubbles to trap fish. They also herd fish and then lunge through the school, and they are often in attendance with other marine predators such as seabirds, seals, and other whales. While pursuing prey, they can dive 300 m (984 ft), but they rarely raise their tails proceeding dives, which suggests that most of their dives are not very deep. When swimming in pursuit of prey or traveling, their average speed is 7 kph (4 mph), but 20 to 25 kph (12 to 16 mph) is the fastest speed reported.

REPRODUCTION: Much is assumed and little is known about the reproduction of Bryde's Whales. As with many tropical species,

Bryde's Whales may breed year-round, and consequently could calve any month of the year, but researchers have noted that calving occurs mostly in winter. Their gestation is 11 to 12 months, and females nurse calves for 6 months. This allows for a period of 6 months to recover and then breed again. Consequently, females give birth every other year.

FOOD: Fish are the main staple of Bryde's Whales, but they also eat cephalopods and small crustaceans such as copepods and krill. This generalist diet no doubt enables the species to feed year-round in tropical waters, in contrast with other baleen whales. Their fish of choice are schooling anchovy, sardine, mackerel, and herring, but occasionally they eat myctophids, lantern fish. They eat krill of the family Euphausia and pelagic red crabs, which are abundant in the subtropics and tropics. Of the cephalopod group, Bryde's Whales feed on cuttlefish, squid, and octopus. One species of squid found in their diet, *Lycoteuthis,* has bioluminescent organs, and is also targeted by seabirds. Bryde's Whales that feed inshore tend to prey on fish, and those offshore tend to feed on copepods and krill, feeding alongside tuna in the northern Pacific, presumably feeding on the same prey.

The amount of food that a Bryde's Whale eats in a day has been estimated at 4 percent of its body weight [660 kg (1,768 lbs)], which is far more than the average estimated consumption of a Sei Whale in a single day, of 277 kg (742 lbs). This consumption estimate is based on feeding amounts during the peak feeding season, but because Bryde's Whales likely feed year-round, these may be overestimates, and those of Sei Whales underestimates, because they likely do not feed while migrating between latitudes.

MORTALITY: Natural sources for mortality have not been documented because the species rarely strands onshore, although sharks and Killer Whales are likely predators.

DISTRIBUTION: Bryde's Whales are closely tied to the warm waters of warm-temperate, subtropical, and tropical zones in the temperature range of 15 to 20 degrees C (59 to 68 degrees F), straddling the equator between southern California and Chile, and east of longitude 150° W in the eastern Pacific. Bryde's Whales do not migrate to polar regions in winter as other baleen whales do, and it is extremely rare to see one off Oregon or Washington, with only one sighting having occurred in 10 years off these states during NOAA surveys. They are the least migratory of rorquals;

BRYDE'S WHALE

Range

nevertheless, there is an observed movement to the equator in winter and away from the equator in summer by whales that occur in warm-temperate waters. The population in the Gulf of California is resident year-round, however, and whales in other areas may also be resident. Presumably, whales are migrating south to warmer waters to calve in the winter, but calving grounds are unknown. Migration likely is also related to the location of predictable abundant prey, and there is a strong positive association between abundance of the whales and sardines in the Gulf of California. There also was a strong positive relationship between abundance and the warm water 1998 ENSO.

VIEWING: Bryde's Whales are rarely seen in the California Current, but they can be observed more commonly in the Gulf of California. When inshore, they are associated with upwelling zones and are found near islands and continental slopes.

CONSERVATION: A survey of records of shore-based whaling stations from the 19th century contains no record for Bryde's Whales, but in less than 15 years, between 1968 and 1981, around 5,000 were killed commercially in the eastern Pacific. Because Bryde's Whales are the smallest of the great whales, they were not hunted much until the 1970s, and there has been no hunting since the 1987 whaling moratorium was implemented. Japan continues to kill the species for "scientific research," but this is likely limited to the western Pacific. Records of mortality from human activities are rare, though some deaths are likely associated with gillnet fisheries for sharks and Swordfish in the eastern Pacific. With nets extending like long fences over several kilometers, it is inevitable that some whales will become entangled in them.

Agreement	Conservation Status
U.S. ESA	Not listed
U.S. MMPA	No special status
COSEWIC	Not listed
IUCN	Data deficient

Very little is known about the species, and that is why the IUCN lists it as data-deficient. The IWC recognizes seven stocks in the Pacific Ocean, three of which occur in the northern hemisphere and one that straddles the equator. The population in the California Current is part of the larger population of Bryde's Whales that occurs in the eastern tropical Pacific and includes the Gulf of California. These stocks may eventually become separate species, but for now, there is insufficient knowledge to determine differences.

POPULATION: Since whaling days, the population has increased with regulated management practices implemented by the IWC, and Bryde's Whales are the most abundant whale species observed on research cruises in the tropics. In 1993, a population estimate for the entire Pacific was 55,000. For the eastern Pacific, the population estimate is 13,000, but it could be as many as 20,000 or as few as 9,000. Nevertheless, during research surveys off California between 1991 and 2005, there was only one confirmed sighting, and the estimated population in the California Current may be as few as 12 animals.

COMMENTS: Bryde's Whales and Sei Whales are very closely related because their genetic separation occurred less than

300,000 years ago. With so little genetic separation, perhaps Sei and Bryde's Whales also understand each other's calls. If calling between individuals helps in finding food dispersed across a vast ocean, then interspecies communication would increase feeding opportunities.

REFERENCES: Kato 2002, Nowak 2003, Oleson et al. 2003, Read et al. 2009, Yoshida and Kato 1999.

COMMON MINKE WHALE   *Balaenoptera acutorostrata*
Pls. 3, 4; Fig. 57.

> Adult male average length is 8 m (26 ft). Adult female average length is 8.7 m (29 ft). Calf average length is 3 m (10 ft). Yearling average length is 2.5 to 3 m (26 to 27 ft); average weight is 450 kg (1,000 lbs).

SCIENTIFIC AND COMMON NAMES: Common Minke Whales are baleen whales of the family Balaenoptera (fig. 57). The Latin name *balaena*, for whale, and the Greek word *pteron*, for fin, are sources for the genus name, and the species name *acutorostrata* comes from the Greek words *acutus,* for "sharp" or "pointed," and *rostrom,* for "snout."

Classification of Common Minke Whales was revised in the late 1990s—from one species, with worldwide distribution, into two species, northern and southern. The southern species *(Balaenoptera bonaerensis)* occurs in Antarctic waters and is a distinct species based on newly discovered morphological and genetic differences. The Common, or Northern, Minke Whale is further segregated into three subspecies, based on location; the northern Pacific Minke Whale *(B. a. scammoni)*, named for the whaler-turned-conservationist Charles Scammon; the North Atlantic Minke Whale *(B. a. acutorostrata)*; and the still-to-be-named Dwarf Minke Whale. The northern Pacific subspecies was previously called *B. a. davidsoni,* a name given by Captain Scammon in 1872. Many common names allude to its similarity but smaller size relative to the Finback Whale, including Sharp-snouted Fin Whale, Lesser Rorqual, Least Finner Whale, Pigmy Whale, Bay

Figure 57. Common Minke Whale.

Whale, and Little Piked Whale. The origin of the name "Minke," however, is drawn from Norwegian lore, which tells of a whaler named Meincke who misidentified a minke whale as a Blue Whale; small baleen whales subsequently were called "Meinke's" whales. A more recent acquired name is Stinky Minke, arising from the smell of the whales' breath, evident when they surface near boats.

DESCRIPTION: A first impression of the Common Minke Whale is that of a sleek creature, designed for speed. Its slender profile is accented by a narrow, pointed head and short, trim flippers. For a baleen whale, the Common Minke Whale is remarkably small, measuring only half the length of the similar-looking Fin Whale. Females are the largest individuals of this species, with a maximum length of 10 m (33 ft) [compared to about 9 m (30 ft) for males] and an estimated maximum weight of 10,000 kg (4 to 5 tons).

The rostrum is very narrow and much shorter than that of other baleen whales. When observed from above, the rostrum has a single distinct, narrow ridge extending from the blowhole guard toward the snout. The baleen is mostly yellowish white in the northern Pacific Minke Whale but is grayish or brown in the Dwarf Minke Whale. As many as 230 to 360 baleen plates on each side of the mouth have been counted, and the plates are shorter in length (4 to 5 in.) and have finer filaments than those of other baleen whales. The mouth gape is somewhat limited, with 50 to 70 short throat pleats that extend only as far as the foreflipper. The dorsal fin is fairly pronounced and falcate. Located two-thirds of the way down the back, the fin is easily observed when the whale surfaces and dives. The tail is notched but not scalloped and has a smooth edge.

The back is black to dark gray, but the undersides have white countershading typical of many sea mammals. Occasionally, white mottling extends along the flanks, and often a lighter-colored patch (in chevrons) occurs across the back, behind the head. Common Minke Whales' most distinctive color pattern, however, is a white or light gray band across the foreflippers. These flashing bands, as well as the chevrons, can be clearly distinguished both above and below the surface.

FIELD MARKS AND SIMILAR SPECIES: The Common Minke Whale is the smallest baleen whale in the California Current—less than half the length of the similar-shaped Fin Whale and Sei Whale. Common Minke Whales are similar in size to Killer Whales but lack the latter's robust muscular shape; the Common Minke Whale's dorsal fin, too, is shorter and more falcate than the Killer Whale's. The flippers are much shorter than in Humpback Whales and slimmer than the paddlelike flippers of Gray Whales and Killer Whales. A key mark for Common Minke Whales is the distinct white or gray band across the flippers and, often, the presence of chevrons across the back. When surfacing to breathe, their blow is short and diffuse, similar to Gray Whales', but the minke is easily identified by its falcate dorsal fin, which the Gray Whale lacks.

NATURAL HISTORY: The Common Minke Whale is the smallest of all rorqual whales in the northern Pacific, and worldwide, it is the second smallest (after the tiny Pygmy Right Whale, *Caperea marginata*, of the Southern Ocean). Because of their diminutive size and quick speed, Common Minke Whales have eluded both whalers and researchers, and only in the past half-century has their natural history been investigated more intensely. Generally, most individuals reach sexual maturity by 8 years of age, but this estimate is down from one put forth in the 1940s, of 12 to 13 years of age: this decrease in age of sexual maturity is thought to be a response to hunting pressures on Common Minke Whales. Their lifespan is estimated to be 30 to 60 years, and individuals followed for the longest time spans were ones identified in 1980 to 1999 in the San Juan Islands.

Common Minke Whales are primarily solitary or occur in groups of two to three in the California Current, but larger groups of a few hundred form where food is superabundant, as in arctic waters. Mother–calf pairs likely comprise the pair group, but the triad usually includes a male, signifying the mating period. During the rest of the year, sexes appear to segregate on the feeding grounds, and in the California Current, females with young are more likely to occur inshore, and males offshore. A study in Washington showed Common Minke Whales grouped into three adjacent but distinct home ranges, but there was no evidence of the adjacent whale groups defending a territory. Each home range was estimated to encompass an area of around 600 km^2 (232 mi^2) and was used for feeding.

Because of their small size, Common Minke Whales can be very acrobatic, and they are sometimes observed leaping completely out of the water. They rarely lobtail or flipper wave, and they distinctly do not display their tail before diving. Their sleek form enables these whales to swim at very high speeds of 20 to 30 kph (12 to 20 mph), both to outswim predators and to chase down fast schooling fish.

The feeding styles of Common Minke Whales are modest, compared to those of Humpback Whales, but they do develop individual feeding strategies. Researchers found some specialists divided according to two feeding strategies: "bird ball feeders" take advantage of feeding flocks of diving seabirds, which force balls of fish to the surface where Common Minke Whales can lunge-feed on them, and breach feeders breach out of the water with a bulging mouth full of fish.

Common Minke Whales do not sing, but they are noted for their underwater "boing" sounds. The sound was first recorded by the U.S. Navy off San Diego but was not identified until 2001, when the Common Minke Whale was proposed as its source. The "boinging" is produced by a solitary whale and, rarely, by whales in duets. The sounds are heard during November through March, apparently during the breeding season. Whales in different regions appear to have different dialects, with those in the eastern Pacific sounding over a longer duration but with a shorter pulse rate. Other Common Minke Whale sounds include low-frequency, downswept calls; high-frequency clicks, whistles, and grunts; and a series of thumps. The thumps, also called "thump trains," have been associated with group feeding.

Although not among the group of "friendly whales" (especially Gray Whales), Common Minke Whales are curious and do approach vessels, surfacing briefly alongside and even pacing boats. This tendency toward friendliness has helped researchers identify individual whales in the Pacific Northwest, around the San Juan Islands, where the species is more abundant. Individuals are identified based on photographic comparisons of the dorsal fin (its shape and coloration), small circular scars, and variations in the pale skin pigmentation along the flanks.

REPRODUCTION: In the northern Pacific, calving occurs throughout the year but peaks from June through August. Most females calve annually at intervals of around 14 months, and in some areas, the pregnancy rate is as high as 85 percent. Consequently, most females ovulate annually 4 to 5 months after giving birth, and some females may ovulate more than once per year if they do not become pregnant the first time. Gestation is 10 to 11 months, and once born, calves grow rapidly and can ingest fish and krill at the age of 6 months. Females may attend their calves for longer than this, however—some for more than 3 years. No one has documented mating in Common Minke Whales, but a trio of whales likely indicates that a male is present with a female and calf.

FOOD: The Common Minke Whales' short baleen and truncated throat pleats reflect their fare of mostly fish. Their diet is diverse, however, and they feed on what is seasonally and locally abundant. Fish food includes small, schooling, "bait"-type fish, such as herring, sardines, anchovies, sand lance, saury, sand eel, salmon, hake, mackerel, and smelt. Invertebrate prey include squid, krill, copepods, pteropods, and mysid shrimp.

Common Minke Whales are effective in chasing down schools of fish because they can match the speed of their prey, diving for 6 to 12 minutes and, rarely, for up to 20 minutes. They lunge while pursing fish, and then gulp their catch. While feeding on the surface, a whale may roll on its side, with its flukes partially exposed out of the water. Common Minke Whales also eat prey that live on the bottom, such as dogfish and mysid shrimp. In coastal bays, they are often observed in mixed-species feeding assemblages of seabirds, seals, and Humpback Whales, and Common Minke Whales may take advantage of the presence of other predators to help catch their common prey. Minke Whales are not known to cooperate among themselves, however.

One estimate of an individual whale's average daily consumption was 277 kg (740 lbs).

MORTALITY: Natural sources for mortality in the Common Minke Whale are rarely identified with certainty, because the species does not commonly strand onshore. In the Pacific Northwest and Alaska, pods of Killer Whales have been observed attacking and feeding on Common Minke Whales, and whalers have noted old Killer Whale tooth scars on flippers and flukes. One observation, in Barkley Sound, Vancouver Island, described a group of three male Killer Whales chasing and cornering a Common Minke Whale in a shallow bay. The observer speculated that the Killer Whales drowned the whale by seizing the flukes and holding it underwater. Observations of Killer Whales chasing Common Minke Whales at speeds of 30 kph (19 mph) for up to 1 hour, and of them ramming the whales, have also been noted. Some Common Minke Whales, however, are able to outdistance the Killer Whales at sustained high speeds. The two species are known to intermingle and coexist while foraging on other prey. Great White Sharks may also kill Common Minke Whales, but there is little evidence of this. Occasionally, Common Minke Whales strand onshore, perhaps out of avoidance of Killer Whales. One young whale ran aground at Point Reyes on a rocky intertidal shoreline, and humans pushed it back into the water as the tide rose, apparently leaving it with only minor scrapes.

DISTRIBUTION: Common Minke Whales are distributed throughout most of the oceans of the world but three separate subspecies have been identified in the Northern Hemisphere. As mentioned above, the northern Pacific subspecies has been further divided into five separate stocks, two in the western Pacific, west of 180° W longitude, and three east of 180° W longitude. The three stocks in the eastern Pacific include the Hawaiian, the Alaskan, and the California–Oregon–Washington. The last of these is associated with the California Current and likely ranges also into Canadian and southern Alaskan waters. Researchers believe that the California Current stock is separated into small, isolated populations that are nonmigratory compared to the Alaskan migratory stock. The migratory stock of Common Minke Whales travels between higher and lower latitudes seasonally, presumably in search of abundant food, and the whales range as far north as the Chukchi Sea and south nearly

Range

COMMON MINKE WHALE

to the equator. Within the California Current, both migratory and resident Common Minke Whales may occur.

Year-round residence has been noted for Common Minke Whales around the San Juan Islands, in Washington–British Columbia, and in Drakes and Monterey bays, in California. Also in the San Juan Islands, individual whales have been identified in multiple years, at summer feeding grounds. One study in the Atlantic identified seasonal distribution of Common Minke Whales in response to shifts in their prey—in spring, to sand eel habitat areas and then, in summer, to herring habitats. No similar distribution shifts have been discovered to date in the Pacific.

During migration, whales segregate, with the adult males arriving first on summer feeding areas, followed by young, and lastly by females. The Common Minke Whale's migratory trip is not as long as that of the Gray Whale, and its average traveling speed is 4 to 13 kph (3 to 8 mph). Females are not accompanied by calves on the summer feeding grounds, and no distinct calving grounds have been identified, as for Gray Whales.

Instead, calving may occur in temperate waters spread along the coast. Newborn calves have washed ashore, for example, in Drakes Bay, in central California.

HABITAT: Throughout their range, Common Minke Whales are associated with areas of consistent upwelling with high primary productivity, and where the seabed sediments are gravelly sand or mud. At areas with these types of sediments, shoaling fish such as herring and sand lance often occur. Common Minke Whales also occur along and on the continental shelf and around islands, reefs, and seamounts. Inshore, they are present in bays and estuaries, at large river mouths, and patrolling along the periphery of kelp beds. They rarely occur further than 160 km (100 mi) from shore and are usually found within 20 km (12 mi) of shore. Most sightings reported in several studies of Common Minke Whales have been in waters less than 90 m (295 ft) deep; in another study, whales occurred in waters averaging 425 m (1,394 ft) deep, with a maximum depth of 2,000 m (6,600 ft).

VIEWING: Although it is a coastal species, you will rarely see a Common Minke Whale from shore; indeed, the best way to see them is by boat. Because of their curiosity, they often will come up to vessels, providing easy viewing. Common Minke Whales are often observed around Vancouver Island, the San Juan Islands, Puget Sound, Depoe Bay, Cape Foulweather, Point Reyes Peninsula, Monterey Bay, Santa Barbara Channel, and the Channel Islands.

CONSERVATION: There is documentation of hunting Common Minke Whales with harpoons in Norway from as early as the Middle Ages, in the 11th century. Scammon reported that native peoples of Cape Flattery and the Strait of Juan de Fuca hunted Common Minke Whales annually, but the numbers killed were likely low since this species would be difficult to catch because of its speed. Along with eluding early whaling pressures, Common Minke Whales may have benefited indirectly from whalers' removal of large baleen whales, making more prey available to them. Before the 1960s, the species was considered abundant along the eastern Pacific, but it was not preferred by whalers because of its small size, low yield of oil, habit of not occurring in large groups, and difficulty to capture. After the 1960s, when numbers of larger rorqual whales began to wane, Common Minke Whales became the target of whalers. Presently, Common Minke Whales are one of only a handful of species legally whaled under the IWC, but not in the eastern Pacific. (Japan has killed these whales under a "scientific" permit in the northern Pacific since 1994.) The commercial value of a Minke Whale was estimated to be $20,000 in 1994.

Agreement	Conservation Status
U.S. ESA	Not listed
U.S. MMPA	No special status
COSEWIC	Not at risk
IUCN	Least concern

As a coastal species, Common Minke Whales face modern threats associated with interactions with humans in fisheries and shipping traffic. Mortality and injury from human activities includes entanglement in fishing gear in Puget Sound and off California, and boat strikes (although only one ship strike has been documented for a Common Minke Whale in 30 years). Amplified ocean noise may affect communication of males singing. Because Common Minke Whales occur mostly inshore, it is not surprising that pollutants such as PCBs, DDTs, chlordane compounds (CHLs), and hexachlorobenzene (HCB) have been discovered in the blubber of Common Minke Whales.

POPULATION: Worldwide, Common Minke Whales are still considered the most abundant baleen whale, with numbers estimated in the millions. For the California Current in U.S. waters, however, the current population estimate is only 900 to 1,000 whales, based on surveys from ships. This survey method may underestimate the true population size, because the Common Minke Whale is hard to see from a vessel. An alternative method may be hydroacoustic surveys, because the distinctive vocalization of the species is easy to detect. With no reliable population estimates, there can be no estimate of trends in the population.

COMMENTS: Biologists speculate that divergence of the Antarctic Minke Whale and the Common Minke Whale occurred around 5 mya and was triggered by global warming, which reduced upwelling areas and caused species isolation and restriction to narrower areas of upwelling. Common Minke Whales may be especially vulnerable to global climate change because of their reliance on upwelling centers for food.

REFERENCES: Best 1982, Carretta et al. 2007, Dorsey 1983, Estes et al. 2007, Ford et al. 2006, Hoelzel and Stern 2000, Macleod et al. 2004, Pastene et al. 2007, Rankin and Barlow 2005, Scattergood 1949, Whitehead and Carlson 1985.

HUMPBACK WHALE   *Megaptera novaeangliae*
Pls. 1, 2; Figs. 8, 13, 37, 41, 42, 58.

> Adult male average length is 12 m (39 ft); average weight is 30,000 kg (66,000 lbs). Adult female average length is 13 m (42 ft). Calf average length is 4.5 m (15 ft); average weight is 1,350 kg (3,000 lbs).

SCIENTIFIC AND COMMON NAMES: Humpback Whales *(Megaptera novaeangliae)* belong to the Balaenopterid family (fig. 58). The genus name, *Megaptera,* was coined by British naturalist John Gray in 1846 to describe the large pectoral flippers of the whale as a "giant wing." The species name, *novaeangliae,* is a Latin derivation from the French name *baleine de la Novelle Angleterre,* given by naturalist Mathurin Brisson, in 1756, from early sightings off New England.

Figure 58. Humpback Whales.

Whalers coined the common name "humpback" for the whale's humping motion, or distinct arched back, while diving. Other common names include Hunchbacked Whale and Hump Whale, the French *baleine a bosse,* and the Spanish *ballena jorobada.*

DESCRIPTION: Winglike pectoral fins distinguish Humpback Whales from any other baleen whale or, indeed, any other whale. The long, narrow fins are about one-third the length of the body. Knoblike bumps, called tubercles, along the front edge create a scalloped appearance. These bumps, from which hairs emerge, also cover the lower lip and the rostrum. The baleen, which is mostly black or gray, consists of 270 to 400 baleen plates on each side; these are short to moderate in length, reflecting a generalist diet. The rostrum is narrower than that of other baleen whales, but not enough so to easily distinguish between species in the field. Observed from above, the rostrum has a single narrow ridge running from the blowhole toward the snout. The mouth cavity expands when feeding, with 14 to 35 throat pleats (half the number that Blue Whales have) extending all the way to the navel. The dorsal fin varies from falcate to rounded or slightly bent. The wide scalloped tail has a serrated rather than a smooth rear edge that, in width, is about one-third the length of the body.

The back is black to dark gray, but the undersides range from pale white through mottled to all-black. The color variegation of the flukes is very distinctive, and researchers use these patterns to identify individuals as they raise their tails out of the water during dives. The pectoral fins of Humpback Whales are brightly white below. Coloration varies, too, because of barnacle infestations around the head, flippers, and flukes. Linear scarring on the back and fins is mostly present in males and is likely related to male competition.

Females are 1 to 1.5 m (3 to 5 ft) longer than males, but identifying the sexes in the field is based mostly on behaviors, such as a female with calf, or a male singing. The primary external physical distinction between the sexes is a round lobe at the tail end of the genital opening of the female. The longest female (18.6 m; 61 ft) ever recorded was from a whaling station in California in the early 20th century, but in recent times, the longest recorded female documented was only 15.5 m (51 ft). The oldest recorded female was estimated to be 48, based on analysis of ear bones; individuals over 75 years old are reported, however, and new aging techniques may reveal even older animals.

FIELD MARKS AND SIMILAR SPECIES: Genetically, the Humpback Whale's closest relatives are Gray Whales and Blue Whales, but the Humpback Whale is unlike either of these whales. The long flippers (30 percent of body length) of Humpback Whales contrast with the shorter flippers (10 to 15 percent of body length) of other baleen whales and also with the broad, paddlelike ones of Gray Whales. When exposed, the tail is scallop-shaped, with serrations along the edge that are lacking in all other large baleen whales. Humpback Whales and Common Minke Whales both may have mottled undersides, but Common Minke Whales lack white mottling around the head and often have distinct chevrons on the front flippers. The Humpback Whale blow is moderate to low at 3 m (10 ft) and may appear V- or heart-shaped.

NATURAL HISTORY: Humpback Whales are charismatic, with alluring songs, extraordinary feats of agility and migration, and surprising originality in feeding styles. Their migration is one of the longest of any mammal, but their migratory destinations have yet to be completely mapped out. They migrate annually, with a strong seasonal pattern that revolves around feeding and calving in separate and discrete coastal waters.

Except between a female and calf, groups of whales are not based on family relations, and associations among individuals differ between feeding and breeding areas. Group size is usually small—three to nine individuals—although large groups of up to 150 form when food is superabundant. On the feeding grounds, whales forage cooperatively. The feeding groups sometimes remain intact over an entire summer, and a few have been reported persisting over several years. Males are social in these feeding areas, but on winter breeding grounds, they are competitive.

For centuries, humans have marveled at the unique and spectacular behaviors of Humpback Whales, the most acrobatic of the large whales. They commonly breach, spy hop, lob tail, flipper wave, and tail wave. A breach often is transformed into a full somersault, with the whole belly gracefully exposed, and more breaches commonly follow in quick succession. We are not certain why whales breach, but because it occurs year-round, there are likely multiple reasons, including communicating underwater, stunning fish, shedding parasites, playing (especially calves), or just zest for life.

These acrobatic maneuvers are enhanced by the shape of the humpback's pectoral fins. Researchers have discovered that not just the wing shape but also the bumps on the outer edges of the flippers actually give this large, cumbersome whale a lift in tight maneuvers. Simulated models, tested in a wind tunnel, confirmed that the lumpy wing shape increases lift and reduces drag by one-third. By planting the long flipper in the water away from the body, the bulky whale actually rotates as quickly and agilely as a ballerina.

Humpback Whales are the only truly singing whales. Like a listing ship, the whale sings while floating with its head tilted down, flippers balanced out to the sides, adrift in around 15 m (50 ft) of water. Because whales have no vocal chords and release no bubbles during singing, researchers are not sure how they produce these haunting sounds, but the droopy posture may help the whale move air through its nasal passages. The warbling whales create a sequence of moans, groans, whistles, cries, "whos," "wos," "yups," "eees," "ooos," and snores that are linked to form the song. The song is long and complex, weaving several themes.

A single song is sung by the males of each population, and the ballad changes over the years as themes are modified. These dialects are one way to differentiate populations. The song may last more than half an hour, and a singer can repeat the song for hours and days. Males are solitary when singing and do not sing in groups, but the sound travels far—32 km (20 mi)—and so they can hear each other. The length of time a singer remains submerged may tell females that he is a good choice for mating. If bulls can dive longer and surface for less time between dives, then they are in good condition and have greater stamina. And indeed, females have been observed joining singers. Simple playback of a whale song, however, has led to both avoidance and nonresponse from other whales nearby. Mate choice may not be the only reason that whales sing, and studies continue to decipher the context and language of the songs.

Along with songs, Humpback Whales also may use echolocation, based on mega clicks, for locating prey or for orientation. The clicks are very similar to the sound a bat makes with its sonar before capturing prey, with a series of clicks ending in a high-pitched buzz.

Originality in the Humpback Whale is also reflected in the species' unique feeding methods, first described in 1929. Alone or in groups, whales create a curtain of bubbles [13 to 15 m (43 to 50 ft) in diameter] when swimming, clockwise, around a school of fish, by releasing air from their mouths or blowholes. The curtain captures the fish as efficiently as a purse-seine net, and the whale or whales then lunge toward the surface to gulp the trapped fish. As the whale herds its prey, the length and shape of its long flippers may complement the rising bubble curtain. To ensure precise timing of the complicated maneuver, for a successful catch, coordination among the whales is necessary, and there is some evidence that a leader acoustically directs the action. Vocalization appears to vary with the targeted prey; for example, a buzzing sound is associated with herding herring.

Another method the humpback has for confusing and concentrating prey is the formation of a large [4 to 7 m (13 to 23 ft) diameter] cloudburst of bubbles from beneath a school of fish or krill. The whales expel a burst of air and then slowly lunge through the cloud. A single whale or groups of two to three might repeat this feeding strategy several times in a row. A third method involves a whale lying on its side at the ocean's surface and swimming rapidly in a circle, lashing the surface with tail and flukes and forming a ring of foam that isolates the prey. Slapping the fluke on the surface stuns or concentrates prey. Bubble formation is not isolated to feeding, however: bubble curtains are also created by escort males in attendance with females and calves, to confuse potential predators, and by whales of all ages while playing.

"Friendly whale" once was a term applied only to Gray Whales, but beginning in the late 1990s, friendly Humpback Whales began approaching and interacting with people and vessels. One encounter involved a group of fishermen and biologists who released a whale that was hopelessly entangled in crab pot lines off San Francisco. The rescuers reported that they were in the water with the whale, while cutting away the tangled lines, and upon release, the whale turned around and momentarily faced each of its rescuers before slowly swimming away. Other encounters have involved Humpback Whales approaching and even rubbing against boats. Humpback whales also are friendly with other species; recently one was documented safeguarding a seal fleeing a pod of predatory Killer Whales in Antarctica. The Humpback Whale lay on its back and lifted the seal out of the water on its upraised belly, out of reach of the Killer Whales.

Maternal-like behavior is common in marine mammals, but not when species are extremely different, as in this example.

REPRODUCTION: Humpback Whales attain sexual maturity in adolescence, at age 5 to 7 years (this is true for both sexes), but they are not fully mature until 8 to 12 years of age. Once females begin giving birth, the calving season takes place annually in winter (October to March) and peaks in early February. Females often skip 1 or 2 years before mating again, and the annual calving rate has been estimated at around 0.46. Many individual females have been documented calving several years in a row, and females ovulate annually. Gestation is 11 to 12 months, and this rapid fetal development and the large demands of nursing force females to seek superabundant food in areas such as at higher latitudes. Calves grow rapidly, nursing for 1 year, but can ingest fish and squid at age 6 months. They are dependent on their mothers for 1 year, and some will nurse for up to 2 years. Pink milk has been described, perhaps related to a diet of krill.

Although no births have been observed, a fresh placenta was collected from a Humpback Whale calf within 15 minutes of birth off Hawaii. The newborn calf, unfolded from the womb but still with curled flukes, was estimated at 3.6 m (12 ft) long. There are no reliable observations of mating, likely because it occurs further offshore or in deeper waters. Mating probably occurs in midwinter, on the calving grounds, because that is when male testosterone levels peak, and where and when males display various mating behaviors, including singing, aggression, and escorting. Singing males do not sing from a single location but will move around, in no apparent pattern; females appear to approach some singing males. Females without calves may form stable groups, but after a female gives birth, the mother–calf pair will separate from the group. The mother and calf are then often joined by an adult male that acts as an escort, fending off other males. Male whales aggressively head-butt, tail-slash, broadside, and ram each other, sometimes drawing blood, to gain access to females. Some males even form coalitions to supplant a dominant male. The mating system has been described as a "floating lek," because there is no distinct resource, except the female, around which the males rally to compete. Females may mate with multiple males within a single year and have been documented, through genetics, to mate with different males in successive years. Male escorting could ensure that a female does not mate with other males, particularly if she ovulates several times within a season.

FOOD: From May through October, Humpback Whales feed almost continuously, but less so during winter. By gulping prey in the water column and sometimes bottom-feeding, Humpback Whales exploit a wide range of invertebrates and fish. They feed on krill, but their main food is often composed of schooling fish, such as anchovies, sardines, herring, juvenile salmon, pollock, sand lance, and mackerel. Occasionally, their diet includes copepods, pteropods, and cephalopods. Preferred prey can vary by location and season, with sardines preferred in Monterey Bay and krill in Trinidad Bay.

As with Gray Whales, Humpback Whales wear marks on one side of the head, indicating a preference on the part of some Humpback Whales for foraging on their right or left side when feeding on the seafloor. Their unique feeding strategy using bubble nets enables Humpback Whales to capture swift schooling fish, such as sardines and mackerel, that other whales may have difficulty in pursuing. Alone or in small groups, Humpback Whales rely on technique rather than speed to capture prey.

MORTALITY: A few Humpback Whale carcasses wash ashore each year, and in at least some instances, researchers can determine cause of death. The most common identified cause in California is ship strike that fractures their skulls. Rarely, Killer Whales have been observed attacking Humpback Whales, with death likely limited to calves. Killer Whale teeth marks, however, have been noted on the tails and bodies of some 20 percent of adult whales in central California and 26 percent off winter calving grounds in Mexico, suggesting that predation may be higher. Great White Sharks probably do not kill Humpback Whales, but sharks do feed on floating carcasses. Mortality rates are low for both calves and adults, and examples of multiple deaths are rare. (One such event occurred off California, when several whales died within a few days of each other. Researchers speculated that they died from eating fish contaminated with a biotoxin, domoic acid, from a harmful algal bloom.)

DISTRIBUTION: Distribution is better understood for Humpback Whales than for many large whales, because they live largely in coastal waters, where researchers can more easily learn about them. They are cosmopolitan—found in all the world's oceans.

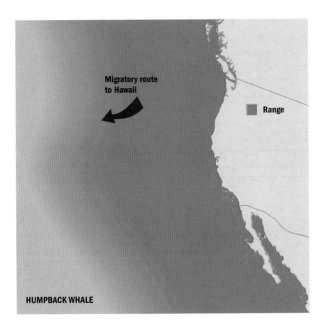

HUMPBACK WHALE

Migratory route to Hawaii

Range

In the Pacific Ocean, three separate populations of the species are recognized, based on distinct winter-to-spring movements: the eastern north Pacific stock (Washington to Baja); the central north Pacific stock (Alaska and British Columbia to Hawaii); and the western north Pacific stock (Bering Sea to Japan–Philippines). All three populations occur along the California Current, but the dominant one is the eastern north Pacific population, which spends spring, summer, and fall feeding in coastal waters from Washington to California, particularly in the superabundant waters of the Gulf of the Farallones and the Southern California Bight. The central north Pacific and western north Pacific populations, which forage also off British Columbia and Washington, occur mostly on summer foraging grounds in Alaska.

Some whales divert from the primary migratory paths and venture to other feeding and breeding areas. One such group of whales, which breeds off Central America, and is separate from the eastern north Pacific stock, currently holds the record for the longest recorded migration of any whale [8,750 km (5,437 mi)]: this group crossed the equator to feed in Antarctica. Another group of Humpback Whales has been identified around the offshore islands of Mexico; the winter grounds of this group are as yet unknown. The separation of the dominant populations, confirmed by genetic analysis, is reinforced by generations of females returning to breed where they were born.

Calves found in the California Current are born and nursed in winter in the warmer subtropical waters off Mexico (around Isabel Island, Tres Marias Island, and the Revillagigedo Archipelago) and Costa Rica. By late winter, these Humpback Whales begin one of the longest migrations of any mammal, swimming north to their primary feeding grounds in Alaska and elsewhere, a distance of more than 8,000 km (5,000 mi). Pregnant females, nonpregnant females, and males depart first, followed last of all by females with young. Some females do not migrate north, apparently finding sufficient feed around the islands to oversummer; if food is sufficient, individual whales may linger off California in any month of the year. Mostly, however, they are present during April through October, when prey abundance is greatest. Then, in autumn, Humpback Whales of the California Current begin the southward migration, led first by females with weaned but dependent calves, followed by juveniles, then by adult males, and finally by pregnant, slow-moving females.

Along their migratory routes, these whales may traverse deep waters, and some individuals also travel across the Pacific to Hawaii, or further north to Glacier Bay and the Bering Sea in Alaska. Individuals also have been seen in both British Columbia and Japan, showing that transoceanic movements, although not common, are possible.

Swim speeds for this large whale are moderate during migration, ranging from 8 to 15 kph (5 to 9 mph), and slow when feeding, at 2 to 6 kph (1 to 3 mph). A whale fleeing a whaling boat, according to one report, was clocked at 27 kph (17 mph).

HABITAT: Humpback Whales are associated throughout their range with the shallow continental shelf and shelf break, with islands such as the Channel Islands, with offshore banks and seamounts such as Cordell Bank and the Santa Rosa–Cortez Ridge, and with reefs. These underwater features are important in creating upwelling areas and convergence zones, where prey is concentrated. In winter calving grounds, whales mostly occur in water 200 m (656 ft) deep or less and warmer than 21 degrees C (70 degrees F).

VIEWING: Although a coastal species, Humpback Whales are rarely seen near shore, and viewing by boat is the best way to see them. Whales are easily observed in season around Vancouver Island, Puget Sound, the Olympic Peninsula, Depoe Bay, Cape Foulweather, Point Reyes, the Farallon Islands, Cordell Bank, Monterey Bay, Santa Barbara Channel, San Pedro Channel, and the Channel Islands.

CONSERVATION: Native Americans hunted Humpback Whales, but intensive commercial whaling began only in 1889. Between 1905 and 1960, an estimated 23,000 Humpback Whales were killed in the northern Pacific, and between 1960 and 1965, an additional 5,000 were killed, mostly off California, and towed to whaling stations such as Trinidad Head, Moss Landing, and San Francisco Bay.

Agreement	Conservation Status
U.S. ESA	Endangered
U.S. MMPA	Depleted
COSEWIC	Threatened
IUCN	Least concern

Because of their haunting songs and graceful breaches, Humpback Whales became a symbol of the early conservation movement that eventually led to full protection for whales in the United States. More recently, the Commission for Environmental Cooperation (CEC) among Canada, Mexico, and the United States selected the Humpback Whale as a marine species of common conservation concern. The group identified ship strikes and entanglement in fishing gear as the major causes of mortality, followed by noise disturbance, loss of prey and prey habitat, and climate change. Humpback Whales are the whale that is second most commonly killed by strikes from ships.

Amplified ocean noise from ships affected the presence and feeding behavior of Humpback Whales in Alaska, resulting in the creation of the first whale sanctuary in Glacier Bay National

Park, where ship traffic was restricted. Ocean noise may also affect Humpback Whales' communication, namely males' singing on the breeding grounds. Because whales are dependent on krill, they are also vulnerable to changes in climate affecting upwelling centers, such as the Gulf of the Farallones, as well as by the prevalence of harmful algal blooms.

POPULATION: A prewhaling population estimate for the northern Pacific was 15,000, which may have been reduced by commercial whaling to only 1,000. The current population for the northern Pacific is 6,000 to 8,000, and off California, 1,400. There was a drop in the population off California during the 1998 ENSO, presumably because of starvation, but precise estimates of this decline are unreliable because the whales range so widely. The reproductive rate of Humpback Whales off California—that is, the number of calves per total whales—is also lower than for other humpback populations.

Most researchers agree, however, that this population has increased during the past 20 years. A population estimate is also calculated from the proportion of known individuals observed during surveys. Researchers identify the photographs of as many as 400 individuals in a year, and they share photographic records of more than 1,500 whales in the Pacific, in a database called SPLASH (Structure of Populations, Levels of Abundance, and Status of Humpbacks).

COMMENTS: Famous individual Humpback Whales, given the names Delta and Dawn (a mother and calf) and Humphrey, inspired people who rescued them from narrow sloughs some 70 miles up the Sacramento River; these whales had wandered upstream from the Golden Gate through San Francisco Bay and apparently could not get out. In 1985, Humphrey spent a month in San Francisco Bay before departing with guidance by boats with human acoustics; he returned in 1990, stranding briefly onshore near Candlestick Park, and was seen in several subsequent years in nearby waters. Delta and Dawn had wounds typical of propeller injuries, and wildlife specialists treated them—darting them with antibiotics—before the two whales left San Francisco Bay after 3 weeks in its upper reaches.

REFERENCES: Calambokidis et al. 2004, Clapham 2002, Clapham and Mead 1999, Payne and McVay 1971, Peacock and Bradley 2008, Perry et al. 1999, Pitman and Durban 2009, Rasmussen et al. 2007, Steiger and Calambokidis 2000.

GRAY WHALE

Eschrichtius robustus

Pls. 1, 2; Figs. 17, 36, 40, 59, 60.

> Adult male average length is 12 to 14 m (42 to 45 ft). Adult female average length is 14 to 17 m (45 to 55 ft). Average weight for both is 27,000 to 36,000 kg (30 to 40 tons). Calf average length is 4 to 5 m (12 to 15 ft); average weight is 900 kg (1 ton).

SCIENTIFIC AND COMMON NAMES: Gray Whales *(Eschrichtius robustus)* are the only species of the family Eschrichtiidae of the Mysticeti suborder (fig. 59). The genus name, *Eschrichtius,* was given by British naturalist John Gray to honor Daniel Eschricht, a Danish zoologist. The species name, *robustus,* is Latin for strong, robust, made of oak. Gray studied subfossils of the species in the 19th century, after their discovery by William Lilljeborg in 1861. By that time, the species had been extirpated from Europe for more than 300 years, although in the 17th and 18th centuries, there was strong evidence of live sightings in Iceland. Lilljeborg named the European extinct species *Balaenoptera robusta,* and shortly thereafter, in 1869, a living specimen was described from the Pacific Ocean. Common names include Grayback Whale, Gray, Scrag, Mussel Digger, Hardhead, Ripsack, and Devil Fish (the names Hardhead and Devil Fish were acquired after Gray Whales smashed whaling vessels). A Spanish name is *ballena gris.*

DESCRIPTION: First impressions are of a stout, muscular animal as compared with other trim and swifter-swimming baleen whales. The head is blunt, narrow, and triangular in shape and is hooked slightly downward. The mouth is generally shorter than that of other baleen whales, and the throat is marked with only two to seven throat groves, and no pleats. Consequently, the shape and size of the head and throat significantly limit the opening of the mouth. The cream- or pearl-colored baleen consists of 130 to 180 plates attached to the upper jaw, and it is the very shortest and thickest of all baleen whales, evolved from the Gray Whale's distinctive method of eating. A number of hairs are scattered around the snout and lower jaw.

Gray Whales are one of only a few great whales, baleen or toothed, that lack a dorsal fin. Instead, a series of six to 12 bumps

Figure 59. Gray Whale, spyhopping.

is visible when the whale arches before diving; called peduncles, these bumps are part of the tail stock. The first peduncle, or knuckle, is the largest and is easily noted when the whale arches. The flippers are broad and paddlelike. The flukes are scalloped, with a deep notch in the middle, and are often scarred with barnacles or Killer Whale tooth marks. At the base of the tail is a bulge, similar in structure to the scent gland of land mammals, but its function as yet remains a mystery.

Their coloring is mostly dark gray, overlaid with molted white, yellow, and even orange resulting from the attachment of and scarring by ectoparasites. At birth and before parasite infestation, calves are uniformly dark or slate gray. The ectoparasites include one species of barnacle of the family *Cryptolepas* and three species of lice of the family *Cyamus,* two of which only occur on Gray Whales (fig. 60). In the warmer waters of the calving lagoons, parasites are sloughed off with old skin, but they leave scars that provide a mottled appearance. The motley pattern is mostly around the head and tail but varies considerably among individuals and with age.

Females are slightly larger than males, and identifying the sexes in the field is only possible based on behavior, such as a female swimming with a calf. Sexual maturity is reached at age 5 to 12 years, with an average for both species of 8. Gray Whales

Figure 60. Community of *Cryptolepas* barnacles and *Cyamus* whale lice on skin of Gray Whales.

are estimated to live at least 60 years, and the oldest recorded female was estimated to be 75 to 80.

FIELD MARKS AND SIMILAR SPECIES: The Gray Whale's head is shorter, blunter, and narrower than that of other baleen whales. When it dives, its arching back lacks a dorsal fin, and its exposed tail is scallop-shaped with a deep medial notch. When surfacing to breathe, its blows are straight up, short [3 to 4 m (9 to 12 ft)], broad, and bushy. Humpback Whales and Common Minke Whales are the only other large nearshore whales with which Gray Whales might be confused. Both Gray and Humpback Whales have white mottling around the head and tail, but the flippers of Gray Whales are short, whereas those of Humpback Whales are long and oarlike. Common Minke Whales have a dorsal fin and lack white mottling around the head but often have distinct white chevrons on their front flippers. Gray Whales spy hop or tail wag with their heads or tails exposed, at which time the head and tail can be clearly distinguished from those of other whales.

NATURAL HISTORY: Of all the great whales, Gray Whales stand out as the ambassadors of ocean mammals. They inspire awe by their astonishing migration and quirky feeding style and engender affection owing to their "friendly" nature. Within a single year,

Gray Whales undertake one of the greatest single migrations of any mammal. The round trip is 15,000 to 21,000 km (9,000 to 13,000 mi), and they lose a third of their body weight while mostly fasting on the journey. Their round trip begins from their highly abundant feeding grounds in the subarctic waters of Alaska, when changes in day length and the buildup of sea ice stir them to depart. They slowly swim south to give birth in the warmer southern waters of California and Baja, and then return northward in spring. They repeat this voyage year after year, from infancy into old age.

Of the great whales, Gray Whales are the only entirely coastal species, and they rarely venture more than 10 km (6 mi) from shore, except when navigating around offshore islands along their route. Notably, they also are the only large whales to give birth and breed in shallow waters and coastal lagoons. Since the 1990s, researchers have also discovered that some 20 percent of females give birth on their migration south rather than in the lagoons of Baja, many in the warmer waters of the Southern California Bight. The southern waters provide warmth to newborn calves that lack insulating blubber, and the lagoons furnish protection from predatory sharks.

"Ecosystem engineer" describes the Gray Whale, which dramatically transforms the seafloor communities where it forages, and "mud sucker" is an apt description of its specialized feeding technique, unique among all whales. It sucks mouthfuls of soft bottom sediments that are laden with millions of small crustaceans and polychaete worms. Its feeding excavations physically leave large furrowed swaths on the bottom, 3 to 6 m (9 to 20 ft) long, and a cloud of sediment follows in its wake as it plows back and forth for hours. One researcher estimated that each whale tilled around 0.4 km² (100 ac) per summer and calculated that the entire Gray Whale population plowed around 1.2×10^8 m³ (1.6×10^8 yd³) of mud, or more than 30 percent of the Beaufort Sea floor, in the same time period. The disturbed mud attracts other benthic animals, such as fish and crustaceans that eat the debris left in the wake of the whales, and the diversity and abundance of species in these disturbed patches are much higher than in areas where no foraging occurs. The suspension of sediments and the nutrient recycling likely contribute to a greater overall biological abundance in the Bering Sea. This effect is direct, upon sea floor communities, and secondary, due to crustaceans brought to the surface and eaten by other animals. Several sea-bird species, including phalaropes, auklets, and gulls, feed in the muddy plumes.

Gray Whales are social and intelligent, but the only enduring relationship is that between mother and calf. Even this relationship, however, ends at weaning. Gray Whales usually gather in groups of three to seven whales, but groups of 40 to 400 whales, and even of more than 1,000, can congregate in the calving lagoons. Large groups of whales are seldom observed on feeding grounds, because whales scatter to plow their own acre of sea bottom. Cooperation of any kind is rare with Gray Whales, except when defending young. Nevertheless, the same individual whales have been recorded over successive summers off Oregon and elsewhere.

While patiently waiting for whales to surface, researchers have studied their watches, counting the minutes whales remain underwater. During migration, whales dive for 3 to 5 minutes, surfacing for only a half a minute to take a few breaths. Usually, they dive for 7 to 10 minutes when foraging, but when resting in lagoons, they can hold their breath for 30 minutes. While underwater, a whale leaves "fluke" prints across the ocean surface, where its tail swish produces a blemish on the surface and a trail for researchers to follow. When swimming, they are not fast, averaging 10 kph (6 mph); however, when being chased by Killer Whales, they can briefly attain speeds of 16 kph (10 mph).

Gray Whales were long considered silent, because the sounds they emitted were confused with ocean background noises such as knocks, roars, pops, and grunts. They have a complex repertoire, however, consisting of sonarlike clicks, moans, and metallic sounds such as "bongs." Their low-frequency clicks are likely useful for detecting coastal features with sonarlike bounce back, as an aid in navigation. Navigation is tricky for Gray Whales, because their route can be within meters of coastal promontories and nearshore islands and is often through opaque waters.

Gray Whales commonly breach, fluke, spy hop, flipper wave, and tail wave, in most months of the year. They also like to rub against objects such as rocks and even human structures such as piers and boats. Rubbing may be a response to skin irritation caused by ectoparasites, but whatever the reason, the activity has led to people interacting with whales rubbing against boats with proffered hands.

"Friendly whales" is a term applied to Gray Whales, which were attracted to boats in the calving lagoons of Baja beginning in the early 1970s. Females with curious calves approached whale-watching boats and skiffs. A generation of friendly whales now approach, follow, and rub against boats and other objects, even allowing extended human hands to touch their young. Adults would have fiercely defended these same young from whalers in the previous century, accounting for the name Devil Fish.

REPRODUCTION: The calving season occurs annually between December and March, and the peak is mid-January to mid-February. Many newborn calves have been reported off central and northern California since the 1970s, as the population has grown and as sea temperatures have warmed. More recently, newborn calves have been seen as early as December 25 off Carmel and have stranded ashore live as far north as British Columbia, including at Bolinas Lagoon and the Smith River in California. During the 1998 El Niño–Southern Oscillation (ENSO), the northernmost sighting of a female with a newborn calf was in Discovery Bay, Washington.

Individual females give birth every other year or every third year, and in any given year, around half of the adult females are pregnant. The other half are in estrus and are receptive to mating, although some females, when they cannot eat enough to sustain a pregnancy, skip 1 year before breeding again. Gestation is 11 to 13 months, and calves nurse on rich, oozy milk for 6 to 7 months. Gray Whales are one of the only species of whale that have been observed birthing. The head emerges first, and the cow promptly supports the calf at the surface after delivery. Most births occur in shallow water, away from predators.

Females are attentive mothers, fiercely protecting their young against predators and, historically, against whalers. Mother and calf are inseparable, with the calf swimming alongside the female at her lower right or left quarter. This swimming configuration, in the slip stream created by the female's wake, likely assists the calf so that it expends less energy. Very young calves will swim and rest on the back of the female. Calves ingest around 190 L (50 gal) of milk per day from the female's two nipples, which protrude from slits on the belly when nudged by the calf. Once calves have gained sufficient weight to keep warm against colder waters, the pair begins its long migration. They leave late in the season and linger along the coast as they nurse and feed along the way. Cow–calf pairs are most often seen migrating north in May but also as early as March and as late as June. Along their route, cow–calf pairs will often linger at protected bays or around river mouths, where they likely nurse, rest, and feed. Examples of way stations along their route include San Clemente Island, Drakes Bay, the Russian River mouth, the Klamath River mouth, and Trinidad Bay. Once in arctic foraging grounds, the bond is continued into August and September for a total of 6 to 8 months after birth. By that time, the calf likely has learned how and where to forage, achieving a length of around 8.5 m (28 ft).

After weaning and while migrating south, females go into estrus from November through January, and peak ovulation is in January. Mating primarily occurs in or around the lagoons and coastal areas of Baja but is not limited to there; it also has been observed all along California at locations such as the Channel Islands, Santa Cruz, Drakes Bay, the Russian River mouth, Big Lagoon, and Trinidad Bay.

Neither males nor females hold territories on calving or feeding grounds, and there apparently is little male rivalry. Nevertheless, dominant males appear to occupy the Baja lagoons, maximizing their opportunities to mate, while younger males may mate with females migrating north from lagoons after they are inseminated by the dominant males, or with those migrating south when they are just coming into estrus. Researchers have long noted threesome mating groups that include a secondary male that was thought to assist; however, researchers have since observed courtship involving multiple pairs with both sexes mating with several partners. Consequently, the mating system has been classified as promiscuous. Male Gray Whales have enlarged testes compared to other large whales, except the Northern Right Whale, which also suggests that males may compete by volume of semen rather than by fighting or singing.

FOOD: Gray Whales need to sustain or increase weight in a short time to make up for their winter fast, so they seek out areas where food is superabundant, feeding continuously from May through October. Until recently, researchers thought that Gray Whales did not eat during migration, but since the population has rebounded, they are commonly seen feeding in bays and river mouths along the northward migration, including in Monterey Bay, San Francisco Bay, Drakes Bay, Tomales Bay, Trinidad Bay, and Puget Sound. Once on the feeding grounds, whales feed almost continuously and rarely rest. One fisherman watched a Gray Whale feeding constantly for several hours near Tiburon in San Francisco Bay.

The most common eating method is plowing the sea floor for invertebrates embedded in mud or swarming just above the surface of the mud. The whale tilts on its side, usually the right side, and then plows along the bottom, forming short furrows as it sucks up mud or sand. The invertebrates living in the mud are then entangled in the baleen as the whale pushes out the sediment with its large muscular tongue. Gray Whales also skim surface waters with mouth agape, sucking in krill, red crabs, or larval crabs, and engulf prey just below the surface, eating schooling fish or krill.

By gulping prey on the bottom, in the midwater, and at the surface, Gray Whales are able to exploit more than 80 species of invertebrates and fish. They feed on benthic organisms such as copepods, amphipods, and worms, but also on swarming mysid shrimp and krill, dense concentrations of larval crab, red crabs, and small schooling fish such as anchovies. In Monterey Bay, they feed in and around kelp beds, and sea grasses are in the stomachs of dead whales, suggesting that they eat crustaceans attached to the grasses. In Alaska, they forage mostly on amphipods embedded in soft mud at concentrations of $1,022/m^2$ ($11,000/ft^2$). Off Oregon, however, they forage in summer mostly on swarming mysid shrimp. One Gray Whale can consume around 170 kg (375 lbs) of invertebrates per day, and a calf raised in captivity was fed more than 800 kg (1,800 lbs) of squid per day.

MORTALITY: Because they hug the coast, their carcasses often wash ashore in fairly fresh condition, and researchers can often decipher the cause of death. Natural mortality is almost exclusively due to starvation and predation by Killer Whales. Until recently, Killer Whale predation was merely surmised, based on bite marks and characteristic removal of the tongue in carcasses, but in the 1990s, many researchers and whale-watching boats observed with a mixture of awe and horror as Killer Whales attacked mothers and calves in Monterey Bay. The attacks occur in deep water as the pair traverses the wide and deep Monterey Bay canyon, and perhaps those that hug the shore elude Killer Whale sonar detection. Great White Sharks also prey on calves, but more die from starvation, as was the case in 1999 to 2000, following the 1998 ENSO. Rarely, Gray Whales are trapped in sea ice and drown.

DISTRIBUTION: Because Gray Whales spend their lives in coastal waters, their geographic distribution is better understood than that of most cetaceans. They range from the Beaufort and Chukchi Seas in Alaska down to Baja California and mainland Mexico. In the Beaufort Sea, they occur west into the Siberian Sea and north to the edge of the pack ice, presently at latitude 69° N, but this line continues to retreat with global warming.

They are attracted by superabundant food and spend May to October on the feeding grounds until they are plumped up enough to undertake the long southward migration to calving grounds, give birth, and migrate back to Alaska. Whales migrate southward as early as August, and a distinct peak occurs in early January (34 percent of migratory animals) that stretches through early March. Pregnant females initiate the exodus south, followed by nonpregnant females, males, and juveniles. After gorging in Alaskan waters, whales apparently require no

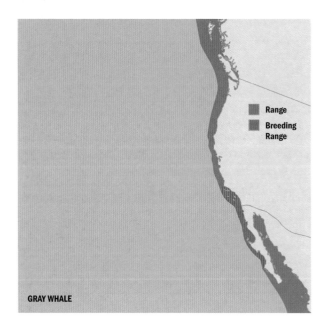

Range
Breeding Range

GRAY WHALE

food on the southward migration and travel directly, reaching their destination within 49 to 55 days. Swimming speeds average only 1.3 to 2.2 kph (3 to 5 mph), but whales traverse 90 to 185 km (55 to 115 mi) per day, traveling continuously with short stopovers. They will stack up at some promontories along the route, presumably to rest or perhaps wait for other migrants.

The migratory pathway is generally close to shore, following a distinct route less than 8 km (5 mi) offshore, which bifurcates off Point Conception, with some whales continuing along the mainland shore and others following a route through the Channel Islands to eventually rejoin the coastal group off Baja. Pregnant females and those with small calves born along the route arrive first at their southern destination, followed by courting males and nonpregnant females. The numbers of whales within the lagoons peak in early March, and then the whales begin their return journey north.

Unlike any other large whales, most females give birth within several lagoons and bays along the western coast of Baja California and the southern waters of the Gulf of California. The core calving lagoons are Guerrero Negro, Scammon's, San Ignacio, and Magdalena (the largest). Until the 1980s, a small group of whales also calved in lagoons in the northwest coast of mainland Mexico, in Bahia Santa Maria, but development and industrial fishing on the coastline caused the whales to abandon the site.

Since the 1990s, many females also have been calving around the Channel Islands in southern California; the species likely has done so prehistorically. They also may have calved off Washington, because the Makah native peoples named the January moon in honor of the Gray Whale with its young.

The northward migration begins as early as December and extends through September, but the departure date varies among groups, with newly pregnant females leaving first, followed by adult males, then by juveniles, and lastly by females with calves. Most are gone from Baja in February, and numbers peak off southern California in March and off northern and central California in April to May. The protracted northward migration reflects the slower movements of mothers with calves. Whales can be seen from shore just beyond the surf, and they likely developed this strategy to avoid sonar detection by Killer Whales. Such a prolonged northward and southward migration means that, in certain time periods, individuals going both directions overlap: females with calves may still be slowly traveling north while other individuals are beginning to travel south.

Adding to the confusion of direction, more and more individual whales are not migrating all the way to Alaska but are spending the entire summer at locations along the way. This group, now called the "Pacific Coast Feeding Aggregation," includes resident whales that have been studied at Vancouver Island, Puget Sound, Whidbey Island, Depoe Bay, the Farallon Islands, Tomales Bay, Monterey Bay, and the Southern Channel Islands.

Gray Whales are strongly associated with the shallow continental shelf throughout their range. The primary migratory pathway north and south is within 2 km (1.2 mi) of shore but may range as far as 200 km (125 mi) offshore. Most whales, however, migrate within 5 km (3 mi) of shore, and the route is strongly influenced by topography and bathymetry. In northern California, where the continental shelf is narrow, whales may migrate within a mile of shore, but in southern California they range further offshore around the Channel Islands. The route of females with calves was not known for a long time, because researchers surveying whales by plane did so slightly offshore, thereby missing those that closely hugged the coast within 10 to 15 m (30 to 50 ft) of the surf line.

These whales are associated, for the most part, with shallow depths of 3 to 122 m (10 to 400 ft). Their feeding habitats are composed of soft bottom substrates of mud or sand, which account for vast areas of the coastal shelves of the Bering and Chukchi Seas, and for the floors of numerous bays from Puget Sound to San Francisco. Soft sediments are also dominant at river mouths such as the Columbia, Klamath, and Russian.

VIEWING: Coastal promontories all along the migratory route have long been acclaimed as the land lubber's alternative to seeing Gray Whales from a boat. Several state and national parks provide special viewing sites: Lime Kiln Point State Park on San Juan Island, Olympic National Park, Redwood National and State Park, Point Reyes National Seashore, Año Nuevo State Park, and Channel Islands National Park. Other excellent locations include Granite Canyon near Carmel, Piedras Blancas along Big Sur, Coal Oil Point in Goleta, Point Vicente on the Palos Verdes Peninsula, and Point Loma in San Diego.

By boat, whales are easily observed around Vancouver Island, Puget Sound, and the Olympic Peninsula; Depoe Bay and Cape Foulweather in Oregon; and Point Reyes, Monterey Bay, Santa Barbara Channel, and the Channel Islands in California. The most exquisite experience, however, is to see hundreds of whales gathered with calves in the lagoons of Baja California, by boat or kayak.

CONSERVATION: Aboriginal people hunted Gray Whales for thousands of years in the eastern Pacific, and European settlers practiced shore-based whaling in the 19th century, until the 1880s, on the site of what is now the University of California Santa Barbara in Goleta. Whaling intensified in the California Current region after the whaler Charles Scammon discovered the calving grounds in Baja in 1857, and in a brief 30-year period, more than 11,000 were slain. Scammon became an advocate for the whale's protection in the late 19th century, but by the 20th century, the species was nearly extinct. In 1931, Gray Whales received some protection under a convention on whaling, but not until 1946 were they truly protected: in that year, many countries joined the IWC and agreed to ban commercial hunting, allowing only aboriginal hunting and collection for scientific research. The Mexican government protected breeding lagoons in Baja, creating the first whale sanctuaries in the world in 1972.

Agreement	Conservation Status
U.S. ESA	Listed endangered 1969; delisted 1994
U.S. MMPA	No special status
COSEWIC	Special concern
IUCN	Least concern

Modern concerns are mostly associated with the coastal migration of Gray Whales. Along the nearshore route, they are exposed to several threats—oil rigs and boat traffic off southern California, intense shipping out of San Francisco, a forest of fishing gear lines from crab pot lines and nets, and polluted sediment and waters near urban centers. Mortality and serious injury from human activities include entanglement in various fishing gear, such as salmon, thresher shark, and swordfish set gillnets, and crab and lobster pot lines. One study estimated that 25 percent of the dead Gray Whales in British Columbia were incidental to gillnet fisheries, longlines, trap, and seine, and another study found that 20 percent of whales had scars from entanglement or ship strikes.

Native American subsistence hunters in Alaska and Washington kill a small number of whales each year, averaging around 97 over a 5 year period in the late 1990s; Russian aboriginals hunted the great majority of these, with the Makah of Washington only hunting around four. Hunters may lose harpooned whales: a whale with a harpoon protruding from its back was seen in 2000 at Shelter Cove, California.

Whales have long been noted to avoid shipping areas and oil drilling sites in southern California, and they are sensitive to continuous sounds that exceed 120 dB^2. Routinely, most Gray Whales avoid San Francisco Bay along their migratory path, and they have altered their migratory routes in Mexico to avoid whale-watching boats. Consequently, the U.S. and Mexican governments promote guidelines for whale-watching (see Marine Mammal Watching). Gray Whales also are more vulnerable to pollution because they reside mostly nearshore, and in developed areas, they may ingest polluted sediments and invertebrates when feeding.

Changes in available food can lead to starvation—amphipod densities in the Chukchi and Bering seas are sensitive to changes in ocean temperature. A large number of Gray Whale deaths in 1999 and 2000 was likely a result of starvation; many died along the migratory route north, and many females did not calve that year. In Oregon, resident whales had poor body condition during a year when upwelling was delayed, causing mysid shrimp, their primary food there, not to reproduce. Changes in available food or in the number of whales also may explain why more are observed foraging along the coast from California to Washington in recent years. The warming trend of waters in arctic and subarctic waters may cause less food for whales in the future, but in the short term, Gray Whales have extended their feeding range in the Arctic as the pack ice has retreated.

POPULATION: Two surviving stocks of Gray Whales flank the Pacific Ocean, the western north Pacific (Korean–Okhotsk) and the eastern north Pacific (California–Chukchi). The current population estimate for the eastern stock ranges from 18,000 to 29,000, with an average of 22,000. The western Pacific stock is critically endangered, with a population estimated at a mere 100, making it one of the most endangered of whales.

Pre-whaling population estimates range from 34,000 to as high as 118,000, with an average of 96,000. The higher estimate is based on an analysis of DNA variability for Gray Whales that form the western and eastern stocks, and it is controversial. In recent times, researchers have estimated the population of the eastern north Pacific stock by counting the number of whales passing points of land in Carmel, Los Angeles, and San Diego, during their southward migration. Between 1967 and 1988, the population increased by 3 to 4 percent per year, and their recovery was so successful that the species was taken off the endangered species list in 1994. Over a 2-year period from 1999 to 2000, however, over 600 whales stranded dead along the coast, compared to the previous average annual rate of 38.

COMMENTS: Gray Whales are the only species of great whale that people have tried to raise in captivity. These mighty efforts involved two Gray Whale calves that were successfully treated in captivity and released—Gigi in the 1970s and J.J. in 1997. J.J. was released after doubling in length to 9 m (29 ft) over 1 year. Whether they lived long or not is not known, but the vast efforts to raise them are a tribute to their role as ambassadors of the sea.

REFERENCES: Angliss and Allen 2008, Alter et al. 2007, Baird et al. 2002, Buckland and Breiwick 2002, Calambokidis et al. 2002, Daily et al. 2000, Gulland et al. 2005, Heckel et al. 2001, Jones et al. 1984, Jones and Swartz 2002, Le Boeuf et al. 2000, Mate and Urban-Ramirez 2003, Moore and Clarke 2002, Newell and Cowles 2006, Oliver and Slattery 1985, Reilly et al. 2008, Rice and Wolman 1971, Shelden et al. 2004, Smith 2005, Sullivan et al. 1983.

SPERM WHALE   *Physeter macrocephalus*
Pls. 1, 2; Figs. 38, 61.

Adult male average length is 16 m (53 ft). Adult female average length is 11 m (36 ft). Average weight for both is 45,000 kg (50 tons). Newborn calf average length is 4 m (13 ft).

Figure 61. Sperm Whale.

SCIENTIFIC NAMES: The scientific term *Physeter macrocephalus* is quite apt. Linnaeus chose the name *Physeter* from the Greek for "bubbler" or "blower." The Latinized Greek adjective *macrocephalus* means "large" or "big-headed." There is a scholarly synonym for *P. macrocephalus*; some scientists call the Sperm Whale *Physeter catadon,* the species term deriving from the Greek for "lower tooth," referring to the great teeth on the lower jaw of the whale (fig. 61). Which name has priority has been long debated, but the International Code of Zoological Nomenclature's Article 24(a) designated *macrocephalus* the preferred name.

COMMON NAMES: The name Sperm Whale comes from 18th-century whalers' belief that the spermaceti found in this whale's head was its sperm. Spermaceti was a valuable, white, waxy substance with many uses, such as in candles, ointments, and cosmetics. Spermaceti derives from Latin, meaning "sperm of the whale," which in turn can be traced back to Greek roots. The term Sperm Whale is the shortened form of the 19th-century Spermaceti Whale. The name Cachalot was an alternate term in most of the whaling period; it persists in English and is the standard term in French. The word originated in the Portuguese *cachalote,* which means "head" or "big head." *Cachalote* was adopted by the Spanish for Sperm Whale as well.

DESCRIPTION: The Sperm Whale is the largest of the odontocetes, or toothed whales, as well as the most powerful. Indeed, it is the most massive predator on Earth. We do not know the maximum size a bull Sperm Whale can attain because whaling in the 18th and 19th century selectively took the biggest males. A few specimens recorded or preserved in New England, however, suggest that bulls were more than 24 m (80 ft) in length. Today, males may be more than 18 m (60 ft) in length, and females may reach 14 m (46 ft). The 41,000 kg (45 ton) bulls are about three times the mass of the 13,600 kg (15 ton) females, which makes the Sperm Whale one of the most highly dimorphic cetaceans.

The most striking feature of the Sperm Whale is its massive boxcarlike head. The spermaceti organ and the rest of the nasal complex account for most of the head's bulk. The whale's head may be up to one-third of its length (Whitehead 2005) and constitutes one-quarter of its mass. The waxy spermaceti fills a cone-shaped chamber that extends from the tip of the snout to the skull. Below the organ is the "junk," a whaler's term for the denser form of spermaceti, which lies in a series of chambers above the mouth. The skull is dish-shaped, like a parabolic reflector for sounds. Within the skull lies the great brain of the Sperm Whale. This brain is larger than that of all living and extinct species of animals, averaging 7.8 kg (15 lbs) and occasionally exceeding 9 kg (20 lbs) in weight.

Below its broad head, the Sperm Whale's narrow jaw alone can be one-fourth of its body length. The lower jaw holds 16 to 30 pairs of teeth; these fit into sockets in the upper jaw. The conical teeth are 8 to 20 cm (3 to 8 in.) in length and can weigh as much as 1 kg (2 lbs) each. The skull is the most asymmetrical of any whale, to accommodate the complex nasal passages, which terminate in the S-shaped, single blowhole on the far left tip of the snout.

The skin of the Sperm Whale's head is smooth, but its back is deeply wrinkled in rows stretching from head to tail. Its color is black, blue-gray, or dark gray, with a pale patch around the genital area and a pale line along the lower jaw. With age, its color lightens, and white patches appear around the head. The skin can also be marred by sucker scars from cephalopods. The back is graced with a low, rounded dorsal fin that precedes a series of bumps extending along the spine to the tail stock. In females and some young males, the dorsal fin has calluses, the cause of which is unknown. The flippers are unremarkable; they are proportionately small and paddle-shaped. The tail has a notched and fairly straight edge, but the trailing edge may be irregular because of nicks and bite marks. When raised out of the water before diving, the tail appears triangular-shaped.

FIELD MARKS AND SIMILAR SPECIES: The Sperm Whale's square head is iconic and distinguishes it from any other whale if well seen. The bumps along the back may be confused with those of a Gray Whale, but upon closer examination, the head shape, blow, and tail are distinctly different. The Sperm Whale's bushy blows are low and directed leftward, because the single blowhole near the tip of its head faces forward and to its left; it is the only whale with such an asymmetrical blowhole.

NATURAL HISTORY: Sperm Whales are among the deepest divers of marine mammals, and males regularly dive to 2 km (1.2 mi), remaining underwater for around 1 hour. The deepest dive recorded was more than 2,100 m (6,890 ft) for 1 hour, but this depth of dive has not yet been recorded using newer technology. The smaller female does not dive as deep or for as long, but she still rivals most other marine mammals. Researchers have even attached video cameras on whales with suction cups and watched their diving, resting, and socializing underwater. One whale eyed the camera up close, perhaps out of curiosity.

The Sperm Whale may sometimes be heard as well as seen. Indeed, for centuries, sailors heard mysterious knocking through the hulls of their ships, as if crews of carpenters sometimes worked at the bottom of the sea. Only in the 1950s were Sperm Whales identified as the source. Recent research shows that Sperm Whales are one of the loudest animals on Earth—they generate clicks rated as high as 230 decibels underwater. The clicks emanate from the whale's single nostril, travel backward through the spermaceti organ, and then rebound from the air sac in front of the skull and travel forward through the junk out into the ocean. In fact, the nose complex of this whale is the largest on Earth, acting as a massive sound amplifier that can beam clicks like a searchlight before it. Researchers have recorded returning echoes of clicks from the sea surface and sea bottom, which may let the whales orient in the blackness of the 400 to 900 m (1,300 to 3,000 ft) depths they most often visit. At diving depths, trains of steady clicks may accelerate into squeals. The squeals suggest that the whale may be closing in on its prey, fine-tuning its position. The characteristic clicks of Sperm Whales serve in field identification during research cruises. Hydrophones towed behind survey vessels detect the clicks, and whales can be located, even though they are deep below the surface.

The social structure of Sperm Whales is complex and has been intensely studied, providing us with some insights into the life of this leviathan of the very deepest waters of the Pacific. Female and young Sperm Whales live in small, stable "units" of nine to 12 whales that are usually found in tropical or temperate seas. Units of Sperm Whales signal with distinctive sets of codas, which are stereotyped sequences of clicks. The units often join with other units, forming "nursery schools" of 20 to 30 individuals in transitory groups as the whales forage or socialize near the surface (Perry et al. 1999). Groups can also rest together head down at midwater depths. In these clusters, Sperm Whales may touch and caress one another with their flippers and jaws.

Units belong to "clans," which are much larger groups of Sperm Whales with similar sets of codas. Clans are represented over vast areas of the scale of 10,000 nautical miles. Clans may overlap with others, but units seem to almost always socialize within their own clans. The distinctive coda characteristics of the units and clans have suggested that Sperm Whales are cultural creatures. Sperm Whale units and clans may share ecological wisdom as well as codas, training the young regarding where and how to hunt.

Young male Sperm Whales leave their units when they are about 6, and then travel to higher, colder latitudes and form bachelor schools that are composed of mixed ages of males of up to 20 years of age. After males reach sexual maturity at 20 years, they seem to live a relatively solitary life and rarely rest close to others like whales in their maternal units. Nevertheless, male whales can be found in very loose aggregations, which may reflect concentrations of prey or undersea geography.

When the males reach sexual maturity, they turn southward, temporarily joining breeding groups and units to mate. There may be violent contests between sexually mature bulls, called "schoolmaster bulls," on the breeding grounds, which are suggested by scars on some of their bodies. Although such fights were described by whalers in the days of sail, there are no recent accounts of these rare events.

Females are sexually mature around 9 years of age, after which growth slows, finally stopping about age 30 years, when they are 11 m (34 ft) long. Puberty is extended in males, however, from 10 to as late as 36 years by some reports, but it occurs at an average age of 25 years. They continue to grow until about 50 years of age, reaching about 16 m (52 ft) in length. With such a long time to reach sexual maturity, it is not inconceivable that their life may span 100 years.

REPRODUCTION: Sperm Whales have an exceptionally long interval of 5 to 7 years between births. Part of the reason for this protracted period is because the pregnancy itself lasts 15 to 17 months. When born between June and November, the single calf is 4 m (13 ft) long. The calf can eat solid food before age 1 but may suckle for several years, one of the longest periods for any cetacean. This extended care-giving further adds to the long time between births. Similarly, the small "nursery school" units point to their long-term care extending to puberty, with several adults contributing to their "training."

There are no known mating grounds, but mating likely occurs when loose aggregations of males join "nursery schools" in lower latitudes from April to August; however, it has never been observed. Similarly, there are no known calving grounds, but occasionally newborn calves have washed ashore, and a few births have been witnessed. One extraordinary story described a birth off Central America in October of 1983, after which several whales came to interact with the calf, apparently assisting. A researcher entered the water after the birth took place to observe the interactions, and was herself approached to within 10 m (32 ft) by the calf, which was then slowly intercepted by the mother.

FOOD: Sperm Whales eat a variety of deepwater squid, octopus, and fish. Bulls tend to pursue larger prey on average, but both females and males are known to capture both giant and colossal squids, and also the smaller Humboldt Squid. Over 40 species of cephalopods have been identified in their diet throughout the world, but off California, 24 species were recognized from stomach contents of harvested whales, including six dominant ones of the species *Moroteuthis, Histioteuthis* (two spp.), *Galiteuthis,* and *Gonatopsis borealis.* The skin of some Sperm Whales bears sucker marks from these squid, testimony to underwater struggles between these great vertebrate and invertebrate predators. Fish account for only a small proportion of their diet in the Pacific, but deep-dwelling rockfish are examples. In the Atlantic, however, several midwater and bottom fish were commonly eaten, such as lumpsucker, rockfish, cod, and Blue Whiting. Sperm Whales may eat 907 kg (1 ton) of prey, whether gelatinous or bony, each day.

Sperm Whales hold the records for the deepest dives by any marine mammal, although these dives may be exceptional and not the rule. A bull whale killed in South Africa after a nearly 2-hour dive had in its stomach two dogfish of a strictly bottom-dwelling species. The depth of the sea floor there exceeded 3,000 m (9,849 ft).

Sperm Whales are often silent at the surface but click when diving, and cameras attached to Sperm Whales have recorded them opening and closing their brightly lined mouths. Whether Sperm Whales primarily find prey by echolocation or by visual means is an open question. The whales may see prey silhouetted by dim down-welling light or may perceive them through the bioluminescence common to many deep-dwelling invertebrates.

MORTALITY: Mass stranding of Sperm Whales is rare, but one such event occurred at the mouth of a river near Florence in Oregon in June 1979, when a school of 41 whales stranded in a line along the beach. About one-third of the group consisted of young males, and the rest were females ranging in age from 11 to 60 years. This extraordinary event provided an exceptional opportunity to study the social structure of a herd, but alas, there was no obvious reason for their deaths at the time.

Killer Whales prey on Sperm Whales, and smaller groups of Sperm Whales may array themselves in a "marguerite" pattern if Killer Whales are nearby, heads inward and tails outward. In one well-described case off the coast of Point Conception in southern California, waves of 25 Killer Whales attacked such a Sperm Whale rosette in a "wound and withdraw" pattern for 4 hours. Sperm Whales left the rosette to lead wounded companions back into formation, despite being attacked themselves. The Killer Whales finally killed an isolated Sperm Whale, and then carried off the carcass, leaving the rest with grievous wounds. Larger groups of Sperm Whales may gather from nearby waters to form "tight flotillas." In one case off central California, more than 50 Sperm Whales joined together in one of these close formations; no kills were observed. Sperm Whales may fall into a marguerite formation when encircled and harassed by False Killer Whales and pilot whales as well.

Curiously, about 6 million years ago, a species of Sperm Whales with large teeth on both lower and upper jaws cruised ocean waters. These extinct cetaceans may have played the master predator role that Orcas do today. "Killer Sperm Whale" is the name that has been suggested for these grand Miocene predators.

DISTRIBUTION: The Sperm Whale is a cosmopolitan species, found in all the world's oceans, in temperate and tropical waters. Only the Killer Whale may have a wider range. In the northern Pacific, males range as far north as latitude 70° N, and older solitary bulls even venture right to the edge of the pack ice. Females and young, however, venture only as far as 50° N latitude. Off the eastern Pacific coast, Sperm Whales favor the deeper waters near or beyond the continental break.

Typical of most large whales, Sperm Whales migrate to northern latitudes in summer and to lower latitudes in winter. In summer, mature males may migrate as far north as the Bering Sea, the Aleutian Islands, and the Gulf of Alaska, in search of various deepwater prey. In winter, all sex and

SPERM WHALE

Range

age classes mingle in tropical waters and rarely venture north of latitude 40° N. Sperm Whales can be observed any month of the year in California and Baja, and all months but December through February off Oregon and Washington. These seasonal patterns may change in response to El Niño–Southern Oscillation (ENSO) events or other large oceanic changes.

Off central California, abundance is higher in mid-May and mid-September, a pattern that is likely related to the whales' migration along the California Current; however, in winter, breeding whales are also seen there over the continental slope.

Sperm Whales rarely venture close to shore, and instead occur in waters deeper than 300 m (1,000 ft) around deepwater features such as islands and seamounts and the continental shelf and shelf break. Deep marine canyons are the exception, where upwelling occurs and food abundance may be higher. Similarly, females with calves are often constrained to sea temperatures of 10 to 20 degrees C (50 to 68 degrees F).

VIEWING: On an 8,000 km (5,000 mi) research cruise between the mainland United States and Hawaii in the 1990s, viewers counted only eight groups of Sperm Whales. However, hydrophones towed behind the vessel picked up the distinctive calls and codas of 28 groups. Extrapolating from the width of the vessel's probable visual or acoustic detection, experts estimated there were about 30,000 Sperm Whales in this broad area. Expect to see Sperm Whales only rarely, even if you spend many hours scanning the sea.

CONSERVATION: Sperm Whales were long favored by whalers because of their valued spermaceti oil, which could be used to create a smokeless candle. The first Sperm Whale killed by shore-based whalers off California was landed by the Carmel Whaling Company in 1875. Whalers sought Sperm Whales specifically because of the quality of the oil produced by the spermaceti and because they also extracted ambergris, which they found in the intestine. Europeans used ambergris in perfumes, but the Sperm Whale's need for this substance is unknown. This fatty material is black in a living whale but bluish white when refined. Judging from whaling logs and other sources of information, more than 400,000 were harvested from the North Pacific alone from 1800 to the end of most commercial whaling in 1987, and 258,000 of these were killed between 1947 and 1987 (stock estimate).

The post–World War II Japanese and Russian factory ships killed a high of over 6,000 whales per year in less than 10 years in the eastern Pacific during the 1960s. Rogue whalers may still take Sperm Whales, and there is a small indigenous Sperm Whale hunting tradition in Indonesia.

Agreement	Conservation Status
U.S. ESA	Endangered
U.S. MMPA	Depleted
COSEWIC	Candidate
IUCN	Vulnerable

Because they feed at great depth, Sperm Whales do not compete with most commercial fisheries. Nevertheless, Sperm Whales do interact with driftnet fisheries off California and longline fisheries for halibut and sablefish in Alaska. Indeed, several male whales that washed ashore in California died from massive balls of fish nets that they had ingested. The fish nets were from fisheries around the Pacific, and theories abound regarding how the netting ended up in their stomachs. One theory suggests that the whales are getting a mouthful of net when they grab food from fishing gear hanging in the water, and another has whales confusing ghost nets for prey. Some whales also become entangled in nets, and a couple of whales are estimated to die each year due to entanglement in the California drift gillnet fishery. Despite these interactions, the collapse of many world fisheries may have a minimal impact on Sperm Whales because their deepwater prey are not usually harvested by commercial fishermen.

Even Sperm Whales get the bends. As Sperm Whales age, their bones seem to slowly suffer from expanding nitrogen bubble damage as they ascend from the depths. The impact of military low-frequency sonar (LFA) on deep-diving whales may accelerate such damage, if the whales surface too rapidly in response to sudden or painful blasts of sound. Evidence is not clear on the effects of LFA on Sperm Whales, but related Dwarf Sperm Whales may have beached because of sonar experiments.

POPULATION: Unlike Northern Bowhead Whales and right whales, Sperm Whales have recovered in some seas, although the effects of whaling persist in the smaller sizes and lowered numbers of bulls. Sperm Whale populations may still be shifting to regions where earlier populations were exterminated. The effects of earlier whaling persist in part because of pre-1987 IWC recommendations that larger bulls be preferentially taken, which assumed they could be replaced by younger bulls. Younger bulls wait to breed, and the loss of many mature bulls meant that many females simply were not impregnated, and so recovery of the species was retarded.

World population estimates vary widely. An estimate of 360,000 has been recently made by projecting numbers for the rest of the world ocean from those in systematically surveyed areas. The original population of Sperm Whales may have been 1,100,000. The population of Sperm Whales of the California Current may be 1,400 individuals or more. Although larger populations of Sperm Whales exist to the west and south of the California Current, there is little evidence of local mixing with these animals. It is likely that California Current Sperm Whales visit northern British Columbia, but there are no population estimates there.

COMMENTS: Unlike most other cetacean species, Sperm Whale bulls were known to fight whalers. One famous 19th-century incident was the sinking of the whale ship *Essex*. The *Essex* sailed from Nantucket in Massachusetts in 1819 and rounded Cape Horn, but it only captured eight "spermaceti whales" off the South American coast. Sailing northeast into the tropical Pacific, the *Essex* finally found a "shoal" of Sperm Whales on the surface. The *Essex* launched its three light open whale-catching boats. The mate, Owen Chase, was harpooner in one of the whale boats, and sighted a whale. The approximately 26 m (85 ft) bull whale rammed the *Essex*, bringing her 216,000 kg (238 tons) to a full stop, as if she had hit a reef. The whale hesitated, then rammed the *Essex* again, going at 6 knots this time, and stove in her bow, and she capsized in 10 minutes. What had happened? Was the bull defending the wounded Sperm Whales attacked by the men in the whale boats? Some experts believe that the immense head of the Sperm Whale evolved as a battering ram from the foreheadlike melon common to the toothed whales. Indeed, among the odontocetes, the greater the extent of sexual dimorphism, the greater the relative size of the melon, with the Sperm Whale representing the extreme case.

REFERENCES: Arnbom et al. 1987, Fiscus et al. 1989, Gosho et al. 1984, Moore and Early 2004, Philbrick 2000, Rice et al. 1986, Watkins et al. 1993, Watkins et al. 2002, Watwood et al. 2006, Weilgart and Whitehead 1986, Whitehead 1998, Whitehead 2002a, Whitehead 2002b, Whitehead 2003.

PYGMY SPERM WHALE   *Kogia breviceps*
Pl. 16; Fig. 62.

> Adult average length is 4 m (13 ft); average weight is 450 kg
> (1,000 lbs). Newborn average length is a bit over 1 m (4 ft).

SCIENTIFIC AND COMMON NAMES: According to one account, the term *Kogia* was a 19th-century jest. This small whale seems to have reminded J. E. Gray, long-time keeper of zoology at the British Museum, of an old "codger," a word that Gray then disguised in a similar-sounding word that he coined as its genus name. Many of his contemporaries were not pleased with his pun; the famed zoologist Richard Owen vainly hoped there would be a "burial and oblivion of [such] barbarous and undefined generic names with which the fair edifice begun by Linnaeus has been defaced."

Brevi comes from "short" in Latin, and *ceps* derives from the Greek word for "head." Thus, *breviceps* signifies "short-headed" and refers to the roundish skull of this genus, which is so unlike the long skulls of the rest of the cetacea (fig. 62). All small sperm whales were once considered to belong to the species *Kogia breviceps,* but in the 1960s, the Dwarf Sperm Whale *Kogia sima* became widely recognized as a species separate from *breviceps.* More recently, genetic evidence suggests that there may be a third *Kogia* species: that *Kogia sima* may be divided into Atlantic Dwarf Sperm Whales and Indopacific Dwarf Sperm Whales.

In English, Pygmy Sperm Whale is the standard name for *Kogia breviceps.* The French and Spanish names are similar: *cachalot pygmée* and *cachalote pigmeo,* respectively.

DESCRIPTION: The Pygmy Sperm Whale is a larger edition of the Dwarf Sperm Whale. Both species of small sperm whale are

Figure 62. Pygmy Sperm Whale; note false gill.

often compared to porpoises and sharks. Their bodies are big headed and robust, making them reminiscent of porpoises—but at the same time, their underslung jaws, dark eyes, and bluntly pointed heads suggest sharks.

The forehead of the Pygmy Sperm Whale contains chambers filled with spermaceti, the valuable oily or waxy substance prized by whalers in the 19th century. The spermaceti helps focus the beam of sound the Pygmy Sperm Whale can broadcast into the sea. The sounds originate in a single pair of "monkey lips," internal lips within the forehead. These lips resonate other structures whose vibrations are directed forward through the spermaceti chambers, exiting the head above the mouth. The foreheads of all species of sperm whale are specialized and complex, with structures sometimes lying mostly to the right or left. The shortened skulls of *Kogia* whales, which house the acoustic organs, are the most asymmetrical of all cetaceans, and perhaps of all mammals.

As in the Dwarf Sperm Whale, the dark eyes are large and prominent, well adapted for vision in dim light. The backs of the eyes reflect light, as in many nocturnal birds and mammals. The double pass of the light through the photoreceptors increases the eyes' sensitivity to any source of brightness in the dark environment. *Kogia* possess two kinds of reflective cells—those that shine back blue-green light and those that shine back blue light.

Light near the surface is predominately blue-green, whereas only blue light penetrates below 91 m (300 ft); thus, vision should be enhanced in both ranges.

Pygmy Sperm Whales are countershaded: bluish gray or brownish black above, and lighter cream or pinkish below. Behind the eyes on either side of the head is a "false gill," a bracketlike marking that is light with a dark border. Below these false gills lie the rounded flippers. The low, curved dorsal fin is less than 5 percent of body length and lies after the midsection. The tailstock is relatively long and is laterally compressed. The flukes are deeply notched.

The blowhole lies slightly forward of the eyes on the back of the head, lying somewhat to the left. The inconspicuous blow projects forward.

In its lower jaw, the Pygmy Sperm Whale has 12 to 16 pairs (sometimes only 10 or 11 pairs) of sharp, fanglike teeth that fit into sockets in the upper jaw. The Pygmy Sperm Whale has no teeth in its upper jaw, unlike the Dwarf Sperm Whale (which can have as many as three pairs).

Like the Dwarf Sperm Whale, the Pygmy Sperm Whale possesses a special intestinal sac that can release a cloud of reddish liquid when the whale is harassed. This colorful inking may account for one of its Japanese names, *tsunabi,* which means "firecracker."

FIELD MARKS AND SIMILAR SPECIES: *Kogia* sperm whales' sharklike heads distinguish them from any other cetacean except for a baby Sperm Whale—it might even be possible to mistake this mammal for a fish. It would be rare to see this whale species under the conditions necessary for a positive identification—calm water and close proximity. Even under ideal conditions, Pygmy Sperm Whales and Dwarf Sperm Whales are difficult to tell apart. The smaller dorsal fin of the Pygmy Sperm Whale is further toward the rear; the taller fin of the Dwarf Sperm Whale lies near the center of the back. The Pygmy Sperm Whale has a squarish head when compared with the more conical snout of the Dwarf Sperm Whale. Pygmy Sperm Whales have been reported to rest lower in the water than their smaller cousin.

NATURAL HISTORY: Little is known about Pygmy Sperm Whales, whether about their life cycle or their social life. When seen on the surface, the whales rest quietly in small pods of perhaps up to seven individuals. Like Dwarf Sperm Whales, they rise, breathe, and dive quietly. The whales may rest motionless at the surface for some time, creating the basis for a historic Japanese name: *uli-kujira*, the "floating whale."

Pygmy Sperm Whale mothers and calves sometimes strand together when the mother is hurt or sick, a sad illustration of the strong bond between the two. Pygmy Sperm Whales sometimes also strand in larger groups, reflecting the social nature of this (as well as many other) cetacean species.

Pygmy Sperm Whales do not whistle like most cetaceans. Their clicks are unusual, being emitted at a high frequency in a narrow band, and being comparatively long and weak. The clicks of Harbor Porpoises, Sperm Whales, and two other genera share similar characteristics, and they are thought to be difficult for Killer Whales to hear. Pygmy Sperm Whales are also inconspicuous and silent on the surface, making detection by predators less likely.

REPRODUCTION: Virtually nothing is known about reproduction in this rare and elusive whale. An annual birthing cycle is suggested by a gestation period of 9 to 11 months.

FOOD: Mid- and deepwater squid dominate the diet of Pygmy Sperm Whales, according to worldwide data. Their prey are found at depths of 500 to 1,300 m (1,700 to 4,300 ft), and their mouth shape implies that they may feed on the bottom. About 13 families of squid have been identified from stomach contents of stranded whales. *Kogia* whales likely use echolocation to locate prey, and their mode of eating likely involves suction, similar to that of other squid-eating whales. Some fish, as well as crustaceans such as shrimp, may also be eaten.

Argument exists about whether sperm whales of all kinds feed right-side-up or right-side-down. If a sperm whale fed with its back toward the bottom, then its prey might be backlit by the down-welling light from the sky above. If a sperm whale fed with its back toward the surface, then it might more easily see the many bioluminescent creatures against the darkness of the depths. But it may be that sperm whales hunt both ways.

MORTALITY: Killer Whales and sharks prey on Pygmy Sperm Whales, and account for some strandings. Some whales bear healed scars, suggesting that predation may be more common than is realized. Lamprey or cookie-cutter shark scars have also been identified on stranded whales. Surprisingly, heart disease has figured in the deaths of stranded Pygmy Sperm Whales in the southeastern United States. Various forms of stress may lead to this pathology.

DISTRIBUTION: The Pygmy Sperm Whale has a worldwide distribution in tropical and temperate waters. However, this inconspicuous whale is rarely seen or identified at sea, even though it may be a common stranded species in some areas. Pygmy Sperm Whales were only identified five times during NMFS survey cruises off the American west coast from 1991 to 2001. All of these whales were over deep waters, more than 280 km (150 mi) offshore of northern and central California. In five other sightings, the identity of the *Kogia* species was not clear. No sightings have been confirmed off British Columbia. Nine sightings of Pygmy Sperm Whales have been reported in the open Pacific off Baja California and in the Sea of Cortez.

The Pygmy Sperm Whale's preferred prey suggests that this species occurs in deeper oceanic waters than does the Dwarf Sperm Whale, which may be more associated with the continental shelf and slope.

Range

PYGMY SPERM WHALE

VIEWING: The Pygmy Sperm Whale seems quite rare and appears well offshore along the northeastern Pacific coast, but it is more common in the warmer waters off Baja California.

CONSERVATION: There is such scant information about this species that threats to its conservation are hard to identify. Historically, it is possible that traditional land-based whalers hunted Pygmy Sperm Whales in the eastern Pacific, as they did—and still do—in Indonesia, Japan, and the Antilles. The impact of this whaling on regional populations is unknown.

Agreement	Conservation Status
U.S. ESA	Not listed
U.S. MMPA	No special status
COSEWIC	Not at risk
IUCN	Data deficient

In the recent past, Pygmy Sperm Whales have been taken in American gillnet fisheries off the west coast of North America, but there are no such captures known since a campaign featuring improvements in gear, the use of pingers, and skipper education. Drift gillnet fisheries off Baja California may still take some Pygmy Sperm Whales.

Rescue centers in the southern United States have failed to keep a single *Kogia* alive long enough for it to be released. New techniques at one rescue center, however, helped calves survive longer in captivity. Because of their short mouths and underslung jaws, the playful calves learn to be fed by tube instead of bottle. Nonetheless, all deaths at the center were ultimately due to intestinal obstruction from constipation. Because *Kogia* may swallow floating plastics, this common form of marine pollution may represent a particular hazard for these whales if it blocks the intestine. A *Kogia breviceps* that stranded and died in Texas had two stomach compartments jammed with plastic bags.

POPULATION: By extrapolating from sightings of Pygmy Sperm Whales during annual cruises from 1991 to 2001, NMFS estimates a population of 250 of Pygmy Sperm Whales in U.S. waters, but there is no such estimate for Baja California. Because these quiet whales are only identifiable under ideal sea conditions, their numbers may be underestimated.

COMMENTS: Imagine a luminous world far below the surface of the sea. Almost all the species dwelling in the near total darkness faintly glow—an array of wraithlike, semi-transparent creatures that sway or move through the cold waters. This is the hunting ground of the large-headed whales, with their specialized reflecting eyes and their biosonar.

REFERENCES: Beatson 2007, Bloodworth and Odell 2008, McAlpine 2002, Young et al. 1998.

DWARF SPERM WHALE *Kogia sima*

Pl. 16; Fig. 63.

> Adult average length is 3 m (9 ft); average weight is up to 270 kg (600 lbs). Newborn average length is 1 m (3 ft); average weight is 40 to 50 kg (90 to 110 lbs).

SCIENTIFIC AND COMMON NAMES: For the derivation of the generic name *Kogia*, please see the account for Pygmy Sperm Whale. The feminine adjective *sima*, meaning "snub-nosed," agrees in gender with the noun. Until the 1960s, the Dwarf Sperm Whale (fig. 63) was thought to be a subspecies or race of the Pygmy Sperm Whale, so the older literature confuses the two species. To complicate matters, populations of *K. sima* in the Atlantic and the Pacific–Indian Oceans are genetically quite distinctive and may represent two species, one of which may bear a new scientific name in the future.

In English, Dwarf Sperm Whale is the conventional common name. Historically, in English, *Kogia* whales have been called Blackfish, Snub-nosed Cachalots, and even Rat Porpoises, the last name having been suggested by their snaggle teeth. In French, the term *cachalot nain* has been coined for the species, and in Spanish, the term is *cachalote enano;* both of these terms mean Dwarf Sperm Whale.

DESCRIPTION: The Dwarf Sperm Whale is sharklike in appearance. The body is robust and compact, with small broad flippers and fins and a stubby, beakless head. Like its larger, squared-headed relative, the Dwarf Sperm Whale's "nose" is dominated by spermaceti-filled chambers, giving it a distinctive

Figure 63. Dwarf Sperm Whale.

countenance. The blunt head shape is thought to be well adapted for bottom-feeding, much as in sharks. Spermaceti is an oily or waxy substance that transmits sounds in this group of whales. The nose of the Dwarf Sperm Whale differs from its more massive cousin's, in that the source of its sounds comes from the back of the nose, not the front. A pair of internal lips vibrates a membrane that resonates other structures, some of which resemble human vocal cords. The resulting sounds travel outward through a horn-shaped spermaceti organ to the melon at the front of the nose. The sounds then pass through the skin into the water via an "oval window" in the blubber. One expert believes that *Kogia*'s sound generator works like a single-reed Scottish bagpipe.

The skull of *Kogia* is the most asymmetrical of all cetaceans and perhaps of all mammals, making it easily identifiable on a beach or a laboratory bench. This asymmetry is the consequence of the evolution of the snout, with breathing and sound-generating structures shifted to the left. The right nasal passage is diverted for sound production, and the left for breathing.

The eyes of *Kogia* are well adapted for seeing in low-light conditions. Its eyes have a large aperture as well as a reflective layer, or tapitum, behind the retina, the cells of which contain highly refractive crystals. This reflective layer passes light back through its photoreceptors, increasing the eye sensitivity. All cetaceans possess tapida, but *Kogia* has a dual system, unusual among mammals. The lower part of the tapitum has its peak reflection in green-blue light, and the upper part in blue light. The green-blue peak reflectance matches the light in near-surface waters, and the blue sensitivity matches that of the down-welling light at great depths.

Dwarf Sperm Whales are countershaded, from dark bluish or brownish gray above to whitish or pinkish white below. Both species of *Kogia* possess "false gills" that are whitish "brackets" at the rear of the head.

The prominent fin of this whale, although variable in form, is usually hooked and rounded, and it tends to lie near midback. The flippers are hooked, and the flukes notched. The Dwarf Sperm Whale carries its short-jawed mouth on its underside. In combination, the bluntly pointed head, large eyes, false gills, and underslung jaw give the Dwarf Sperm Whale a sharklike visage. Unlike sharks, however, in the Dwarf Sperm Whale, the teeth are unimpressive—sometimes one to three pairs of slender, incurving teeth in the upper jaw, and seven to 12 pairs in the lower. The teeth in the lower jaw are less than 30 mm (1.2 in.) long and 4.5 mm (0.2 in.) wide.

Both species of *Kogia* possess a special sac in their intestine that is filled with reddish brown fluid that looks somewhat like chocolate syrup. When startled or harassed, the whales may expel a reddish cloud of this murky liquid, which could serve to conceal the whales or repel a predator.

FIELD MARKS AND SIMILAR SPECIES: A stranded Dwarf Sperm Whale reminds some beachgoers of a shark, but *Kogia* has a blowhole, no real gill slits, and only a few pointed teeth, not jaws lined with triangular blades.

At sea, Dwarf Sperm Whales suggest porpoises to some, but if seen well, their head shape is distinctive, as is the blowhole position on the side of the forehead. Dwarf Sperm Whales and Pygmy Sperm Whales are difficult to distinguish, however. The Dwarf Sperm Whale is slightly smaller in length and mass, and it has a larger dorsal fin near the midpoint of its back. The Dwarf Sperm Whale also has a shorter, more pointed snout than the Pygmy Sperm Whale, giving its head a conical look rather than the squarish look of the Pygmy Sperm Whale. In the distance, *Kogia* might resemble many small cetaceans.

If the Atlantic Dwarf Sperm Whale and Pacific–Indian Dwarf Sperm Whale prove to be different species, then one should be able to identify them simply by range. There may also prove to be behavioral differences, as are found in closely related and very similar bird species in their calls or feeding strategies.

NATURAL HISTORY: Observers encounter either individual Dwarf Sperm Whales or groups that may range up to 10. When diving, Dwarf Sperm Whales do not pitch forward or raise their tailstock like other cetaceans; instead, they sink vertically. The whales are often reported to be shy at sea, which may account for the infrequent sightings of this species. In observations at the Gulf of California, the whales were reported to lie quietly at the surface for about a minute, simply disappear, and then reappear within 3 minutes for another surface rest. Whales there were hard to approach, retreating when approached within 200 m (650 ft). Photographs show distinctive notches and marks in the dorsal fin, making it theoretically possible to track individuals.

The large, dark eyes and distinctive forehead of the Dwarf Sperm Whale hint of its life beneath the surface. Both vision and biosonar might help this whale detect and capture prey, likely at great depths of perhaps to 500 to 1,300 m (1,700 to 4,300 ft). Remember that the depths visited by these cetaceans are not completely dark. During the day, as reported by humans in submersibles, a faint light toward the surface can be seen down to 800 m (2,600 ft) in clear waters. At any time, a background glow from the many bioluminescent species of plankton, squid, and fish illuminates the deep. When deep water is stirred, the light of the many species intensifies. Squid can be attracted to light and could even be drawn to the movements of the Dwarf Sperm Whale.

Because *Kogia* has proved to be difficult to age, its life history is hard to document from stranded specimens. Males are reported to reach puberty at 2 m (6.5 ft) in length, and to become sexually active at 2.1 to 2.2 m (6.3 to 7.2 ft).

REPRODUCTION: There is scant information about reproduction in this rare whale. An annual birthing cycle is suggested by an estimated gestation period of 9 to 11 months, but other researchers do not agree that either species of *Kogia* has a distinct annual breeding season. Young calves are born with a rounded head, but as they mature, the snout becomes blunt like the adults'.

FOOD: Like beaked whales, *Kogia* can feed by suction, drawing in squid and other prey in less than half a second by rapidly depressing their tongues. The Dwarf Sperm Whale captures midwater squid in continental slope waters, although fish may also be taken. There is a suggestion than younger Dwarf Sperm Whales may hunt more along the continental shelf.

From examining stranded whales, researchers have identified in their diet squid from 13 families from around the world, the most common of which are associated with the continental shelf. Whales collected in Japan had eaten identifiable fish from 18 families, with lanternfish and other deep-sea and bottom fish being the dominant species. All fish species found in the Dwarf Sperm Whale's diet inhabit depth ranges greater than 250 m (800 ft).

MORTALITY: Predators of Dwarf Sperm Whales are likely to include Killer Whales and sharks. It has been suggested that *Kogia*'s resemblance to a shark is not due to chance but may have evolved to confuse or deter visually hunting predators.

DISTRIBUTION: Dwarf Sperm Whales are found in tropical and warm-temperate seas the world over. The frequency of sightings

Range

DWARF SPERM WHALE

varies greatly throughout this range—reflecting the number of observers as well as the number of whales. Stranded whales, however, provide some insight into the species' range. Over the last 50 years, five strandings of Dwarf Sperm Whales have been reported on the west coast of North America, from La Paz in Mexico to Vancouver Island in British Columbia. Other strandings have occurred at Point Reyes and Morro Bay.

Dwarf Sperm Whales have been counted in study areas in waters from 247 m (810 ft) to 1,565 m (5,136 ft) deep. In the Bahamas, Dwarf Sperm Whales concentrate on the upper slopes of the continental shelf in winter but split their time between the slope and deeper waters in summer. More studies should help us tell about seasonal movements between habitats in this species.

VIEWING: Considering the rarity of this species in the eastern Pacific, many of us would be more likely to encounter dwarf species while on holiday in the islands of Kauai or Hawaii. Dwarf Sperm Whales are the sixth most commonly sighted cetacean in Hawaiian waters. Dwarf Sperm Whales also frequent the protected waters of the southwestern Gulf of California.

CONSERVATION: Small whaling catches of Dwarf Sperm Whales have been reported in the West Indies, Sri Lanka, and Japan. It is hard to gauge the impact of this whaling in such a reclusive species. Incidental take of Dwarf Sperm Whales in gillnets is possible, but none has been observed in any fishery on the west coast of North America.

Agreement	Conservation Status
U.S. ESA	Not listed
U.S. MMPA	No special status
COSEWIC	Data deficient
IUCN	Data deficient

Plastics, including bags and sheets, have been implicated in the deaths of several stranded Dwarf Sperm Whales. Perhaps some floating plastics, in their transparency or movements below water, resemble squid or other prey and are actively ingested by these small whales.

Attempts to rehabilitate stranding *Kogia* calves have failed because of constipation in the animals. The contents of the digestive tract are fluid in *Kogia,* unlike in all other mammals. Constipation can create an array of ruptures, blockages, and other devastating consequences, and in a major veterinary hospital, it has led to the death of all rescued calves. This suggests that plastic debris may be a critical threat to *Kogia* at sea.

POPULATION: Along the west coast of the United States, no identifiable Dwarf Sperm Whales were sighted in 10 years of vessel surveys. There were five reports of the closely related Pygmy Sperm Whale, and four of small sperm whales that could not be confirmed as to species. Given the lack of data, there is no official estimate of population for the west coast of North American Dwarf Sperm Whales, nor is there an estimate for the world population.

COMMENTS: Dwarf Sperm Whales descend from the earliest family of toothed whales, which appeared at the end of the Oligocene epoch some 30 mya. Their foreheads represent the great evolutionary step taken by this venerable family—biosonar. Biosonar meant that toothed whales could "see" in the dark. Their voices, senses, and brains evolved to "draw" an interior image of the depths—to illuminate the dark waters from within. The family is relict now, with only two lineages left from its more triumphal times. In the Miocene, Sperm whales once even occupied the Killer Whale niche and were masters of the world ocean. Now we are left with this handful of species.

REFERENCES: Chivers et al. 2005, Clarke 2003, McAlpine 2002, Nagorsen 1985.

BAIRD'S BEAKED WHALE   *Berardius bairdii*
Pl. 5; Figs. 64, 65.

> Adult male average length is 10.4 m (34 ft). Adult female average length is 10.5 m (35 ft). Average weight for both is 12,000 kg (13 tons). Newborn average length is 4.5 m (15 ft).

SCIENTIFIC NAMES: The young Norwegian immigrant Leonhard Stejneger of the Smithsonian Institution found an unusual four-toothed skull on Bering Island in 1882. One year later, Stejneger described a new Bering Sea whale that was closely related to

Berardius arnuxii, Arnoux's Beaked Whale, of the southern Pacific seas. This southern whale had been named for Auguste Bérard, the captain of the boat that carried a skull of this species from New Zealand to Paris, where it was studied. Arnoux was the ship's doctor. Stejneger thought the northern form belonged to the same genus as the southern form, *Berardius.* However, he believed that the northern whales represented a different species, which he named *bairdii* in honor of Spencer Fullerton Baird, the Secretary of the Smithsonian, and his boss. Thereafter followed more than a century of argument as to whether the northern and southern species were separate species or not. If the consensus becomes that both forms represent the same species, then the scientific name for Baird's Beaked Whale may become that of the subspecies *Berardius arnuxii bairdii.*

COMMON NAMES: In English, the north Pacific form is most often known as Baird's Beaked Whale, and sometimes as the Northern Four-toothed Whale (fig. 64). Both the northern and southern forms together or separately are called Giant Beaked Whales. In French, the common name has been rendered as *baleine à bec de Baird* and *bérardie de Baird,* meaning Baird's Beaked Whale or Baird's Berardius. In Spanish, they are called the *ballena de pico de Baird* or *zifio de Baird,* again meaning Baird's Beaked Whale.

DESCRIPTION: Baird's Beaked Whale is the largest of the beaked whale family, Ziphiidae. The southern form of *Berardius* is

Figure 64. Baird's Beaked Whales.

generally shorter. The longest Baird's male recorded was 11.9 m (39 ft) long, and the longest female was 12.8 m (42 ft). The long body of Baird's Beaked Whale is tubelike, tapering at either end. The head is proportionately small and is about one-eighth of its total length. The well-rounded melon is high and prominent and is broader and more bulbous in males. A visible depression shows in front of the melon. The blowhole is crescent-shaped with the rounded edge facing forward. The Baird's beak is thin and tipped with two visible teeth in the lower jaws of both males and females. Two lesser teeth lie hidden within the mouth behind the larger triangular visible teeth. The jutting front teeth are about 9 cm (3.5 in.) long, and the back teeth are 5 cm (2.4 in.) long. The lower jaw extends about 10 cm (4 in.) beyond the upper jaw. The mouth line curves gently up into a low arc.

The flippers are well forward, small, and slightly rounded at the tips. The dorsal fin lies back about two-thirds down the whale's length. The low fin may be triangular or hooked and has a rounded tip. The fluke's width is about one-quarter of the whale's length. The flukes are slightly rounded and have no center notch.

The color of the Baird's Beaked Whale may be brown, blue-gray, or black on its upper or dorsal side. White patches mark its lower or ventral side, especially on the throat, between the flippers, and near the umbilicus.

FIELD MARKS AND SIMILAR SPECIES: Baird's Beaked Whales can be positively identified at sea, unlike many individuals from other beaked whale species. When surfacing to breathe, Baird's Beaked Whales bring their head and beak steeply out of the water, clearly showing their long beak and prominent melon. The pair of triangular front teeth often flash in sunlight as the whale surfaces. The size and length of these beaked whales can help confirm their identification. Baird's at a distance might be confused with other species of beaked whales, particularly Hubbs' Beaked Whale or Stejneger's Beaked Whale. Both of these species are smaller and have shorter beaks and less conspicuous melons. The teeth of the mature male Hubbs' Beaked Whale and Stejneger's Beaked Whale lie back further in their lower jaws.

Baird's Beaked Whales may also be confused with Northern Bottlenose Whales, but the latter only occur in the Atlantic Ocean, not in the eastern Pacific Ocean. When a pod of Baird's Beaked Whales lies at the surface, it may resemble a similar

Figure 65. Baird's Beaked Whales traveling in tight formation.

group of small Sperm Whales; however, the dorsal fin of the
Baird's Beaked Whale is distinctive once it shows.

NATURAL HISTORY: The large size of Baird's Beaked Whales suggests
a deep oceanic life surpassing that of other beaked whales. They
are often seen in tight groups of perhaps three to 30 or more
individuals (fig. 65). Groups may surface and breathe together,
hinting at well-coordinated movements beneath the waters.
Individuals may stay close to one another, with snouts showing
as the whales slide over one another's backs. Japanese data based
on the growth layers in teeth indicate that adult males can live
up to 84 years, and females up to 54 years. This long life would
reinforce strong relationships among members of a group.

Recordings off Hecata Bank near Cape Blanco, Oregon, and
off Cedros Island in Baja California show that Baird's Beaked
Whales whistle and emit tonelike sounds as well as regular
echolocation clicks. However, the whales also click in short
and erratic bursts, unlike most other whales but like transient
Killer Whales.

REPRODUCTION: As in other beaked whales, the extensive linear
scarring on males hints at underwater contests over social
dominance, but little else is known regarding the reproduction

for this whale. Males become sexually mature at 9.3 to 9.6 m (30 to 31.5 ft) at about age 30, and females 10 to 10.3 m (33 to 34 ft) at about age 10 to 15. Gestation had been estimated as from 10 to 17 months. In Japan, calving peaks in March and April, but no information exists for those in the California Current. With such a long gestation, calving likely occurs every other year, or with even longer intervals, given the longevity of the species.

FOOD: Researchers attached a data logger to a Baird's Beaked Whale off the coast of Japan. Dives recorded were deep [over 1,000 m (3,200 ft)], intermediate [100 to 1,000 m, (320 to 3,200 ft)], or shallow [less than 100 m (320 ft)], with intermediate dives tending to follow deep dives, and shallow dives to follow intermediate dives. Deep dives appeared to reach the sea floor.

Although many marine species are known to swallow stomach stones, few whales have been documented to carry such gastroliths. The pebbles found in stomachs of some Baird's Beaked Whales are exceptional and may reflect bottom-feeding.

The Japanese have long hunted local Baird's Beaked Whales from the Boso Peninsula near Tokyo and from the island of Hokkaido. Much of our knowledge of Baird's Beaked Whale biology stems from this tradition of small-boat coastal whaling. A study on the Pacific coast of Japan showed that Baird's Beaked Whales caught prey similar to that from bottom trawl nets set at depths of more than 1,000 m (3,200 ft). Another study showed that these Baird's Beaked Whales preyed on many fish species as well as squid. In contrast, a study off Hokkaido, Japan, showed that 87 percent of the prey there were squid.

MORTALITY: Killer Whales are one of the few possible predators of Baird's Beaked Whales. One wonders if the resemblance of Baird's Beaked Whales' erratic clicks to those of the mammal-hunting Killer Whales plays some protective role.

DISTRIBUTION: Baird's Beaked Whale is a northern Pacific endemic, ranging from the Sea of Okhotsk and the Sea of Japan in the west to the offshore waters of Baja California in the east, and northward as far as the Pribilof Islands.

NMFS surveys indicate a seasonal presence of Baird's Beaked Whales off the west coast of the United States. Most sightings are in summer and fall along the continental slope. It may be that these whales migrate further offshore in winter. Baird's Beaked Whale is now rarely seen in Canadian waters.

BAIRD'S BEAKED WHALE

Range

Baird's Beaked Whales also appear to move northward in fall off Japan. Whether any whales migrate over longer distances between the western and eastern Pacific is unknown. The western population may be the larger of the two, because it seems to sustain the continued shore whaling tradition off Japan, while eastern sightings are relatively rare.

VIEWING: Baird's Beaked Whales can sometimes be seen in Monterey Bay in summer and fall. The whales dive along the slopes of deep marine Monterey Canyon, which comes very close to shore. They can dive for as long at 40 minutes yet come up in the same spot. On the surface of the Monterey Bay, Baird's Beaked Whales travel in small groups of less than 20.

CONSERVATION: Knowledge about the threats to the preservation of this species is limited, but they include whaling and fisheries interactions. The traditional Japanese shore whalers are limited to 62 Baird's Beaked Whales per year today, which may have no impact on the population of the west coast of North America, unless there are long-distance migrations in this species. The defunct coastal whalers of California and British Columbia harvested Baird's Beaked Whales before the 1970s, which may have some lingering effect on current numbers, given the comparative rarity of the species in the eastern Pacific. It is possible that gillnets may trap Baird's Beaked Whales, but no drownings have been reported by fisheries observers in the United States since the introduction of pingers, improved nets, and a skipper education program by NMFS in 1997.

Agreement	Conservation Status
U.S. ESA	Not listed
U.S. MMPA	No special status
COSEWIC	Not at risk
IUCN	Data deficient

POPULATION: The stock assessment population estimate for Baird's Beaked Whales in 2007 was a mere 313 for California, Oregon, and Washington. No data exist for Baja or British Columbia.

COMMENTS: There are no records of live strandings of Baird's Beaked Whales. When carcasses float ashore, the whales are impressive in death. Their long bodies and their great melons hint at their lives as secretive and swift divers in the darkness of middepth to deep waters. Their landed bodies are treasure houses of knowledge for researchers, who may travel long distances to measure and photograph specimens.

In 2003, a 12.6 m (41.5 ft) male Baird's Beaked Whale stranded on Ocean Beach in San Francisco. This stranding was one of three on record on the central California coast. Researchers from the California Academy of Sciences collected the skull and preserved tissue samples before the whale was buried. Specialists from around the world inquired about obtaining specimens pertinent to their interests. If you are fortunate enough to find a beached beaked whale, take photographs and contact the NOAA stranding website. In Canada, contact the Vancouver Public Aquarium to obtain report forms.

A Baird's Beaked Whale stranding in Monterey Bay, California, in 1925 was the source of the legend of the Moore's Beach monster. Locals were mystified by the unusual carcass, and speculations about its identity grew quite imaginative, continuing to this day. Sightings of live Baird's Beaked Whales in the bay may have inspired some of the other fanciful yarns of sea monsters told on the Monterey waterfront years ago.

REFERENCES: Kasuya 2002, Minamikawa et al. 2007, Walker et al. 2002.

CUVIER'S BEAKED WHALE

Ziphius cavirostris

Pl. 5; Fig. 66.

Adult average length is 5 to 7 m (16 to 23 ft); average weight is 1,845 to 3,090 kg (4,000 to 6,800 lbs). Newborn average length is 2 to 2.7 m (5.5 to 6 ft); average weight is 250 to 300 kg (550 to 660 lbs).

SCIENTIFIC AND COMMON NAMES: *Ziphius* derives from the Greek for "sword" or "swordfish," referring to the narrow beak of some members of this family of whales. The specific name *cavirostris* comes from Latin, with *cavus* meaning "hollow" and *rostris* meaning "face" or "snout." The term refers to the dished or hollowed face of this whale.

Baron Cuvier, the famous French zoologist of the early 19th century, described this whale in 1823, from a find made in 1804. Because the skull of that specimen appeared petrified, Cuvier mistakenly thought it was an extinct fossil species. Cuvier's Beaked Whale or Cuvier's Whale are the ordinary English names, but this cetacean is also commonly called the Goosebeak (or Goosebeaked) Whale because of its short, gooselike beak (fig. 66). French common names include the *Baleine de Cuvier* or *Baleine à bec de Cuvier,* which translate as Cuvier's Whale or Cuvier's Beaked Whale. Sometimes the French call this whale by its scientific name, *Ziphius.* In Spanish, the names *ballena de Cuvier* or *zifios* may be heard.

DESCRIPTION: This whale's features, which very lucky whale-watchers might see, are its relatively small head, which is perhaps 10 percent of its overall length, its "smile," and its 40 cm (16 in.) dorsal fin. The body is spindle-shaped and robust. The concavely shaped head is blunt, with a short and indistinct beak, followed

Figure 66. *Upper:* Cuvier's Beaked Whale, head with slight smile. *Lower:* Cuvier's Beaked Whale back, with reddish color.

by a gently sloped forehead. The mouth turns upward in a fixed "smile." In adult males, the head is whitish, and it may be scored with linear tooth-rake scars. The heads of females can also be lighter in color than their bodies. Both sexes sometimes have dark eye patches. As in all beaked whales, the throat is pleated with two pairs of converging grooves. There is a depression in the animal's back, behind the blowhole. Body color varies widely and may be gray, brown, near-black, or reddish brown. The reddish brown or burnished color is due to an infestation of diatoms and algae. The skin of Cuvier's Beaked Whales may bear the circular scars of cookie-cutter shark bites. The dorsal fin lies two-thirds of the distance down the whale's length. The flukes are broad—about a quarter as wide as the whale's body. The flippers are narrow, rounded at the end, and small, and they can fit neatly in a depressed pocket in the whale's side, a characteristic of beaked whales.

Adult males have two erupted teeth, about 8 cm (3 in.) in length, that project beyond the slightly jutting lower jaw. The conical teeth are exposed when the mouth is closed and may be covered in barnacles. These teeth never break the gum in females. As many as 34 unerupted vestigial teeth have been found in the upper jaw in stranded animals, although this is rare.

FIELD MARKS AND SIMILAR SPECIES: A male Cuvier's Beaked Whale is a distinctive cetacean, with its whitish, short head that can be sometimes seen at the surface, and with its diagnostic teeth projecting from the lower jaw. However, females and immatures might be confused with similar-sized mesoplodont (another genus of beaked) whales. Mesoplodonts generally have better defined beaks and, in males, teeth that generally erupt further toward the rear of the lower jaw. Larger beaked whales may be distinguished by their size, bulbous foreheads, and longer tubelike beaks.

NATURAL HISTORY: The Cuvier's Beaked Whale is the poster child for the revelations gained from research using new recording and tagging techniques, developed in the last decade. We now know a great deal about how Cuvier's Beaked Whales move under the sea and how they hunt prey. In one remarkable study, researchers attached D-tags to two Cuvier's Beaked Whales at the same time. D-tag is shorthand for digital tag, a device to record the sound, depth, and orientation of whales. D-tags are about the size of a sandal and attach to the whale with four suction cups. Later, they drop off, bob to the surface, and broadcast their location so that they can be retrieved. The two instrumented Cuvier's Beaked Whales dove together, kept close to one another, moved on similar paths through the entire deep-dive sequence, and then finally surfaced near one another. The whales had hunted as a pair.

The social life of Cuvier's Beaked Whales, on the other hand, remains almost entirely unknown to scientists. Observers report sighting lone individuals or small groups of whales, usually fewer than 10. The linear scars on males suggest male contests

THE DEEPEST DIVER?

The dives of the Cuvier's Beaked Whale have proved to be deeper than the dives of other cetaceans studied thus far, with a maximum of 1,888 m (6,194 ft). The longest dive recorded lasted 87 minutes. The average dive was 1,070 m (3,500 ft) with a duration of 58 minutes, which exceeds the average for any other air-breathing diver in the ocean. Only Sperm Whales may dive deeper; research in this field persists as sensor technology shrinks and research capability expands. After deep dives, Cuvier's Beaked Whales spent extended periods of rest, averaging 1 hour, near the surface. It is possible that these deep-diving whales switch from aerobic to anaerobic metabolism, which necessitates long rests near the surface between dives. In contrast, Northern Elephant Seals and Blue Whales dive without long breaks, but not as deeply.

for females or dominance in social groups, but no one has ever witnessed such a struggle. The relationships among members of a group may be long, because these whales possibly live to 60 years of age.

Many cetologists believe that whales hear through their jaws. Sounds pass through a thin window in the bones of cetacean lower jaws and then travel along a fatty pathway to the ear. Whales are thus said to have evolved "jaw hearing." To explore how this worked, researchers performed a medical scan on the head of two Cuvier's Beaked Whales that had stranded and died. With a large industrial scanner, the scientists managed to map the heads; they then determined their physical properties in a careful necropsy (i.e. an animal autopsy). Using modeling techniques originally developed to understand how vibrations act on bridges and other structures, the researchers simulated the passage of sound through the Cuvier's Beaked Whale's head. They discovered a second likely passage for entering sounds—below and between the lower jaws, then through an opening in the bones behind the throat to reach the ears. This throat pathway may be ancient and may have been the original underwater acoustic route for cetacean hearing.

REPRODUCTION: Cuvier's Beaked Whales become mature when they reach 5.5 m to 6.1 m (18 to 20 ft) for males and 6.1 m (20 ft) for females, which equates to between 7 and 12 years of age. Cuvier's Beaked Whales calve year-round, but most often in spring. Gestation takes 1 year, and the calving interval is 2 to 3 years.

FOOD: Cuvier's Beaked Whales' general lack of teeth implies that they are suction feeders, consistent with their predation on soft-bodied squid. Cuvier's Beaked Whales feed primarily on deepwater squid and sometimes on small fishes and crustaceans. There has been a suggestion that some whales bottom-feed on fauna at deep, cold, methane-rich ocean seeps.

MORTALITY: Off Hawaii, a Cuvier's Beaked Whale was photographed bearing a kiss-shaped scar from a shark attack. The healed wound may have been the work of a Tiger Shark, Galapagos Shark, or Great White Shark. Like Blainville's Beaked Whales, Cuvier's Beaked Whales only echolocate at depth, and their silence at the surface may help them evade sharks or another likely predator, the Killer Whale.

DISTRIBUTION: Cuvier's Beaked Whales are seen infrequently but are distributed widely on the high seas, except in polar waters. These whales tend to be relatively common over certain submarine features such as deep underwater canyons (Monterey Canyon, for example), and especially near islands; they also associate with continental shelf breaks. It may be that different species of beaked whales prefer waters of varying depths. Cuvier's Beaked Whales dive deeper, and in deeper waters, than Blainville's Beaked Whales, the only other cosmopolitan beaked whale.

Cuvier's Beaked Whales have been sighted rarely in Canadian waters. In recent NMFS surveys, all sightings along the U.S. coast were offshore, many more than 185 km (100 mi) from land. Based on NMFS surveys in the eastern tropical Pacific from 1986 through 1993, models of population density predict that relatively high numbers of Cuvier's Beaked Whales should be found in the Gulf of California and along the west coast of the Baja Peninsula.

CONSERVATION: Threats to Cuvier's Beaked Whale conservation include fisheries interactions, plastic pollution, and naval sonar. Japanese whaling that targets Baird's Beaked Whales takes low numbers of Cuvier's, as well. The American drift gillnet fishery

CUVIER'S BEAKED WHALE

Range

also caused bycatch of Cuvier's Beaked Whales, until the NMFS launched a take reduction program. After this program began educating skippers and installing net pingers, bycatch of Cuvier's Beaked Whale fell to no known drownings after 1997. Mexican drift gillnet fisheries for swordfish may still bycatch Cuvier's Beaked Whales.

Plastic pollution may be of special concern. One estimate is that more than half the stomachs of Cuvier's Beaked Whales that beached in Europe contained plastics. There is no comparable information for the U.S. west coast. It may be that plastic sheets or bags in the water resemble squid, or that prey shelter in such drifting plastics and that Cuvier's Beaked Whales ingest the debris while suction feeding.

There are few underwater observations of the impact of boat noise on whales. In one published case, a close ship approach apparently disrupted a deep foraging dive by a D-tagged Cuvier's Beaked Whale, perhaps due to the boat masking echolocation sounds, prey echoes, or social signals. More research in this

Agreement	Conservation Status
U.S. ESA	Not listed
U.S. MMPA	No special status
COSEWIC	Candidate
IUCN	Least concern

difficult area of study will be vital to understanding the impact of shipping noise on sea life.

Naval sonar is now recognized as potentially significantly affecting Cuvier's Beaked Whales. This discovery occurred when a team of researchers in 2000 started a long-term study of the elusive Cuvier's Beaked Whale in the Bahamas. The shy whales seemed relatively approachable in the clear, tropical waters. The scientists began a photo catalog that recorded the distinct markings of the individuals that frequented the island waters. But their work changed abruptly with a mass stranding of 16 whales, nine of them Cuvier's Beaked Whales, over a wide area. Six beaked whales died—five Cuvier's and one Blainville's. Encounter rates for groups of whales in the area dropped from a dozen or more times a year to a single sighting of a pair of previously unidentified whales in a year-long period, and none of the 35 Cuvier's Beaked Whales identified in the photo catalog was ever seen again. The cause for the stranding was a U.S. Navy battle group that cruised through the islands using active sonar. Investigators found that the dead Cuvier's Beaked Whales had bled internally around their ears and brains. No one is certain how many whales were actually killed by the active sonar, because other injured animals, instead of stranding, may have died at sea, been scavenged, or sunk. However, the disappearance of all recognizable whales in the wake of this naval exercise suggests that high-intensity sonar caused a regional extirpation of Cuvier's Beaked Whales.

In 2002, there was a mass stranding of a group of 14 whales in the Canary Islands. As in the Bahamas, strandings occurred across a broad area. Nine of the group were Cuvier's Beaked Whales, and they had suffered tissue damage from bubbles, like what happens with the bends. Sudden decompression, such as human scuba divers might experience if they ascend too abruptly, can cause nitrogen in tissues to outgas and bubble, and the bubbles in turn can cause an array of symptoms from pain to paralysis and death. In the case of the beached whales, one theory proposed was that the strong vibrations of the high-intensity sonar had somehow directly released nitrogen gas bubbles in the whales' bodies. Another theory was that the whales had panicked and, in a flight to escape the sounds, had ascended or descended in some way that resulted in fatal decompression sickness. There are many more instances of Cuvier's Beaked Whales beaching during or after naval high-intensity sonar, but the aftermath usually is not documented so thoroughly as that for the Bahamas and Canary Islands strandings.

For as long as records have been kept, Cuvier's Beaked Whales have occasionally beached, but strandings became more frequent after 1960, and more than a single whale often stranded. The increase coincided with the U.S. Navy's adoption of active sonar during the Vietnam War and subsequent naval tests and drills. Since that time, one-third of all strandings of beaked whales have been linked to naval exercises, though not conclusively, and perhaps 80 percent of the whales so stranded were Cuvier's Beaked Whales.

POPULATION: Despite difficulties in finding and identifying this elusive whale, surveyors came up with a provisional population estimate of 2,220 Cuvier's Beaked Whales off Washington, Oregon, and California. No estimate of population trends was possible, given the rarity of the species. There is no Canadian or Baja west coast estimate, but more than 20 strandings are recorded along the British Columbian coast.

COMMENTS: There are sensitive topics in the scientific whale literature—subjects rarely addressed. One is the suffering of cetaceans at the hands of people. In the literature, one can sometimes read a few detailed descriptions and see photographs of the widespread body damage endured by stranding, dying beaked whales after high-intensity sonar episodes, but no one mentions their pain. Another issue often ignored is the extent to which many species of cetacean may be affected by high-intensity active sonar.

REFERENCES: Dalebout et al. 2004, Heyning 2002, Heyning and Mead 1996, MacLeod 2003, Tyack et al. 2006.

BLAINVILLE'S BEAKED WHALE *Mesoplodon densirostris*
Pls. 7, 8; Fig. 67.

> Adult average length is 4.5 m to 6 m (15 to 20 ft); average
> weight is 820 kg to 1,030 kg (1,800 to 2,300 lbs). Newborn calf
> average length is 1.9 to 2.6 m (6 to 8.5 ft); average weight is
> 60 kg (130 lbs).

SCIENTIFIC NAMES: *Mesoplodon* derives from the Greek *mesos* for "middle," *hopla* for "weapon," and *odon* for "tooth," meaning "armed with a middle tooth." The generic term refers to the two exposed teeth of this species of whales, which protrude from the middle of the lower jaw in adult males (fig. 67). The specific name *densirostris* comes from Latin. *Densi* signifies dense, and *rostris* means face or snout. *Densirostris* refers to the especially heavy weight of the rostrum in this species.

COMMON NAMES: The famous French zoologist and anatomist Henri de Blainville formally named this species in 1817 on the basis of bone from its snout, which was the heaviest bone he had ever seen, denser than elephant ivory. The snout or rostrum was all that was known of the species at that time.

Figure 67. Blainville's Beaked Whale cow with calf.

Many common names in English, French, and Spanish refer to Henri de Blainville. The species may be called Blainville's Beaked Whale, or simply Blainville's Whale, in English, *Baleine àbec de Blainville* or *Mésoplodon de Blainville* in French, and *Ballena de Blainville, Ballena de pico de Blainville,* or *Ballena picuda de Blainville* in Spanish, with identical meanings. In Spanish, *zifio de Blainville* may be used, with *zifio* meaning Beaked Whale (see Cuvier's Beaked Whale). A second pair of English common names refers to the heaviness of the snout bones: the Dense-beaked, or Densebeak Whale.

DESCRIPTION: Blainville's Beaked Whale has a deep, robust body that is laterally compressed, especially near its tail. The dark dorsal fin is small and triangular or hooked and is set two-thirds of the way down the whale's body. The pectoral fins tuck into pocketlike depressions, as in other beaked whales, which must reduce drag in ascent and descent. Base colors may be a warm brown or blue-gray, which then countershade abruptly to a light gray or whitish belly. The eyes are surrounded with a dark patch. Overall, males are darker than females, and females develop white lower and upper jaws as they mature.

The head of the male Blainville's Beaked Whale is distinctive when clearly seen. The lower jaws arch above the longish, tubular beak, and each jaw carries a single erupted tooth or tusk up to 20 cm (8 in.) in length, which is angled forward about 45 degrees. Only the tip of each tooth shows, but they may be encrusted in barnacles, which makes them resemble a pair of "dark pompoms." The flattened melon sits low behind the toothed jaws. Males bear many whitish tooth rakes, surely inflicted in combat with one another, although no one has witnessed such a fight. These typically linear scars do not repigment, and so the whale's skin is a history of aggressive encounters. The pair of large teeth does not break the skin in females or immatures. However, the adult female's jaw is also somewhat arched, but not as much as the male's. Oval scars on the bodies of both sexes are probably the work of cookie-cutter sharks.

The Blainville's middle snout bone may be the densest and hardest on Earth. It features heavy, stiff, rodlike structures, which may be unique to this genus. The rods develop as animals mature, particularly in males, and are somewhat brittle. Possibly, the hard bone reinforces the snout in fights with rivals, but it would probably not serve well in blows, given its brittleness.

Another theory is that the heavy bone serves as ballast for this deep-diving species.

The throat of this species is pleated, as in other beaked whales, presumably helping it to suck in squid during its deep dives. The gape of the mouth is small, compared with that in other families of whales, perhaps also to facilitate the intake of soft-bodied prey.

FIELD MARKS AND SIMILAR SPECIES: Identifying beaked whales can be difficult. One study in the South Pacific showed that collectors misidentified 20 percent of a sample of dead beached beaked whales, which should have been easy to inspect visually. This sample included the first record for a Blainville's Beaked Whale in New Zealand. On the other hand, mature Blainville's Beaked Whale males are unmistakable with their high-arched jaw and tusks, if seen surfacing in the sea nearby—a rare event along the west coast of North America. Younger Blainville's Beaked Whales and those at a distance might be mistaken for other Mesoplodon species.

NATURAL HISTORY: Of all the beaked whales, Blainville's Beaked Whale has proven easiest to study. In fact, photographic identification of individuals has been possible in the Bahamas, the Canary Islands, and the Mediterranean since the whales seem to tolerate boats and divers to some degree in these areas. Observations there indicate that some individuals repeatedly visit certain underwater features and thus may be residents. This research has also shown that associations can exist between individual Blainville's Beaked Whales, including stable mother offspring relationships that can persist up to 3 years. Group sizes are small, typically three or four, ranging from two to 11. Studies of the social life of Blainville's Beaked Whales in the eastern Pacific have been limited, and maximum lifespan is unknown.

D-tags, which record sound and orientation, have been attached to Blainville's Beaked Whales, allowing researchers to track the depth and timing of dives. Like Cuvier's Beaked Whales, Blainville's Beaked Whales dive long and deep to forage. These dives occur about every 2 hours and can reach a depth greater than 1,400 m (4,800 ft), making these whales among the deepest divers of any marine mammal. Deep dives exceed the aerobic dive limit, and the whales must have an oxygen debt by the time they surface. This would explain the long intervals between the deep dives, during which the whales rest near the surface or take shallow dives. As with Cuvier's Beaked Whales, their descent is swift but their ascent is slower, involving active swimming. The reason for the slow ascent is uncertain, but it might help explain the damage caused by low-frequency sonar. A whale might make a panicked ascent because of fear or pain during sonar emissions. The quick ascent might create some form of decompression sickness, which would account for the gas and fat bubbles seen in their injuries.

REPRODUCTION: Information on reproduction for this beaked whale, as for others, is limited. Sexual maturity for one female examined from a stranding was reported at age 9. Females give birth to one calf, but gestation and lactation data are not available.

There is no information on the mating behavior of the species in the eastern Pacific, but in the Atlantic, small polygynous groups occur with a dominant male in attendance with a group of females. Although no one has observed aggression between Blainville's Beaked Whale males, their linear scars suggest they must fight back to back. The greatest concentration of scars is on the head and back of males behind the low melon. Because the points of the Blainville's Beaked Whale's teeth extend upward over the jaw, the only way one male could scar another like this would by swimming "upside down" past its rival. It has been speculated that males could possibly each "charge" by one another back to back.

FOOD: Blainville's Beaked Whales feed primarily on deep- and midwater squid, and sometimes on small fishes and crustaceans. Prey are from depths of greater than 200 m (656 ft), and likely much greater, based on dive data collected from other beaked whales.

This species makes two distinctive types of clicks as it forages. Search clicks are emitted at lower rates and have a chirplike upsweep. The long-duration search pulses are modulated like those of many species of bats, and unlike those of other families of toothed whales. The relatively loud search pulses probably reveal the general location of prey. Buzz clicks are softer bursts, first made from a body length away in the last stages of hunting. As the Blainville's Beaked Whale homes in on a target, there may be 300 or more pulses, each an update on the prey's precise position. The unusual search clicks may be an adaptation to soft-bodied squid, which would have fainter echoes than fish. The softer buzzes may better resolve targets in a cluttered acoustic environment, as has been discovered in bat sonar studies.

MORTALITY: Blainville's Beaked Whales in Hawaii may bear mouth-shaped scars from large shark attacks. Tiger Sharks, Galapagos Sharks, and Great White Sharks are possible predators there. Blainville's Beaked Whales are also regularly bitten by the smaller cookie-cutter sharks. Blainville's Beaked Whale is silent near the surface and only echolocates at depth. It is thought that their silence may help conceal them from predatory sharks as well as from Killer Whales.

DISTRIBUTION: Although Blainville's Beaked Whales are infrequently seen, the species has a worldwide distribution in tropical and warm-temperate seas. These whales tend to be most often sighted over certain deep seabed features, such as slopes, continental breaks, and rises near islands, possibly because deep currents ascending near these features may concentrate prey. Blainville's Beaked Whales may prefer shallower waters than some other beaked whales. They are the most widely distributed

■ Range

BLAINVILLE'S BEAKED WHALE

of the mesoplodonts, and Cuvier's Beaked Whale is the only other cosmopolitan species in the Ziphiidae family, overlapping with the Blainville's Beaked Whale's range but extending further into cold-temperate waters.

CONSERVATION: The primary conservation issues for Blainville's Beaked Whales are plastic pollution and active naval sonar. The stomach of a Blainville's Beaked Whale that beached in Brazil was jammed with bluish plastic threads. The rest of the gut was empty, which suggests that the plastics stopped digestion. Suspended plastics at sea may harbor prey and be incidentally swallowed, or may resemble squid, and thus be mistakenly eaten.

Agreement	Conservation Status
U.S. ESA	Not listed
U.S. MMPA	No special status
COSEWIC	Not at risk
IUCN	Data deficient

This species strands during naval activesonar exercises, although more Cuvier's Beaked Whales beach in these incidents. However, it may be that Blainville's Beaked Whales tend to go out to sea to escape the traumatic sounds, and thus injured or dying animals simply may be less likely to be encountered. Death from high-intensity sonar may also be underestimated when whales sink. The dense snout bone of Blainville's Beaked Whale may hasten its sinking. Whatever the case, unless the whales strand alive or float dead on to shore, injuries from military sonar would not be evident.

POPULATION: Because there were only 26 Mesoplodon sightings in two recent North American west coast surveys, it is impossible to draw any conclusions about population trends for any of these species off the United States. If there were a 50 percent decline in the beaked whale stock over 15 years, then statistically the decline would be undetectable, simply because of the rarity of sightings. Extrapolating from the survey database yielded an estimate of 603 Blainville's Beaked Whales for the North American west coast. Some 2,100 were estimated for the warmer waters off Hawaii.

The unique echolocation search pulses of the Blainville's Beaked Whale can be used to estimate populations, if ample hydrophone arrays are available to listen for signals. In the Tongue of the Ocean off the Bahamas, an array of naval hydrophones picked up small groups of Blainville's Beaked Whales as they moved in a 1,536 km² (588 mi²) area. The number of whales was estimated on the basis of the number of groups observed over an 8 day test. The use of passive acoustic devices like these to estimate populations is being refined for other whale species with distinctive sound signatures.

COMMENTS: We are just beginning to explore the ocean's midwater zone, which lies below 300 m (1,000 ft). We are piloting small research submarines and ROVs to these depths. The diversity and numbers of creatures encountered is astounding, with striking new species reported off the west coast of North America. We need to further open the doors of our perception on this magic world, which sustains Blainville's Beaked Whales and the other members of its ancient family. If we can understand how the midwater's great productivity originates and flows, perhaps we can protect this part of the Earth from unthinking exploitation.

REFERENCES: Johnson et al. 2006, Pitman 2002, Tyack et al. 2006.

GINKGO-TOOTHED BEAKED WHALE

Pls. 6, 8.

Mesoplodon ginkgodens

Adult average length is 4.5 to 4.9 m (14.8 to 16 ft); average weight is 1,500 to 2,000 kg (3,300 to 4,400 lbs). Newborn average length is 2.4 m (8 ft).

SCIENTIFIC NAMES: For the etymology of the genus name *Mesoplodon*, see Blainville's Beaked Whale. The species was named *ginkgodens* by Nishiwaki and Kamiya in 1958, meaning "ginkgo tooth," based on a whale stranded near Tokyo. The two teeth of the adult male resembled the ginkgo leaves familiar in the Orient. The English word *ginkgo* derives from the Japanese *ginkyo*. The first Ginkgo-toothed Beaked Whale specimen was actually found in San Diego, California, but was not recognized as a new species.

COMMON NAMES: Ginkgo-toothed Beaked Whale is the ordinary English name; others are the Japanese Beaked Whale and the Ginkgo-beaked Whale. The name is *mésoplodon de Nishiwaki* in French (Nishiwaki's Mesoplodon), and in Spanish it is known as the *ballena de pico de Nishiwaki* or the *zifio de Nishiwaki,* both meaning Nishiwaki's Beaked Whale.

DESCRIPTION: Ginkgo-toothed Beaked Whales have the typical long spindle shape of their genus. The maximum recorded length for a male is 4.8 m (15.7 ft), and for a female 4.9 m (16 ft). Males are dark but bear 3 to 4 cm (1 to 1.5 in.) diameter white spots below on the rear third of their bodies. The dark color reported is that seen after death and may be the result of postmortem fading. The white spots could be the consequence of parasite attacks and not be natural pigmentation. The short beak may be lighter than the blue-black background color. Unlike other beaked whale species, males are not marked with white linear scratches. Females and juveniles have medium gray backs and light gray bellies.

Ginkgo-toothed Beaked Whales have a modest melon, which underlies a slightly bulged forehead that drops steeply to the beak. The upper jaw is narrow and ends in a sharp point; the lower jaw is highly arched. Adult males carry a barely visible tooth at the front edge of the arch on each side of the jaw. Only the tip of the tooth extends outside the mouth. Hidden within the mouth is the 10 cm (4 in.) broad base of the tooth, which is the widest among the mesoplodonts. It is the tip of this tooth that resembles the stem and the symmetrically curved base that resembles the margin of a ginkgo leaf. The only evidence for this species ever recorded on the Galapagos Islands was a pair of their diagnostic teeth discovered on a beach. The teeth do not break the gum in females and juveniles.

As with other mesoplodonts, the flippers and fins are small. The low dorsal fin sits far back and may be hooked. The flukes are broad, unnotched, and fairly triangular in shape.

FIELD MARKS AND SIMILAR SPECIES: Except for adult males, Ginkgo-toothed Beaked Whales look much like other mesoplodonts. Adult males may be distinguished by their lack of linear white scars, by differences in tooth form and placement, and sometimes by coloration. Hubbs' Beaked Whale males bear large teeth extending above the beak, and have a white-capped melon and white-tipped jaws. Blainville's Beaked Whale males also bear prominent teeth outside their arched mouths, and their heads are lightly colored. Perrin's males are dark but have teeth at the front of the mouth. *Peruvianus* males are small and dark and have short, peglike teeth that erupt at the gentle rise in their lower jaws. Cuvier's Beaked Whale males have a light-colored, blunt head that has an indistinct beak, and an upward-angled mouth with teeth projecting forward.

NATURAL HISTORY: We have no direct knowledge from scientific sightings about the social life of the Ginkgo-toothed Beaked Whale. We might guess that their natural history is like that of other mesoplodonts, with strong social attachments expressed in synchronized diving, surfacing, and resting. The lack of dramatic scarring in stranded specimens might suggest a less combative lifestyle for males than in other beaked whale species. However, the skin of the Ginkgo-toothed Beaked Whale may mostly heal with pigmented tissue, unlike other beaked whale species, so battle scars could be hard to detect. Also, the teeth are barely exposed and may not deeply gouge the skin of rivals in underwater battles.

REPRODUCTION: There is no information on reproduction in this species, but because males are distinctly different from females and have tusks, male competition plays some role in mating strategies.

FOOD: The Ginkgo-toothed Beaked Whale, like other beaked whales, probably eats mostly squid, but there are no data to support this notion.

MORTALITY: There is no information on natural sources for mortality.

DISTRIBUTION: The distribution of Ginkgo-toothed Beaked Whales spans two oceans, the Indian and Pacific, and includes tropical and warm-temperate waters. What little is known of their range has been gleaned from stranded animals in Sri Lanka, Malaya, Japan, Australia, New Zealand, the Galapagos Islands, California, and Baja California, and from fisheries (see below).

CONSERVATION: Little is known about the species, but some issues have been identified that threaten their conservation, including fisheries, pollutants, and noise. There are reports of a small take of Ginkgo-toothed Beaked Whales off Taiwan. Ginkgo-toothed Beaked Whales have been contaminated by toxic organic tin compounds off the coast of Japan. Plastics may be a special problem for them. Plastic fragments ingested by zooplankton, soft-bodied salps, and squid may bioaccumulate as the indigestible polymers pass up the food chain. Beaked whales may suction in

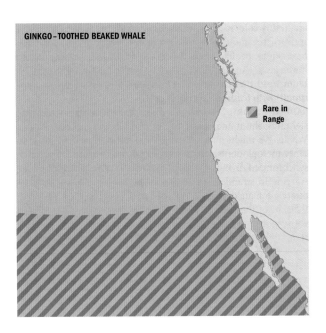

GINKGO-TOOTHED BEAKED WHALE

Rare in Range

the plastic waste directly or indirectly in their food, which can hamper digestion. We know from strandings that many beaked whales swallow plastic refuse such as fishing lines and plastic bags, which sometimes has led to their demise. As with other beaked whales, they suffer from mortality and injury from underwater sound, and Ginkgo-toothed Beaked Whales have mass stranded at the same time that naval sonar exercises have taken place.

Agreement	Conservation Status
U.S. ESA	Not listed
U.S. MMPA	No special status
COSEWIC	Not at risk
IUCN	Data deficient

POPULATION: There is no population estimate for this rarely sighted but widely distributed whale.

COMMENTS: Six species of *Mesoplodon* share the northern Pacific with three other members of the secretive Ziphiidae family—Cuvier's Beaked Whale, Baird's Beaked Whale, and Longman's Beaked Whale. Three species of sperm whale forage in the same offshore waters. All these cetaceans feast on squid and other prey from the deep ocean, which is perhaps the most abundant food source on Earth. But how do they divide up these riches? By latitude or current? By depth? By type of prey? By underwater features? How little we know.

REFERENCES: Pitman 2002, Taylor et al. 2008.

PERRIN'S BEAKED WHALE   *Mesoplodon perrini*
Pls. 7, 8.

> Adult male average length is 3.9 m (13 ft). Adult female average length is 4.4 m (14.8 ft). Calf average length is 2.1 to 2.5 m (6.9 to 7.5 ft).

SCIENTIFIC NAMES: *Mesoplodon* refers to the two teeth often prominent in this genus of whales (see Blainville's Beaked Whale for etymology). The species was named *perrini* after marine mammalogist William Perrin, for his part in collecting the first two specimens, and for his innumerable contributions to the study and conservation of marine mammals.

COMMON NAMES: The English name Perrin's Beaked Whale has been proposed for this recently discovered cetacean, first recognized in 2002. In French, its common name has appeared as *le Baliene à Bec de Perrin* or *mésoplodon de Perrin*. In Spanish, its name has been rendered as *ballena Picuda de Perrin* and as *zifio de Perrin*. The French and Spanish terms can be translated as Perrin's Beaked Whale.

DESCRIPTION: Perrin's Beaked Whale is only known from five specimens stranded along the California coast: an adult male from Carlsbad, a female and calf from Camp Pendleton, a calf from Torrey Pines State Reserve, and another calf from Fisherman's Wharf in Monterey. Because the description of this species is based on a limited sample, it is tentative.

Perrin's Beaked Whale has the characteristic shape of its genus, with a small head, a long body, and a short tail. The rostrum (or snout) is short in Perrin's adults compared to most mesoplodonts, and is even stubbier in calves. The melon is modest, and the crescent-shaped blowhole is broad. The male's back is dark gray and countershades to white below. The tops of the tail flukes are pale gray. The calves are light to dark gray above, and white below. The lower jaw and throat are white. A masklike dark gray patch runs from the corner of the mouth to the eyes and snout. (The female had been dead too long to determine her coloration). The male's flank is marked with linear scars like other beaked whales. The scars on the Carlsbad male seem to have been made with a single tooth. The Carlsbad male and the Monterey calf bear rounded scars typical of cookie-cutter shark bites.

The Carlsbad male has two erupted teeth at the front of its mouth. The teeth are roughly triangular in shape, but are somewhat convex on the sides and flared slightly outward. The teeth are about 64 mm (2.5 in.) in length, with 33 mm (1.3 in.) exposed above the gum line. Three barnacles rode on the teeth of the sole adult male. The mouth line of all specimens is straight. Throat groves are present, 2 m (6.6 ft) in length in the male.

FIELD MARKS AND SIMILAR SPECIES: Mesoplodont whales can be impossible to identify in the field. Perrin's Beaked Whale is a case in point. Initially experts believed that the first four specimens found were Hector's Beaked Whales, a species from the southern Pacific. Only after a genetic examination of the California specimens' tissues was it clear that the whales were not Hector's Beaked Whales but a species new to science. Genetic comparison with other beaked whales suggested that the new species may be most closely related to the Pygmy Beaked Whale that was first identified off Peru in 1991.

Hector's Beaked Whale and Perrin's Beaked Whale share one distinguishing field mark with the larger Baird's and Cuvier's Beaked Whales—two teeth near the tip of the jaw. If further research shows there are no Hector's Beaked Whales in the northern hemisphere, then this field mark might identify male Perrin's Beaked Whales along the west coast of North America. The Baird's Beaked Whale is easily distinguished because of its robust head and body compared to the smaller Perrin's Beaked Whale. In the other possible mesoplodonts of the west coast of North America, the males do not have prominent exposed teeth near the tip of the lower jaw like

Perrin's Beaked Whale. Thus, if you ever have an incredibly lucky close encounter with a stubby-beaked whale at sea, look for its teeth. If they are at the end of its jaw, it may be a Perrin's Beaked Whale male. If you find any beaked whale stranded on a beach, check out its teeth and photograph it. For any female and juvenile mesoplodonts at sea, and for distantly seen animals, identification only to genus is standard for professionals in the field.

NATURAL HISTORY: Because only stranded specimens have been identified with certainty, little can be said about the natural history of Perrin's Beaked Whale. The linear scars on the adult male imply competition between rivals over females, as is likely the case in other mesoplodonts. The Camp Pendleton adult female appeared to be the mother of the Camp Pendleton calf, with the adult stranding about 1 week before the calf, reflecting the maternal bond common to all mammals The fact that three out of the five stranded whales were calves suggests that females may come closer to the coast to give birth. Growth layers in the teeth of the adult male and female suggested they were both about 9 years of age, which at least establishes an age that animals are sexually mature.

REPRODUCTION: No information is available regarding the reproduction of this species, but since males are distinctly different from females and have tusks, male competition likely plays some role in mating strategies.

FOOD: Only two of the beached whales had stomach contents that could be examined. A squid eye was in the stomach of the Torrey Pines calf, and the Camp Pendleton adult female had eaten at least two squid and one vertebrate of some indeterminate kind. A diet of mostly squid would be typical for deep-diving mesoplodont whales.

MORTALITY: There is no evidence regarding the natural sources of mortality of this species.

DISTRIBUTION: All five specimens were found on the California coast from San Diego to Monterey. A possible live sighting has been reported off Baja California. But if positive identification requires genetic verification, then we will have to be patient and wait to collect more tissue samples from living or dead whales to establish the range of this species.

CONSERVATION: Scant information about this species limits our understanding of the potential or historic threats to their conservation. High-intensity military sonar may be troubling

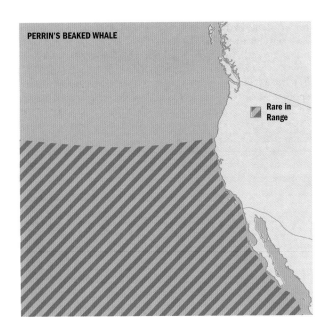

PERRIN'S BEAKED WHALE

Rare in Range

for this rare cetacean, for other beaked whale species are affected, and naval exercises include the California coast.

Agreement	Conservation Status
U.S. ESA	Not listed
U.S. MMPA	No special status
COSEWIC	Not at risk
IUCN	Data deficient

POPULATION: No population estimate exists for Perrin's Beaked Whale. With only five specimens ever recognized, this species may be very rare. Perrin's Beaked Whale also may be somewhat more numerous further offshore in the Pacific, and thus it may only infrequently strand or be sighted.

COMMENTS: The marine mammalogist Kenneth Balcomb wrote that beaked whales were like a race of dinosaurs that have been hiding from us in the deepest oceans. Only in the early 19th century was the first species identified, solely on the basis of some skull bones (see Cuvier's Beaked Whale). From time to time over the following decades, an additional species would be recognized, often on the basis of a few skeletal remains or beached specimens. Now we know that a family of large mammals was lurking in the depths long before we humans sailed the sea. This beaked whale family includes more than one-quarter of all cetacean species in existence.

Perrin's Beaked Whale is currently a California exclusive. We may expect to encounter other Perrin's Beaked Whales along the west coast of North America and possibly elsewhere in the future. We might even discover a new species. In fact, eight of the 21 species of mesoplodonts have been discovered since 1900. It is remarkable that a new species turned up in southern California, one of the best searched and researched coasts on Earth.

REFERENCES: Dalebout et al. 2002.

HUBBS' BEAKED WHALE   *Mesoplodon carlhubbsi*
Pls. 6, 8.

> Adult average length is 4.8 to 5.3 m (15.7 to 17.4 ft); average weight is 1,500 kg (3,300 lbs). Newborn calf average length is 2.5 m (8.2 ft).

SCIENTIFIC NAMES: For the etymology of the genus name *Mesoplodon*, see Blainville's Beaked Whale. The species was named *carlhubbsi* by Joseph Moore in 1963 to honor the revered aquatic biologist Carl Hubbs, who collected the first specimen in 1945 at the foot of the Scripps Pier in La Jolla, California. Knowledge of northern Pacific beaked whales was then in its infancy, and the whale was at first mistaken for the related Andrew's Beaked Whale of the southern Pacific. Twenty-three years later, comparative study showed that it was a new species, whose distinctive identity has recently been confirmed by genetic comparisons.

COMMON NAMES: The English common names for this meso-plodont species are Hubbs' Beaked Whale and the Arch-beaked

Whale. In French, the whale is known as *mésoplodon de Hubbs*, or Hubbs' Mesoplodon. In Spanish, this species is the *ballena de pico de Hubbs*, or the *zifio de Hubbs*, both meaning Hubbs' Beaked Whale.

DESCRIPTION: What could be more striking than an adult male Hubbs' Beaked Whale: long white scars against the background of a black or dark gray body, a white-capped head with white-tipped jaws, and a pair of broad teeth extending like knife blades from the high corners of its mouth. The 12 cm (4.7 in.) high teeth lie mostly below the gum line but their emerged tips reach slightly above the beak. The mouth barely opens, its movements limited by the architecture of the sinuous jaws and the surrounding binding tissue. Two grooves run down its throat, allowing it to expand.

The shape of a Hubbs' Beaked Whale's body resembles that of all mesoplodonts—long, with scaled-down extremities. The 55 cm (22 in.) foreflippers sit well forward and fit into "pockets," or slight depressions, on its side. The pockets are darker than the gray background in adult females. The dorsal fin is about 22 to 23 cm (9 in.) high and sits well back on the body. Its tail's width is a quarter its total length.

Females and young are medium gray above, countershading to white below. Their beaks are somewhat lighter than their heads, and they may have some white on the melon and outer jaws. Their jaws are arched like the males, but not so extremely. The 5 cm (2 in.) teeth of the females and juveniles do not break the gums. The features of juveniles are less distinctive. Calves may bear dark patches around the eyes.

FIELD MARKS AND SIMILAR SPECIES: A male Hubbs' Beaked Whale would be unmistakable if ever seen close-up at sea. The white scars against a uniform dark gray or black, the bulbous white melon or "beanie," and the white outer snout and white outer lower jaw are all field marks that differ from those of adult male Stejneger's Beaked Whales, Perrin's Beaked Whales, and Peruvian Pygmy Beaked Whales. However, Hubbs' Beaked Whale females and immatures resemble other mesoplodonts at sea even when stranded. Experts need to be called in for measurements and tissue samples for genetic confirmation.

NATURAL HISTORY: Several inferences about this elusive species may be drawn from stranded specimens. But other than these clues, little is known of the natural history of this rare species. The scarring and flattened teeth of the males indicate that there may be undersea battles between rivals, perhaps over single females or over groups. The tusklike teeth might defend the males from predators, but more likely they evolved in sexual selection over an extended period.

REPRODUCTION: Virtually nothing is known about the breeding biology of the species; however, newborn calves have stranded from June to August, implying a seasonality in reproduction. The calves are large, having the greatest relative length of any beaked whale, nearly half of the mother's length. Because males are distinctly different from females, male competition plays some role in mating strategies.

FOOD: The stomachs of four beached Hubbs' Beaked Whales each contained squid beaks—one had nearly 1,500, most of which were hook squid. Only one stomach had any fish remains, containing some 40 ear bones representing at least seven species, mostly lanternfish. This limited sample indicates that the Hubbs' Beaked Whale largely catches squid. The limited opening of its mouth fits with the concept of Hubbs' Beaked Whales sucking in squid and small fish in deep oceanic waters.

MORTALITY: Oval scars on Hubbs' Beaked Whales resemble the bites of cookie-cutter sharks, but these parasites do not usually cause mortality.

DISTRIBUTION: Hubbs' Beaked Whale is one of those mysterious mesoplodonts—it has only been identified in a few instances at sea. NMFS surveys reported in 1994 several groupings off Oregon, but none since then during their annual cruises. Thus, this account of the Hubbs' Beaked Whale's distribution is based mostly on stranded specimens from the west coast of North America and from Japan.

The Hubbs' Beaked Whale is endemic to the northern Pacific. In Japan, strandings have been reported at the confluence of the north-flowing warmwater Kurishio current and the south-flowing coldwater Oyashio current. On the west coast of North America, Hubbs' Beaked Whales have beached from Prince Rupert, British Columbia, to southern San Diego, California, largely along the path of the cool California Current. The Hubbs' Beaked Whale's range might extend further south into Baja California, but strandings on this coast are less likely

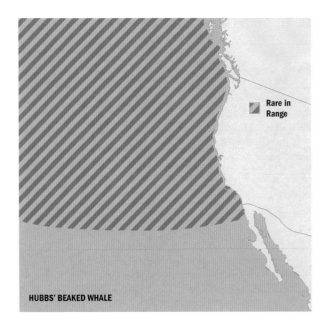

Rare in
Range

HUBBS' BEAKED WHALE

to be noted. Determining the range of any species from strandings depends on the distribution of marine biologists as well as ocean currents.

More strandings of Hubbs' Beaked Whales are reported than of any other mesoplodont along the west coast of North America. Stejneger's strand largely to the north, Perrin's overlap in California, and Pygmy beach to the south. Hubbs' Beaked Whale has been called the dominant species in its stranding range on the west coast of North America, and it may be more closely associated with the California Current than other beaked whales.

CONSERVATION: Because the species is so rare, little is known about past or current threats to its conservation. In the past, Japanese whalers hunted some Hubbs' Beaked Whales, but none have been seen captured in drift gillnets along the American west coast of North America. As with other beaked whales, they are at risk for injury and death from naval sonar, and from consuming plastics.

Agreement	Conservation Status
U.S. ESA	Not listed
U.S. MMPA	No special status
COSEWIC	Not at risk
IUCN	Data deficient

POPULATION: The small range of Hubbs' Beaked Whale suggests that these deepwater whales could be rare creatures. However, if the species were concentrated in the relatively unvisited central northern Pacific, it might escape detection by research cruises or capture by fishing boats. There is no population estimate for the species.

COMMENTS: According to native people in Siberia and Oregon, the meat of beaked whales is inedible; according to a Siberian folk tale, it is suitable only for unwelcome guests. The story of the fate of the first known Hubbs' Beaked Whale, however, is poles apart. Although its skeleton was preserved to become the type or reference specimen of the species, the whale itself was eaten by "locals" near the Scripps Institute pier where it stranded. Carl Hubbs himself testified that the meat "was of good flavor and tender when roasted or fried." Attitudes toward whales were different in the past, and war rationing had undoubtedly heightened people's appetites. But imagine unwittingly eating a specimen that would be named after you!

REFERENCES: Heyning 1984, Hubbs 1946.

PYGMY BEAKED WHALE   *Mesoplodon peruvianus*
Pls. 7, 8; Fig. 68.

Adult male average length is 3.8 m (12.5 ft). Adult female average length is 3.7 m (12.3 ft). Calf average length is 1.6 m (5.3 ft).

SCIENTIFIC AND COMMON NAMES: Almost all *Mesoplodon* whales were first encountered as bones or carcasses found on beaches. But one distinctive species was discovered alive and well on the high seas by marine surveyors in the eastern tropical and

Figure 68. Pygmy Beaked Whale.

temperate Pacific beginning in 1981. For 20 years, this beaked whale lacked a proper Linnean name, because it was sighted but never examined in a laboratory (fig. 68). Its provisional name became *Mesoplodon* Type A.

In 1991, Julio Reyes, a marine biologist, and his colleagues described a new species from whales caught by shark fishermen along the coast of south central Peru. Accordingly, they named the species *Mesoplodon peruvianus*, but they expected this whale to turn up elsewhere. In 2001, the American marine mammalogist, Robert Pitman, argued that *Mesoplodon* Type A and *M. peruvianus* were most probably the same species, and they are treated as the same species in this guide.

English common names for *Mesoplodon peruvianus* include the Peruvian Beaked Whale, Pygmy Beaked Whale, Lesser Beaked Whale, and Small Beaked Whale. The French common names are *Mésoplodon pygmée* and *baleine à bec pygmée*. Spanish names are *zifio menor* or *zifio peruano*.

DESCRIPTION: The Pygmy Beaked Whale is small, the shortest and lightest of its genus. The largest adult documented was 4.1 m (13.5 ft). The pygmy's body form is typical of a mesoplodont, being spindle-shaped and tapering toward its head and tail. Its melon bulges slightly and then slopes smoothly onto a very short brown-tipped beak. The color pattern of all Pygmy Beaked Whales has been summarized as dark gray above and whitish gray below, with light gray below the eye as well. There can be brown spots in the angle of the jaw and in the throat folds.

The jawline of the Pygmy Beaked Whale is curved, rising toward the rear. A pair of peglike teeth rest at the high point of the lower jaws. Only the tips of the teeth pierce the gum in males. The teeth of females and young are unerupted. In sub-adults, the teeth incline 20 to 40 degrees forward; in adults, they become perpendicular to the jaw.

The dorsal fin is low and triangular with a rounded top. The flipper can be 37 cm (14 in.) in length. The flukes can reach 74 cm (29 in.) in breadth and range from 20 to 25 percent of total length.

There are two color morphs of the Pygmy Beaked Whale, which accounts for the confusion as to species: those which are a uniform light brown or gray-brown (thought to be females) and those with a bold white blaze that runs diagonally down both sides and which also bear extensive white linear scars (thought to be males). The diagonal dark background before the white blazes could be fancied as bandoliers, a colorful analogy. The white blazes begin behind the melon and look like a chevron when viewed from aircraft. The area in front of the chevron is a kind of light orange-brown. Behind the chevron lies a contrasting dark black-brown color, which completes a three-color field mark for aerial observers.

FIELD MARKS AND SIMILAR SPECIES: The pale white cape of the males can distinguish them from other male mesoplodonts at sea. The small size and the small, triangular dorsal fin, which is reminiscent of the more northern Harbor Porpoise, might help identify a female or young Pygmy Beaked Whale. The peglike teeth in the short jaws of the males distinguishes them from other male mesoplodonts. The coloration of males is not clear at present—one Latin American illustration of the species has question marks over most of the body. Females and young resemble other mesoplodonts. Photographs of fresh specimens and careful measurements can supplement genetic confirmation of the species' identity.

NATURAL HISTORY: Little is known about the natural history of this rare species. Groups of Pygmy Beaked Whales with males present (the only groups easy to distinguish) swim in small schools averaging 2.3 whales, which range from one to eight individuals. The male morphs bear the linear scars characteristic of *Mesoplodon*, which are relics of conflicts at sea.

REPRODUCTION: The few publications available shed little light on social structure or reproduction, but the scarring on the back of one morph is indicative of male competition similar to other species of the genus.

FOOD: Pygmy Beaked Whales presumably eat similar prey as other members of the family, including midwater oceanic squid and fishes beyond the continental shelf.

MORTALITY: No information exists on the mortality of Pygmy Beaked Whales. Some bear white oval scars consistent with the bites of the cookie-cutter shark.

DISTRIBUTION: Surveyors have sighted the Pygmy Beaked Whale in waters from Peru to Baja California and halfway out to Hawaii. Strandings have been documented in Peru, Baja California, and New Zealand. Although there are few strandings with which to

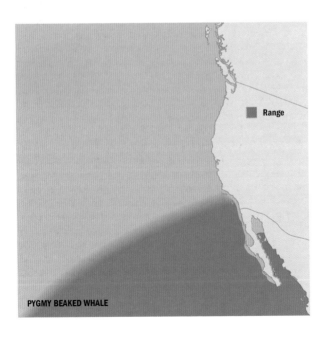

Range

PYGMY BEAKED WHALE

compare, the stranding in New Zealand suggests the species' range may be greater than originally thought.

CONSERVATION: So little is known about this species except from a few sightings and stranding records that threats to its conservation are difficult to deduce. The Pygmy Beaked Whale would be covered by Canadian and American marine mammal legislation, but most of its range lies far to the south and the west. Peruvian Law 26585, meant to conserve small cetaceans, protects six dolphin and porpoise species but no beaked whales.

Agreement	Conservation Status
U.S. ESA	Not listed
U.S. MMPA	No special status
COSEWIC	Not at risk
IUCN	Data deficient

Beginning about 1960, a market for dolphins developed in Peru for muchame, an Italian-style appetizer. This market expanded, and by the 1990s, it was estimated that more than 15,000 small cetaceans were taken each year. A number of laws banning this market were passed, and whale meat is no longer sold in supermarkets. Perhaps more than 3,000 small whales are still taken, which may include Pygmy Beaked Whales. If Pygmy Beaked Whales make long-distance migrations, then the intensive local Peruvian whaling may have impacted their population.

Bycatch in fisheries is a more likely threat. The Pygmy Beaked Whale was first described from specimens caught in the shark driftnet fishery. This is not a fortunate distinction because no other species has been discovered from bycatch. Peruvian gillnet fisheries continue and may still capture small cetaceans, some of which may be sold in the whale meat market. Asian driftnet fisheries for squid and tuna caught many cetaceans before the nets were banned in 1992. Because squid are the primary prey of many beaked whales, the large fishing fleets of the 1980s may have affected many species of these deep-diving offshore cetaceans. We will never know whether their impact was substantial or minor.

POPULATION: No population estimates or trends exist for the Pygmy Beaked Whale.

COMMENTS: At one time, marine mammalogist R.L. Pitman suggested that the *Mesoplodon* Type A beaked whale should have a more suitable name, and playfully dubbed the small cetacean the "bandolero beaked whale" for the X-like pattern on males' backs that suggested a bandolier. In Spanish, one might translate this freely as *El Bandolero,* the one with a bandolier.

REFERENCES: Pitman et al. 1987, Pitman and Lynn 2001, Reyes et al. 1991.

STEJNEGER'S BEAKED WHALE   *Mesoplodon stejnegeri*
Pls. 6, 8.

> Adult average length is 5 to 7 m (16.5 to 23 ft); average weight is 1,900 kg (4,200 lbs). Newborn average length is 2.3 to 2.5 m (7.5 to 8.2 ft); average weight is 80 kg (175 lbs).

SCIENTIFIC NAMES: For the etymology of the genus name *Mesoplodon,* see Blainville's Beaked Whale. The species was named *stejnegeri* in 1885 by F.W. True, one of the first marine mammalogists, in honor of Leonhard Stejneger, who had collected the type skull of this species on an Aleutian beach 3 years earlier. No other specimen of *M. stejnegeri* was reported until 65 years later.

COMMON NAMES: Stejneger's Beaked Whale is the English name, and other names include the Bering Sea Beaked Whale and the Saber-toothed Whale. In French, they are called *mésoplodon de Stejneger,* meaning Stejneger's Mesoplodon, or the *baleine à bec de Stejneger,* meaning Stejneger's Beaked Whale. In Spanish, the name is *zifio de Stejneger* or *ballena de pico de Stejneger,* both meaning Stejneger's Beaked Whale.

DESCRIPTION: Like other mesoplodont whales, Stejneger's Beaked Whales are long and spindle-shaped. Unlike other mesoplodonts, the melon is relatively flattened, so the smallish head more gently slopes to the beak. The jawline curves upward, ending in a high arch. The body of the adult male is mostly gray to black. Females and juveniles are brownish gray on the top but lighter on the bottom. The beak of the males is framed by two prominent teeth, 8 cm (3 in.) wide that emerge from the lower jaw. These 0.8 cm (2.7 in.) thick teeth appear before the apex of the arch and are canted forward and inward. The crowns of the teeth extend above the beak. The 5 cm (2 in.) wide teeth of females, and those of juveniles, do not break the gums.

In many stranded individuals, a dark cap on the head covers both the blowhole and the eyes. Mottling and irregular pigmentation often mark the whales from the throat to the base of the tail. The flipper pocket may be darker than the nearby skin. The small dorsal fin lies two-thirds down the back. The bottom surface of the flukes may bear striking concentric striations, which vary from gray to white and which seem to become brighter with increasing age.

FIELD MARKS AND SIMILAR SPECIES: Stejneger's Beaked Whales most closely resemble other mesoplodont beaked whales. Distinguishing species at sea is only possible for males under ideal conditions. A Stejneger's Beaked Whale male may be black-capped and has a flattened melon, while a Hubbs' Beaked Whale male has a prominent white-capped melon and has white-tipped jaws. The teeth of a Hubbs' Beaked Whale are further back in the lower jaw. Hubbs' Beaked Whales are not known to range north of British Columbia into cold-temperate and subarctic waters, where most Stejneger's Beaked Whales are sighted. Blainville's Beaked Whales bear teeth outside their even more extremely arched mouths, but they lack any dark head cap and tend to range to the south of the Alaskan waters favored by Stejneger's Beaked Whale. Ginkgo-toothed Beaked Whale males show few if any scars, and only the tips of their teeth are visible.

NATURAL HISTORY: Stejneger's Beaked Whales are the only beaked whales known to strand along the Alaskan shores, suggesting that most live mesoplodont species sighted in those waters probably were Stejneger's Beaked Whales. In one study in Aleutian waters, the beaked whales were seen in pods that ranged from five to 15. Water depth ranged from 730 to 1,560 m (2,400 to 5,100 ft). The whales traveled in tight formations, with synchronized surfacing and submerging. In two of the groups, the members appeared to be touching or nearly touching.

Strandings of Stejneger's Beaked Whales in Alaska from 1975 through 1994 confirmed that the whales live or travel in small social groups. Twenty of the 23 beached animals came ashore in groups. Stranding numbers ranged from two to five whales, although some of the beached whales may have been missed. Composition of the groups varied: one was all male, two were all female, and the rest were mixes of males and females. The fact that most strandings included more than a single individual may reflect strong social bonds within groups of these coldwater beaked whales.

REPRODUCTION: Data from Alaska and Japan suggests that birthing may extend from spring to fall. Oddly, the mother's milk of the Stejneger's Beaked Whale is blue-green, like that of a water buffalo. Milk from a stranded Stejneger's Beaked Whale had two to four times more protein than that of Sperm Whales or Beluga Whales but only half as much fat. The tusklike male teeth, useless in taking prey, suggest struggles over mates or dominance, as in other beaked whales. Some specimens of Stejneger's Beaked Whale have been reported with healed fractures of the jaw. Sexual maturity appears to be at about 4.5 m (14.8 ft) for both sexes. The oldest Stejneger's Beaked Whale found in a study of tooth rings was 35 years old.

FOOD: The mouth of the Stejneger's Beaked Whale can only open a few centimeters, which limits its diet to small or soft-bodied

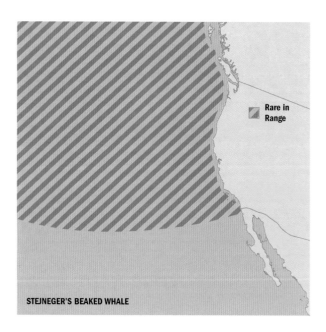

Rare in Range

STEJNEGER'S BEAKED WHALE

prey. Eleven stomachs sampled in Alaska held beaks almost entirely from two families of squid, with armhook species being the most common. The evidence from the stomachs of the stranded Alaskan whales suggests that these *M. stejnegeri* had fed at mesopelagic and bathypelagic depths. Nevertheless, some researchers have noted that this beaked whale pursues salmon off Japan and has been trapped in salmon nets there.

MORTALITY: Stejneger's Beaked Whales bear scars consistent with the bites of cookie-cutter sharks and the parasitic copepod *Penella*. Because these wounds do not readily repigment, the skin of these whales can be a graphic history of such predation. Accordingly, older Stejneger's Beaked Whales appear to bear the most scars.

DISTRIBUTION: Stejneger's Beaked Whales are limited to the northern Pacific. Strandings are concentrated from British Columbia, along the curve of Alaska and the Aleutians, in the Bering Sea, and in the Sea of Japan. In Alaska, strandings suggest that the species frequents the deeper Aleutian basin and the Aleutian trench. The southernmost stranding on the west coast of North America was at Cardiff, California, at latitude 33° N. Most British Columbian strandings are on beaches favored by recreationists along the outer coast of Vancouver Island. If more of the rugged British Columbia coast were surveyed, undoubtedly more specimens of Stejneger's Beaked Whale would be discovered.

They are believed to inhabit the deeper waters along the continental slope. When sighted at sea, they have been in waters 730 to 1,560 m (2,400 to 5,100 ft) above the Aleutian Basin.

CONSERVATION: Some native Siberians and Americans believed that eating beaked whales would cause diarrhea, which probably limited prehistoric whaling. Small numbers are still taken in the Far East.

Agreement	Conservation Status
U.S. ESA	Not listed
U.S. MMPA	No special status
COSEWIC	Not at risk
IUCN	Data deficient

Modern-day threats to the species' conservation include plastic pollution, fisheries interactions, and climate change. Plastic twine up to 200 cm (80 in.) length was found in two Stejneger's

Beaked Whales' stomachs in Alaska, possibly due to incidental ingestion while the whales were suction feeding, or to a misperception of the plastic debris as prey. Because the "biomass" of plastics in the ocean can be greater than the biomass of plankton in some areas, this is hardly a trivial concern. In deepwater or central-ocean species, death from plastic ingestion would be hard to detect.

Some Stejneger's Beaked Whales have been caught in salmon driftnets off Japan, but none in the Alaskan fisheries. A single Stejneger's Beaked Whale was bycaught in the California drift gillnet fishery from 1990 to 1995, although other beaked whales were more frequently taken. Following the introduction of acoustic net pingers the next year, no further beaked whale drownings were reported for that fishery.

Climate change may result in radical changes in oceanic current systems. If the Arctic ice cap vanishes, then currents could flow between the Atlantic and the Pacific Oceans across an open Arctic Ocean. Presently, water enters the Arctic Ocean through the Bering Strait from the Pacific Ocean and in the Norwegian Current from the eastern Atlantic, and then exits through the East Greenland Current. The impact of a melted ice cap is difficult to forecast either for northern latitudes or for the globe. No Stejneger's Beaked Whales have ever been observed in the Arctic Ocean, but warming seas could conceivably shift the range of this species northward.

POPULATION: Japanese scientists estimated there were 7,100 Stejneger's Beaked Whales in waters near Japan in 1998. No current population estimates exist for U.S. waters because of the rarity of sightings.

COMMENTS: Why do whales strand in groups? One reason may be that a number of individuals are fleeing some common factor such as blasts of intense and painful military sonar or a pod of approaching mammal-eating Killer Whales. Another reason might be that the whales act like a herd, following their leader or companions. Cetaceans have strong social bonds, and dolphins or whales will accompany or aid a wounded comrade who might be drifting ashore. Cetaceans also might "stampede" in some form of panic, rushing together toward shallows or shorelines. Stejneger's Beaked Whales seem to be highly bonded, if swimming in tight and synchronized pods is an indication. Close social ties may help explain the many mass strandings seen in Alaska and on the west coast of Japan.

REFERENCES: Arai et al. 2004, Tsuneo and Tadasu 2003, Walker and Hanson 1999.

LONGMAN'S BEAKED WHALE   *Indopacetus pacificus*
Pl. 5; Fig. 69.

> Adult average length is 7 to 9 m (23 to 30 ft). Newborn average length is 2.9 m (9.5 ft).

SCIENTIFIC NAMES: The genus name *Indopacetus* is an invented Latin word meaning "Indo-Pacific whale," referring to the Indian and Pacific oceans. The species name *pacificus* refers to the Pacific Ocean. Thus, the scientific name is somewhat redundant, literally translating as "the Pacific Indo-Pacific whale" (fig. 69). Originally, Longman named the species *Mesoplodon pacificus,* and may have chosen *pacificus* to distinguish this whale from the somewhat similar *Mesoplodon mirus* (True's Beaked Whale) of the Atlantic Ocean. Later, the marine biologist Moore renamed the genus *Indopacetus,* indicating its likely range.

COMMON NAMES: The English common name Longman's Beaked Whale honors the Australian zoologist H.A. Longman, who

Figure 69. Longman's Beaked Whales.

first described the species in 1926. The French and Spanish common names are similar: *baleine a bec de Longman* and *zifio de Longman*. The Tropical Beaked Whale is the name given to a rare cetacean, sighted in the Pacific and Indian Oceans, that is likely identical to Longman's Beaked Whale.

DESCRIPTION: The Longman's Beaked Whale is quite large; Longman's original projection, from a skull found in 1882, was 7.6 m (25 ft). A stranded adult female from Somalia was about 6 m (20 ft) long; a stranded baby male from South Africa was 2.9 m (9.5 ft) in length and 228 kg (502 lbs) in weight; a second South African specimen, a young male, was 3.6 cm (11.9 ft) long and weighed 510 kg (1,124 lbs).

Its overall color has been described as various shades of brown or gray. Two individuals photographed at sea show a dark dorsal fin and fin base that have linear white scars; these may be adult males, because they bear tooth-rake marks, as seen in other beaked whale species. The melon of males seems to meet the beak at a 90 degree angle. Gray-brown whales attending calves show no clear color patterning or rake scars, and probably represent adult females. Younger Longman's Beaked Whales have a black saddle on their backs that begins at the blowhole and extends down over the eye and down toward the flipper. The dark saddle contrasts with a whitish melon.

The dorsal fin of this whale is tall, and usually hooked and pointed. It is positioned far back on the body. The flipper is small and fits into the typical ziphiid "pocket."

Observers at sea have not seen teeth in the mouths of the Longman's Beaked Whales, but the linear rake scars on the putative males suggest their presence. The lower jaws in adults extend slightly beyond the upper. Unerupted conelike teeth have been documented near the tips of the jaw in the stranded specimens. The teeth point forward at a 45 degree angle.

FIELD MARKS AND SIMILAR SPECIES: Longman's Beaked Whale most closely resembles the Southern Bottlenose Whale (*Hyperoodon planifrons*). Both are burly whales with striking melons and prominent dorsal fins. Near the west coast of North America, though, we are well to the north of the range of the Southern Bottlenose Whale.

Possible confusion is with Shepherd's Beaked Whale (*Tasmacetus shepherdi*), but this southerly species does not reside within the California Current. Within the California

Current, Longman's Beaked Whales may be confused with unscarred immature Baird's Beaked Whales and with young Risso's Dolphins that have similar coloring and shape of dorsal fin. The Risso's Dolphin, however, has a blunt head compared to the Longman's Beaked Whale's head.

NATURAL HISTORY: Longman's Beaked Whales, like most cetaceans, join schools. Over many surveys, researchers noted that schools in the Pacific, as a whole, averaged 41 individual Longman's Beaked Whales, with a range of one to 100. Schools in the eastern Pacific were significantly smaller, with an average of nine whales. The whales move vigorously together in their tight schools, with their beaks and melons breaking the water. Longman's Beaked Whales may also breach, and sometimes they school with Short-finned Pilot Whales.

REPRODUCTION: Nothing is known about the breeding biology of Longman's Beaked Whale. The one neonate examined as a beached dead whale suggests that the size of calves at birth may be very large, an estimated one-third the length of an adult female. Because males are apparently different from females, male competition likely plays some role in mating strategies.

FOOD: As with other beaked whales, Longman's Beaked Whales most likely prey on cephalopods, but few have been examined to find out upon what species they prey.

MORTALITY: Some scars seen at sea may have resulted from cookie-cutter shark bites, indicating that the whales were feeding at least at a depth range of 30 to 910 m (300 to 3,000 ft). A long tooth-rake on a South African infant calf may have been made by a large shark.

DISTRIBUTION: The distribution of the Longman's Beaked Whale can only be deduced from rare sightings and strandings. Sightings of this beaked whale are widely distributed in the Indian and Pacific Oceans in tropical and subtropical waters, but they are not usually associated with the California Current. Nevertheless, Longman's Beaked Whales can occur between 40° latitude North and South. Whether the sightings represent different populations or subspecies is unknown.

CONSERVATION: The rarity of the species hinders our understanding of threats to its conservation, but some information can be gleaned from stranded animals. The smallest confirmed Longman's Beaked Whale, the newly born South African calf, carried a load of chemical toxins. Judging by his age, he probably

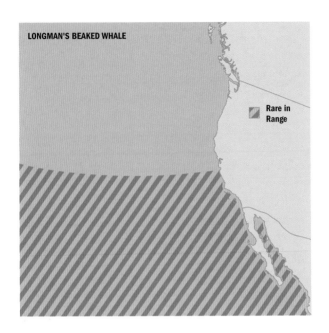

LONGMAN'S BEAKED WHALE

Rare in Range

received this dose in his mother's milk. The contaminants included DDT and related compounds.

Agreement	Conservation Status
U.S. ESA	Not listed
U.S. MMPA	No special status
COSEWIC	Not listed
IUCN	Data deficient

POPULATION: No census estimates exist for this widespread species, whose rarity makes projections of population numbers unreliable. During a series of surveys in the eastern tropical Pacific, totaling 600 sea days, seasoned observers sighted these whales only four times.

COMMENTS: Science is a detective story: a skull, a hunch, circumstantial evidence; bodies found, buried, disinterred, then reburied; samples taken for genetic fingerprints—and at last we are closing in on the identity of the missing living cetacean. Sherlock Holmes would be proud of the generations of scientists who have surveyed the shores and seas to solve the case of Longman's Beaked Whale.

This species was named on the basis of a skull—from a museum collection in Queensland, Australia—that was found in 1882. A second skull was retrieved from a Somali organic fertilizer plant in 1955. Until this century, these two skulls were all that were known for certain of this species. But beginning in the 1970s, large beaked whales of an unknown species were sighted during surveys in warm waters of the Pacific and Indian Oceans. Originally, some observers thought that the mystery whales might be a tropical version of the Southern Bottlenose Whale of the seas surrounding Antarctica. But a second interpretation was that the uncommon whales were actually the missing Longman's Beaked Whale.

From the two original skulls, it was impossible to tell exactly what a living Longman's Beaked Whale would look like. You could see that this cetacean belonged to the family of beaked whales, and you could infer from the size of the skulls that it was one of the most massive members of the family. But how to link the bones to a flesh-and-blood whale? Then, an answer came, from three recently stranded beaked whales from South Africa and the Maldive Islands. DNA samples from the two isolated skulls and from the bodies of the beached whales were compared, and they matched. The bodies resembled those of the mysterious tropical beaked whales. Many scientists concluded that the tropical beaked whale and Longman's Beaked Whale were probably the same cetacean. More sightings and DNA samples should confirm or qualify this conclusion.

REFERENCES: Ballance and Pitman 1998, Dalebout et al. 2003, Pitman et al. 1999, Reyes et al. 1996.

KILLER WHALE — *Orcinus orca*
Pl. 9; Figs. 11, 15, 70, 71, 72.

Adult male average length is 9.8 m (32 ft); average weight is 10,000 kg (22,000 lbs). Adult female average length is 8.5 m (28 ft); average weight is 7,500 kg (16,500 lbs). Newborn calf average length is 2.5 m (8 ft); average weight is 180 kg (400 lbs).

SCIENTIFIC NAMES: The roots of the zoological names *Orcinus* and *orca* go back to antiquity. *Orcinus* derives from Orcus, the Roman god of Hell, and may be taken to mean "hellish." *Orca* stems perhaps ultimately from the same root, and has been linked to the English and French word "ogre." In English *ork* (or *orque*) has signified large whales as well as sea monsters.

COMMON NAMES: Today, most field guides and scholars favor the name Killer Whale. However, the alternate name Orca has been popular since the 1960s. Along the British Columbian coast, Euro-Americans traditionally used the term Blackfish for Killer Whales—including as the translation for many names for this whale species in Native American folklore and art. However, the term Blackfish refers to many species of fishes and to several different species of dark-colored dolphins. Some native words for Killer Whale survive on the Northwest coast: for example, for the Makah, the Orca is *Klasqo'kapix*; for the Haida, *Ska-ana*; and for the Tlingit, *Keet*. French words for Killer Whale include *epaulard,* meaning Swordfish, and *orque.* In Spanish, you may say *orco* or *orca,* the masculine and feminine forms of Orca, or *ballena asesina,* the Assassin Whale.

DESCRIPTION: The greatest of the oceanic dolphins, Killer Whales bear a handsome pattern of white field marks against their black bodies: white chests and sides, and white patches above and behind the eyes (fig. 70). Just behind their conspicuous dorsal fin is a dark gray "saddle patch." Each Killer Whale has an individually shaped and scarred dorsal fin, as well as a unique saddle and eye patch. The 1.8 m (6 ft) tall dorsal fins of males are triangular and upright; the shorter 0.7 m (2.3 ft) dorsal fins of females generally droop. Adult males have larger pectoral fins, tail flukes, and girths than females, as well. Killer Whale flippers are paddle-shaped, broad, and round, and are nearly 1.8 m (6 ft) in length and 0.9 m (3 ft) wide. The flukes are notched and slightly concave, and can reach 2.7 m (9 ft) in breadth. Killer Whale teeth are conical, about 7.6 cm (3 in.) tall and 2.5 cm (1 in.) in diameter, but some can be up to 13 cm (6 in.) long. The teeth are curving, interlocking, and pointed slightly inward and backward—meant for grasping and tearing prey. There are 10 to 13 pairs in each jaw, for a total of 40 to 52 teeth.

Photographic catalogs of Killer Whales have been created on the west coast of North America and in other areas; these records, which enable researchers to identify individuals, are a

Figure 70. Transient Killer Whale female and calf, breaching.

boon to our understanding. In fact, we can identify all individuals in some populations of the west coast of North America, a remarkable feat in wildlife biology.

FIELD MARKS AND SIMILAR SPECIES: Close up, the adult male Killer Whale is unlikely to be confused with any other cetacean. Far away, female or juvenile Orcas might somewhat resemble False Killer Whales, Risso's Dolphins, or similar species. None of these, however, bear the distinctive white-and-black pattern of the Killer Whale. Also, when Orcas are seen in a distant group, the taller dorsal fins of males contrast with the shorter ones of the females and juveniles, which are about half their height; this is unlike the case in other dolphins.

NATURAL HISTORY: Killer Whales are the world's most widely distributed cetacean species. Individual whales, however, are not world travelers, but belong to regional ecological groups (called ecotypes) that have distinctive genetics, calls, social structures, ecological roles, and local ranges. Some experts suspect that at least some ecotypes may actually be different species, or may be species in the making. It could be that in the Killer Whale, we are witnessing a process of evolutionary creation, as the

Figure 71. Southern resident, fish-eating Killer Whales.

single world-dominant cetacean breaks into several new and specialized kinds of whales.

Residents, transients, and offshore Killer Whales are the three major ecotypes found along the California Current. The most commonly encountered is the southern resident Killer Whale ecotype, so named because individuals belong to stable, salmon-eating social groups, called pods, that mostly reside in the inland seas of Washington and British Columbia (where Killer Whales were first intensively studied; fig. 71). Residents swim in family groups called matrilines, each consisting of a matriarch and her descendants. Several matrilines form a pod. Remarkably, no resident Killer Whale female ever leaves her matriline nor abandons her ancestral waters. Each pod and matriline chatters away in its special dialect. The resident pods can be reliably observed in their small summer home ranges of the Pacific Northwest and are perhaps the best known of all cetaceans.

Transient Killer Whales range from southern California to Alaska (fig. 72). They travel alone or in small groups, usually of not more than 10, and hunt marine mammals and sometimes seabirds. Transients do not stay in stable matrilines, although firstborn males tend to stay with their mothers. The transients' movements over their large range tend to be erratic, as they appear here and there in their pursuit of prey. Usually silent,

Figure 72. Transient, mammal-eating Killer Whales, female and calf.

their voices can sometimes be heard after a marine mammal kill. Altogether, they number perhaps 300, and, given the vast seascape that they roam, they are rarely seen.

Offshore Killer Whales are not well known. These smaller Killer Whales were first described offshore of British Columbia and Washington, from whence derives their name. Offshore Killer Whales are sometimes seen at the surface in noisy groups that can number more than 100. These Killer Whales eat fish, and their nicked fins suggest they may also take squid.

A fourth possible west coast North American Killer Whale group has been suggested—the LA Pod, so called because they frequent the waters of Los Angeles. The LA pod is small, but infamous, because a member of this group killed a Great White Shark in plain sight of a whale-watching boat off the Farallon Islands in 1997, with the episode caught on videotape. The LA Pod ranges south to the upper Sea of Cortez in Mexico and north at least to the Gulf of the Farallones.

The dorsal fins of the three major ecotypes vary in shape (see Plate 9). Residents have a rounded dorsal fin that ends in a sharp corner, transients have a pointed dorsal, and offshores have a rounded dorsal. These differences reflect the genetic differences between the ecotypes. Inbreeding between these groups may not have occurred for tens of thousands of years.

If you are fortunate enough to see a Killer Whale, try to take clear photographs and good notes. Researchers can add well-documented observations to the photographic databases that help us understand the whales.

Killer Whales throughout the world generally appear to have different regional cultures. In the Pacific, Killer Whale pods signal in distinct dialects (scientists are not sure what the calls mean). In the North Atlantic, some Killer Whales take herring by thrashing their flukes through schools of fish. In the South Atlantic, other Killer Whales surf out onto beaches to snare young sea lions resting on the sand. In the last case, young Killer Whales have been observed in a sheltered bay being trained by their mothers to ride waves and to roll back into the sea. These kinds of observations suggest to some biologists that Killer Whales are cultural creatures that pass their ecological skills and social behaviors down across generations. Cultural wisdom may give the Killer Whales a considerable advantage for life in the sea.

Their acculturation could also constrain the survival of some whales, if prey should suddenly vanish. Can cultural specialists such as fish-eating Killer Whales adjust when local fisheries collapse, forcing them to either move or switch prey? Recently, numbers of the southern resident pod along the Washington and British Columbia coasts have begun to drop. Failing salmon runs in the Northwest may be one cause for the decline, although there are other possible reasons. To the south, runs of salmon were even more heavily damaged, and many are now extirpated. What were the numbers of North American west coast Killer Whales like before the Columbia and Snake Rivers were dammed, and the California Central Valley Project drained the Sacramento? Were there once resident salmon-eating Killer Whales off San Francisco Bay?

Communication is essential for a highly intelligent and social animal, and Killer Whales are among the most intensively studied of any species. Like the Common Bottlenose Dolphin, the Killer Whale can produce two sounds at once, and it is one of a few species conclusively shown to echolocate in captivity. Its discerning of objects, however, is not as fine as that of the Common Bottlenose Dolphin, because Killer Whales can only detect an object down to the size of a bean (compared to a pin for the dolphin).

REPRODUCTION: Both sexes mature at about 4.5 to 5.4 m (15 to 18 ft), and females average around 4.3 m (14 ft) long when they first give birth. Gestation is around 15 months, and there is no seasonal timing to births. Newborns appear at any point in the year, but less than half live until the next year in some populations. Calves nurse for 2 years, taking solid food at around 1 year; females often calve at 5-year intervals. In captivity they can give birth every 2 years; food availability may affect the calving interval. They can give birth until age 40, producing about five calves in a lifetime. Females may then live into their seventies, and perhaps even into their nineties, becoming matriarchs in their society. In their post-reproductive phase, these "grandmothers" may serve as babysitters, teachers, guides, and keepers of the clan's or pod's traditions. Males serve no such function in the society, and they live only to around 30 years old.

FOOD: The diet of Killer Whales reflects their cosmopolitan distribution and regional preferences. The world over, transient Killer Whales take a great range of prey, including many species of marine mammals in this guide, from fast-swimming Common Minke Whales to diminutive Sea Otters. In Washington and British Columbia, their main prey are Harbor Seals. In Monterey Bay, California, northbound Gray Whale calves are pursued by pods of transient Killer Whales, despite the resistance of their mothers. Killer Whales may repeatedly ram Gray Whale mothers and calves to weaken or separate the pair. In response, the mothers may roll like a log to take the Killer Whales' blows or may lift the calves onto their backs. In the Pacific Northwest, Killer Whales have been documented chasing down and killing Common Minke Whales, holding them underwater until they drown. Off central California, some 20 percent of feeding Humpback Whales have rake marks from Killer Whales, hinting that this predator may be preying heavily on large whales. Many of these hunts entail special, learned techniques, carried out by cooperating Killer Whales.

Other ecotypes take an array of bony fishes, sharks, rays, squid, and octopuses, which vary by region and season. In the Pacific Northwest, the resident Killer Whales feed almost exclusively on salmon. In the Antarctic, they occasionally eat penguins.

When pursing prey, Killer Whales are among the fastest of whales, swimming up to 40 kph (25 mph); their usual travel

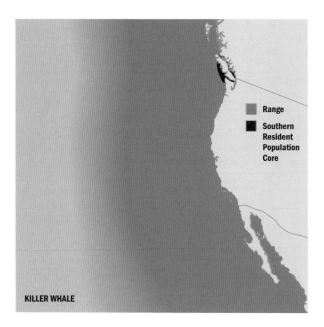

KILLER WHALE

Range

Southern
Resident
Population
Core

speeds, however, are in the range of 6 to 10 kph (4 to 6 mph). If harassed by boats, they may assume a formation with females and calves in the center and males guarding the flanks.

MORTALITY: Predators of juvenile Killer Whales include a few large sharks, but even Great White Sharks are eaten by Killer Whales. However, Killer Whales can die in the act of predation if an intended victim strikes back: a North Pacific Right Whale off Alaska slew a Killer Whale with a single blow of its mighty flukes. Similarly, stingrays may inflict mortal wounds on Killer Whales using their stingers.

DISTRIBUTION: Killer Whales as a species range worldwide and are seen in the furthest reaches of every ocean. Typically, they favor colder, fertile waters near coasts, but the offshore ecotype ranges to unknown oceanic destinations, and even the southern resident ecotype has been sighted in central California. Most sightings are within 800 km (500 mi) of continents. Most individual Killer Whales are not world travelers but belong to regional ecological groups (called ecotypes; see above).

VIEWING: The inland waters of British Columbia and Washington State are the best place to see Killer Whales. The resident pods there are found in relatively small home ranges and swim in protected waters. Note that there is controversy about the impacts of Killer Whale ecotourism in the Northwest. Most Killer Whale-watching vessels are concentrated in the south, in the Haro Strait, the Strait of Juan de Fuca, and by the San Juan Islands. A smaller whale-watching squadron is focused on the Johnstone Strait on the midcoast of British Columbia. Killer Whales can also be seen reliably on whale-watching trips in Monterey Bay, during the northward Gray Whale migration when females attend calves.

CONSERVATION: Killer Whales, like other large whales, were hunted for their oil and meat, and one whale was reported to yield around 750 to 950 kg (1,650 to 2,090 lbs) of high-quality oil. Killer Whales may be the most polluted of any marine species, and when they wash ashore dead, researchers wear protective clothing similar to that required for toxic cleanups. The Northwest coast Killer Whales suffer from high levels of pollutants such as PCBs and PBDEs, flame retardants. PCBs are oily chemicals used as lubricants and coolants in high-heat devices and were banned in North America in the 1970s. Created to resist breakdown in hot brakes and electric transformers, PCBs also resist breakdown in the cold seawater. Flame retardants are a new class of persisting pollutants found in clothing, foams, and plastics. Both pollutants are endocrine disrupters that can act like sex or thyroid hormones and affect reproduction in all mammals.

Agreement	Conservation Status
U.S. ESA	Southern resident endangered and critical habitat
U.S. MMPA	Southern resident depleted; transient group AT1 depleted
COSEWIC	Southern resident endangered; transient and offshore ecotypes threatened
IUCN	Data deficient

Absorbed in the flesh of animals, PCBs are passed up the food chain of the sea, from plankton to fish to mammal. Killer Whales reign at the top level on this marine food pyramid:

they also rank among the most tainted sea mammals in the world. A record-holding transient female Killer Whale that washed up dead at Dungeness Spit in Washington was so toxic that she was off the scale of the scientists' instrument. But perhaps the most contaminated marine mammals of all may be the male transient whales. Mammal-eating transients are more polluted than fish-eating residents, with the transients being one trophic step higher than the residents. Male Killer Whales become more contaminated as they age; females lose much of their toxicity when giving birth, passing on their load of pollutants to their young in milk. As much as 90 percent of the toxins carried by the mother can be transferred to her firstborn. Killer Whales can live 90 years; today's elders were never exposed to endocrine disrupters in their youth. We do not know the consequences of the gross contamination of infant Killer Whales in the sensitive stages of their early development. In Harbor Seals, much lower levels interfere with development and reproduction.

Killer Whales are also affected by sound pollution. The calm inland waters of British Columbia are ideal for raising Atlantic salmon on aquatic farms, but Harbor Seals quickly learned to raid the aquaculture pens. To discourage Harbor Seals, four salmon farms in the Broughton Archipelago near the north end of Vancouver Island installed acoustic harassment devices designed to drive off the seals. The unintended consequence of this Harbor Seal harassment was that Killer Whales avoided the Archipelago for the 6 years when the high-intensity sounds were broadcast. This was a demonstrable but reversible longterm effect due to human-generated sound; the whales returned when the speakers were silenced.

Sounds are also generated by boats. The southern resident pods can be shadowed by more than 20 motorized vessels during the summer season: there can be far more whale-watchers than whales. Whales respond to escorting tour vessels by changing their courses, which means that whales are wasting energy avoiding tourists. "Leap-frogging," or getting ahead of the whales, also drives whales off course. The presence of boats also may have caused Killer Whales to abandon their favored rubbing rocks in the Robson Bight (Michael Bigg) Ecological Reserve. Whether the underwater din from tour boat motors interferes with the whales' ability to echolocate salmon or to hear other whales is not known; however, in theory the sounds made by boats could mask Killer Whale calls and returning echoes.

POPULATION: NMFS identifies four stocks of Killer Whales that occur completely or partly in the California Current, some of which have overlapping distributions. They recognize two stocks of resident Killer Whales, one northern in the range from British Columbia to Alaska, and one southern in the coastal waterways of southern British Columbia and Washington. Photographic catalogs list a maximum of 216 northern resident Killer Whales living at one time. The southern resident stock has ranged from 71 to 96 individuals between 1974 and 2004. Live captures for oceanaria took an estimated 47 whales out of this stock before the earliest population census. For the transient Killer Whale ecotype, the minimum estimate of the stock is 314 individuals. The stock of offshore Killer Whales probably numbers more than 450.

NMFS estimates a total population of 1,214 Killer Whales off the U.S. west coast, excluding Alaska. It is not possible to separate ecotypes reliably on surveys, but a photographic analysis of individuals off California and Oregon identified about two-thirds as transients and one-third as offshore. A British Columbia study identified about 200 offshore Killer Whales, of which 20 were also seen off California.

The southern resident population of Killer Whales has been so depressed that the population was designated as endangered in 2005, and critical habitat was identified throughout much of Puget Sound and associated island waters of Washington in 2006.

COMMENTS: Two hundred years ago, the world was ruled by two species of warm-blooded vertebrates. In the sea that covers most of the Earth, Killer Whales were the ancient and unchallenged masters—apex hunters that fed even on massive baleen whales and sleek deadly sharks. On land, people had reached the last continent, Antarctica, and had become the top consumer of land-based life.

There were curious parallels between these two vertebrates. Killer Whales and people were seemingly both cultural beings, with groups that varied in how they fed themselves, how they communicated, and how they bonded to one another. It was as if there were nations of Killer Whales in the world ocean and nations of people on the continents.

REFERENCES: Au et al. 2004, Baird 2002, Baird et al. 2005, Dahlheim and Heyning 1999, Ford 2002, Goley, and Straley 1994, Heyning and Dahlheim 1988, Pitman 2009, Pitman and Chivers 1999, .

SHORT-FINNED PILOT WHALE

Pl. 10; Figs. 48, 73.

Globicephala macrorhynchus

> Adult male average length is 5.3 to 7.2 m (17 to 24 ft). Adult female average length is 4.1 to 5.1 m (13 to 17 ft). Newborn average length is 1.4 to 1.8 m (4.6 to 6 ft).

SCIENTIFIC NAMES: *Globicephala* derives from the Latin *globus,* for "globe" or "ball," and the Greek *kephale* for "head." "Globe-headed" is an apt name for this genus, as the description below will suggest (fig. 73). However, the species name *macrorhynchus,* which comes from the Greek *macro* for "large" and *rhynchus* for "bill" or "beak," is inappropriate, because this whale's snout is actually short and inconspicuous beneath its domed head.

COMMON NAMES: This species is called the Short-finned Pilot Whale to distinguish it from its close relative, the Long-finned Pilot Whale. One story has it that whales of this genus seemed to be led by individuals that acted as pilots for their pods, whence came its common generic name. Coastal folk often call this whale and other dark dolphins Blackfish. An old English common name is the descriptive Pothead. The French common name is *globicéphale tropical,* meaning the Tropical *Globicephala.* Spanish names include *el calderon negro,* the Black Cauldron, and *el calderon de aletas cortes,* the Short-finned Cauldron.

DESCRIPTION: The Short-finned Pilot Whale has a long, robust body with a heavy tailstock, and it is one of the larger members of the Delphinidae family of whales, and the largest of the oceanic dolphins, after Killer Whales. The head is bulbous, with the prominent melon becoming more exaggerated and squarish as males age. The Short-finned Pilot Whale is highly sexually dimorphic, in both size and look. In mature males,

Figure 73. Short-finned Pilot Whales.

the melon can overhang the short upturned jaws by 10 cm
(4 in.). The jaws hold from seven to nine rows of peglike teeth,
for a total of some 36 teeth at the front of the mouth. The
pointed flippers are sickle-shaped and are about one-sixth of
the body length. The low, wide, hooked dorsal fin lies about
one-third of the way down the back. The flukes are sharply
pointed and notched.

The Short-finned Pilot Whale is largely black or dark gray-
brown, except for its variable lighter markings, which are more
obvious under the water. A whitish teardrop-shaped blaze arches
from behind the eyes to below the fin. Behind the fin, a faint gray
saddle may be evident, but this feature may be absent in pilot
whales off the west coast of North America. On the chest lies a
gray, anchorlike marking that is joined by a thin line to a gray
genital patch.

FIELD MARKS AND SIMILAR SPECIES: The Short-finned Pilot
Whale is so like the Long-finned Pilot Whale that experts can't
reliably tell them apart where the species' ranges overlap. Long-
finned Pilot Whales do tend to have more teeth, a narrower
skull, and longer fins, of course, averaging one-fifth of the
body length; their fins also show more "elbow." Only the skull
measurements are absolutely diagnostic. Because Long-finned
Pilot Whales are antitropical, found now only in the north
Atlantic and the Southern Ocean, northern Pacific whale-
watchers will only see the short-finned species. It should be
noted that skeletal remains from Japan suggest that there was

a northern Pacific population of Long-finned Pilot Whales before the 12th century, which may have been exterminated in early medieval drive fisheries.

At a distance, Short-finned Pilot Whales resemble other "blackfish," particularly the False Killer Whale. The Short-finned Pilot Whale's broad-based but shallow-angled fin can help distinguish it from other far-off dark dolphins.

NATURAL HISTORY: Short-finned Pilot Whales are exceptionally social and highly intelligent, often performing in oceanaria. In the wild, they are found in groups of 20 to 90 or more, but they separate into social clans that average around 20 in number. These clans apparently join others, in a community, to forage or travel. Off the coast of Southern California, pilot whales have been reported in three kinds of social configurations. In the "chorus line" formation, the whales travel or hunt on a broad front up to 3 km (2 mi) wide but only a few whales deep. In an ordinary feeding group, the whales may loosely drift in a given direction but hunt independently. In a loafing group, 12 to 30 whales float together, nearly touching, and may be composed of mostly nursing females. Often, Short-finned Pilot Whales school with other cetaceans, particularly Common Bottlenose Dolphins, as well as with Common and Pacific White-sided Dolphins, and Gray Whales, Fin Whales, and Sperm Whales. But they also harass Sperm Whales on occasion. Sperm Whales respond with counter-aggressive behavior, such as fluke swishes and tail slaps, as well as by assuming defensive formations. Perhaps Sperm Whales are responding to the approaching pilot whales as if they were False Killer Whales, which do attack them.

The pilot whales are polygynous, but the basic social unit is matrilineal: female calves likely stay in their mother's stable pods for life. Female Short-finned Pilot Whales outlive the males and live long past the time when they stop reproducing. In a sample from 27 pods in Japan, the oldest male was 46 years of age; the oldest females were 63. The females no longer bred after 40, but they continued to nurse their last calf and engaged in nonreproductive mating. A substantial fraction of the pods consisted of elder females. These matrilineal pods can temporarily join others, to forage or travel.

An extended lifetime after menopause is rare in nature, but it characterizes Killer Whales and Short-finned Pilot Whales, and possibly also False Killer Whales and Sperm Whales. One explanation is that the long-lived cetacean matriarchs may be valuable clan elders, passing down the successful social or foraging traditions of the group. Another theory, dubbed the "grandmother hypothesis," is that the older females serve as beneficial babysitters for their younger relations. A third reason, at least in the case of Short-finned Pilot Whales, might be the selective payoff resulting from the long nursing of the last calf. Interestingly, Long-finned Pilot Whale females as old as 55 years have been found fertile, and thus were lacking a long post-reproductive life.

When males reach maturity, they leave their natal clan, and only a few adult males are clan members. From genetic studies, researchers in Japan have discovered that a clan is not a harem, but that mature males instead move between clans, seeking mating opportunities. Males compete by biting, head-butting, and tail-slashing, as can be seen in their scars.

Pilot whales maintain strong social networks, and so communication is essential. Vocalizations of Short-finned Pilot Whales are complex and even include individual signatures. The fundamental frequencies of the whistles of pilot whales are low and within the range of human hearing, although some of their overtones may be ultrasonic. Their pulsed calls—some of which may sound to us like rude noises—are also in the audible range. Although we can come to recognize different vocalizations, we do not know their meanings. Advances in audio, video, and computational techniques may help us identify the context of the calls, bringing us closer to the dream of understanding their voices.

REPRODUCTION: Pilot whales have an exceptionally long reproductive cycle for any mammal, but there is no regular seasonal pattern related to the cycle. Nevertheless, they tend to mate in spring and summer, but gestation takes 14 months, and so birthing occurs the following summer or fall. The calves are suckled for a minimum of 2 years and, more often, 3 years or longer. The average calving period is 7 years, but calves can eat solid food as early as 6 months. The last calf to be born to a female may nurse for as long as 15 years, and sons are commonly suckled into the teenage years, but daughters no longer than 7 years.

Males mature at an average age of 17 years, although mating at age 13 to 16 has also been reported. Social maturity, however, is reached when the whales are in their twenties. Females mature much earlier, ranging from 7 to 12 years.

FOOD: Short-finned Pilot Whales may be the oceanic equivalents of cheetahs or peregrine falcons. Suction cup recorders attached to Short-finned Pilot Whales off the island of Tenerife in the Atlantic surprised scientists when they showed the whales racing in the depths. Their expectation was that the whales would move slowly, at 1 to 2 m/second (3 to 6 ft/second), conserving energy and oxygen like beaked whales do. Instead, the pilot whales cruised down to about 500 m (1,640 ft) and then made high-speed dashes to as deep as 1,000 m (3,300 ft), periodically reaching speeds up to 9 m/second (30 ft/second). The whales clicked and buzzed during these sprints, with the rush of water driven by the flukes sounding on the recording multisensor like an approaching train. Finally, the sound of a bump or crash marked the end of the chase, as the whales struck. The whales took large squid, with 1.2 m (4 ft) long tentacles sometimes dangling from their mouths when they regained the surface. Indeed, parts of the arms of the giant squid *Architeuthis dux* have been found floating where the whales come up. Pilot whales eat by sucking, and they have a large tongue that, when depressed, creates a vacuum to suck the prey into their mouths.

Short-finned Pilot Whales primarily forage for squid, secondarily for fish, and occasionally for octopus. Because Short-finned Pilot Whales are sighted in both coastal and offshore waters, they may take a wide variety of prey. Off the west coast of North America, studies of the stomach contents from strandings indicated that local pilot whales favor species from the continental shelf, especially squid of the species *Loligo opalescens,* the opalescent inshore squid. Daily consumption based on captive fed animals has been measured up to 27 kg (60 lbs) per day.

MORTALITY: Natural mortality is virtually unknown except for the examples of mass strandings. Numerous causes for mass strandings have been hypothesized, but there is rarely direct evidence to support any theory. Parasites are common and may contribute to mortality, but the composition of parasites varies by region.

DISTRIBUTION: Short-finned Pilot Whales are nomads that swim in all the world's tropical and warm-temperate seas, but are absent from the Mediterranean. As with other cosmopolitan

Range

SHORT-FINNED PILOT WHALE

cetacean species, there may be distinctive local races. A smaller pilot whale with a squarish head has traditionally been taken in drive fisheries off central Japan, but a larger form with a rounder head and a more pronounced saddle has now been identified to the north. Short-finned Pilot Whales of the California Current may differ morphologically from these two forms, but differences remain unresolved.

For this species in the eastern Pacific, no obvious migratory movements are reported except the onshore–offshore movements characteristic of many species that follow prey, particularly squid, which move inshore in winter and early spring. Such species generally range over areas with complex bottom topography, as well as along the continental slope and shelf break.

Pilot whales particularly are seen in habitats with complex topography, such as seamounts and ridges, and along the continental shelf break and slope.

CONSERVATION: Historically, and up to the present day, pilot whales have been hunted for food, and this has been the primary threat to their conservation. Short-finned Pilot Whales have been hunted in small numbers in several areas within their broad range, with some periods of intensive exploitation occurring in Japan, especially following World War II. Today, Japan's drive fisheries take fewer than 100 pilot whales a year. In the seafaring village of Taiji on coast of Honshu, Japan, pilot whales have been hunted for centuries, and people's long heritage of whaling is a source of pride. Recent reports there of mercury contamination in pilot whales ended local school lunches featuring pilot whale meat, and may lead to changes in traditional attitudes.

Agreement	Conservation Status
U.S. ESA	Not listed
U.S. MMPA	No special status
COSEWIC	Not at risk
IUCN	Data deficient

Today, several human activities result in mortality or behavioral changes of pilot whales. Off the coast of California and Baja California, drift gillnet fisheries may incidentally take Short-finned Pilot Whales. Pingers and other conservation measures may be helping to reduce the rate of bycatch off California now. Historically, Southern California squid purse-seine fisheries also bycaught these whales, but that was before the species became uncommon in the early 1980s.

Pilot whales can react strongly to disturbances from ecotourism. In 1991, a pilot whale in Hawaii seized a snorkel diver by the leg and pulled her 10 to 12 m underwater before letting her free. The circumstances of this attack caused controversy at the time, but was the case exceptional? Small groups of pilot whales often float on the surface like logs, quietly near the surface, and permit whale-watching boats and swimmers to approach closely. Although the whales don't seem disturbed, they actually may be recovering from the especially energetic deepwater dives described above. Charging prey at a 20 knot speed through a medium as viscous as water is arduous. The energy consumed is reckoned as the cube of speed, so a sprinting whale going four times its normal speed may consume 64 times as much energy as usual. The apparent passivity of resting pilot whales may thus result from exhaustion, and frequent interruptions by ecotourists may interfere with this recovery and aggravate the whales.

Pilot whales are also exceptionally vulnerable to injury or mortality from naval sonar, because the bonds between whales in this genus are so strong that groups tend to strand together. Even when beaching pilot whales are escorted by humans back to open sea, the "saved" whales may return to strand again and die. It is as if pilot whales do indeed have a leader that they follow no matter what the consequence. In one stranding, on 5 miles of beach along Cape Hatteras in North Carolina, 15 Short-finned Pilot Whales, six of them pregnant, were found dead, and seven had to be euthanized. Two Dwarf Sperm Whales and a newborn Common Minke Whale beached as well. In this case, the strandings coincided neatly with naval sonar off North Carolina's outer banks. However, mass strandings of pilot whales long preceded the advent of active military sonar, so other factors must play a role, such as confusion regarding coastal topography or illness. Perhaps because of low population levels, mass strandings of Short-finned Pilot Whales are rarely reported along the western coasts of North America.

POPULATION: Pilot whales are found along the west coast of North America from south of Baja California to Alaska; however, estimates of pilot whale abundance off Baja California are lacking. Pilot whales were once seasonally common in the Southern California Bight, and a resident population appeared to be present year-round off Santa Catalina Island. Following the extreme 1982 to 1983 El Niño–Southern Oscillation (ENSO), pilot whales virtually disappeared from this location, perhaps because of changes in squid movements. In survey cruises in 2001 and 2005, only seven pilot whales were counted off the U.S. coast; this generated an official abundance estimate of 245 whales. Short-finned Pilot Whales are only infrequently seen off the coast of British Columbia.

COMMENTS: Pilot whales, Sperm Whales, and Killer Whales often spend their lives in small clans founded by female elders. Despite the vastness of the world ocean, the cetacean clans are tied to traditional territories, routes, and techniques of hunting. Are the matriarchs the mothers of their culture, the keepers of their ecological wisdom?

REFERENCES: Bernard and Reilly 1999, McAuliffe and Whitehead 2005, Olsen and Reilly 2002, Soto et al. 2008.

FALSE KILLER WHALE *Pseudorca crassidens*
Pl. 10; Figs. 74, 75.

Adult male average length is 5.3 m (17 ft); weight can be as much as 2,200 kg (4,800 lbs). Adult female average length is 4.5 m (15 ft); weight can be 1,200 kg (2,600 lbs). Calf average length is 1.5 to 2 m (5 to 7 ft); average weight is 80 kg (175 lbs).

SCIENTIFIC NAMES: *Pseudo* means "false" in Greek, and *orca* is an ancient Latin word for "demon," as well as the modern species term for the Killer Whale, *Orcinus orca. Crassidens* comes from the Latin *crassus,* meaning "thick," and from *dens,* meaning tooth. *Pseudorca crassidens* thus could be translated as "thick-toothed false orca" (fig. 74).

COMMON NAMES: The English common name, False Killer Whale, translates the species' generic name, *Pseudorca,* as do the French name *faux orque* and the Spanish *orca falsa.* In any of these languages, this whale is named thus by negation—simply, "not an Orca." Sometimes, False Killer Whales are also called Blackfish in English, but this term can refer to several other dolphins as well. Less frequently, the False Killer Whale is called the False Pilot Whale in English.

Figure 74. False Killer Whales.

DESCRIPTION: The False Killer Whale is long, slender, and lively; it is one of the largest of the ocean dolphin family. The head is relatively small, narrow, and rounded. The tapering, conical melon extends slightly over the upper jaw in adults, and so the False Killer Whale lacks a beak. Males are larger than females, and the melon overhangs more in adult males. Male False Killer Whales average 5.3 m (17 ft) in length, and females average 4.5 m (15 ft). The dorsal fin is tall and hooked, with a rounded or pointed tip. The small flippers become thin and pointed toward the end, lie forward on the body, and have a distinctive hump on the leading edge. This hump is sometimes called an "elbow," although from a human's point of view, it faces the wrong way. The flukes are notched. The False Killer Whale's body is almost all black. An anchor-shaped gray to near-white patch marks its chest, and there may be a pale gray patch on the sides of its head. Seven to 12 pairs of interlocking conical teeth in both jaws suggest the active predatory life of the False Killer Whale.

FIELD MARKS AND SIMILAR SPECIES: False Killer Whales resemble Killer Whales in their sociality and in their occasional pursuit of other sea mammals. Their teeth, also, are like those of Orcas, which may be the reason why they were first called False Killer Whales in the 19th century. If all a Victorian zoologist had to examine were collections of museum bones, then their skulls could be sorted into two piles, one for Killer Whales and one for "false" Killer Whales, with the false Killer Whales having a wider jaw. In life, the smaller, dark False Killer Whales obviously differ from the more massive, blocky Killer Whales with their striking white markings. Orcas also possess a taller dorsal fin and huge, paddlelike flippers, in contrast to the more modest falcate dorsal fin and narrower tapered flippers of the False Killer Whale.

Genetically, False Killer Whales are not closely related to Killer Whales but instead are related to pilot whales, Pygmy Killer Whales, Melon-headed Whales, and Risso's Dolphins. False Killer Whales resemble pilot whales in size and in their basic black coloration, but they are not thick-bodied like the pilot whales. The head of the pilot whale is bulbous, pushing water forward on the sea surface, in contrast to the narrow head of the False Killer Whale. The pilot whale's thick, low dorsal fin lies far forward, unlike the relatively centered, taller dorsal of the False Killer Whale. The pilot whale's thin flippers lack the diagnostic "elbow" of the False Killer Whale.

False Killer Whales also resemble the largely black Pygmy Killer Whales and Melon-headed Whales, but the latter two species are only about half *Pseudorca*'s length. Also, the Pygmy Killer Whale and the Melon-headed Whale have distinctive white lips. If it can be seen, the "elbow" or hump on the leading margin of the False Killer Whale's flipper again distinguishes this species from its smaller relations.

NATURAL HISTORY: False Killer Whales travel and hunt by day in pods of 10 to 60 or more. Sometimes groups as large as 600 or 700 whales have been seen. False Killer Whales are known for their relatively frequent and sometimes dramatic strandings, in which as many as 835 have beached together. This species holds a record for this extreme form of social behavior, which may seem like mass suicide to those attempting rescues. Despite this, in one mass stranding in Australia, 1,500 volunteers saved the main pod of beaching whales, with only a single death. We usually do not know why these whales strand, although a number of theories have been offered, including animals confusing underwater topographies; experiencing liver dysfunction, disease, general disorientation, or injury from military sonar; or acting based on intense social attachments to injured or dying whales.

At sea, False Killer Whales are exuberant—racing, breaching, splashing on their sides—like some of the spritely, small, toothed whale species. Sometimes, False Killer Whales bow or wave ride, too, putting on a lively show for whale-watchers.

We know little of the social structure of False Killer Whales. Although individuals can be identified by scars in their dorsal fins, few studies have been reported tracing individuals. Thus, we do not understand the makeup of the groups we encounter on the high seas, which, once achieved, could vastly increase our knowledge about this cosmopolitan whale. The oldest ages recorded are 58 years for a male and 63 years for a female. After 45 years, females are no longer reproductive. As with Killer Whales and Sperm Whales, however, the role of menopausal females may be important in social cohesion, and it may indicate a similarly matrilineal social structure.

Genetic data suggests that there may be regionally distinctive populations. For example, there is evidence that Hawaiian inshore and offshore False Killer Whales rarely interbreed and that some Hawaiian offshore False Killer Whales resemble those off the west coast of North America.

Although wide-ranging, open-ocean cetaceans are difficult to track, we do have detailed knowledge about two aspects of *Pseudorcas'* behavior—their sense of sound and their intelligence. False Killer Whales can survive well in captivity, and a few have been long-term subjects in echolocation research in Hawaii.

The False Killer Whale's echolocation pulses have been intensively studied. These animals make single echolocation pulses, longer trains of pulses, and more complex tones. If we listen on hydrophones, we may hear snaps, buzzes, moans, and whistles. What they sound like to these cetaceans is a question. We have learned that the False Killer Whales can hear greater than 100 khz, far above the human 20 khz limit of hearing.

The pulses they make are very brief, less than milliseconds. False Killer Whales beam their echolocation pulses forward through their melons, aiming slightly below the long axis of their bodies. Echoes return from fish or other targets, creating brain waves that researchers can pick up with electrodes mounted in suction cups on the back of the whale's head. We know that, depending on the situation, False Killer Whales can adjust the loudness of pulses. When the whales are out at sea, their pulses are louder than those of whales in pens. One open-ocean study calculated the intensity of these echolocation pulses at 235 decibels, almost as loud as those of the Sperm Whale. The range at which a dolphin-sized prey would be detectable was more than 300 m.

False Killer Whales in oceanaria readily learn tasks. They may be able to recognize themselves in mirrors, like several species of primates and Common Bottlenose Dolphins, although more research will be necessary to support this hypothesis. They also can perceive the elements of complex sound tones such as the ones they produce. But we don't know what their signals mean or whether they have individual signatures.

REPRODUCTION: Males and females are sexually mature at 8 to 14 years, although one report had males maturing at 18 years. Once mature, females ovulate year-round and can do so spontaneously. Calving thus can occur year-round, but there tends to be a peak in winter. In captivity, gestation for one female was 14 months, but estimates range from 11 to 16 months, resulting in a long calving interval of up to 7 years.

FOOD: False Killer Whales pursue fish, squid, octopuses, and, more rarely, marine mammals. Examples of prey include Yellowfin

Tuna, perch, barracuda, and squid. Salmon were observed caught off Vancouver Island, and in oceanariums the whales readily eat mackerel, herring, and smelt. There may be regional variation in their prey preferences. Sometimes, False Killer Whales steal fish from commercial and recreational fishermen, and they may prey on weakened dolphins released from seine nets by tuna fishermen. Other rare examples of *Pseudorca* prey include Sperm Whales and Humpback Whales.

It is not known how deep False Killer Whales dive, but specimens studied in South America had ingested mud (bottom sediment) along with deepwater squid.

Foraging styles likely vary, but one style noted for this species is herding prey using a line formation, called a chorus line, with several hundred participants over an area of several miles (fig. 75). The estimated rate of daily food consumption is 4.5 percent of their body weight.

MORTALITY: There is no evidence of other predators attacking and killing False Killer Whales. These animals do, however, strand and die on beaches frequently, for unknown reasons. Stranded groups have ranged in size up to around 1,000 individuals; one group that stranded in Florida consisted of many members that were ill.

Figure 75. A chorus line of False Killer Whales herding fish.

DISTRIBUTION: False Killer Whales are widely distributed in tropical, semitropical, and warm-temperate oceans, ranging between latitudes 50° North and South. They are reported in colder waters much less often but have been seen as far north as the Aleutians.

False Killer Whales have stranded in British Columbia, with several records there of single individuals. Many of these records may represent the same whale seen over several years, in Barkley Sound on the west coast of Vancouver Island. False Killer Whales are occasionally sighted off the U.S. coast, particularly in warmwater years, from Washington to California. False Killer Whales are regarded as rare seasonal transients in Monterey Bay.

Seasonal movements associated with those of their prey take False Killer Whales in the northern Pacific from warmer water in winter to cooler water in summer. They are most likely to be seen in deep oceans and offshore areas, but they also occur around islands. Sightings are rare, regardless of season or location.

■ Range

FALSE KILLER WHALE

VIEWING: Scientific and whale-watching cruises report False Killer Whales off the coast of Baja California. This may be the most likely area for eastern Pacific False Killer Whale sightings.

CONSERVATION: False Killer Whales were taken in the infamous Iki Island Japanese whale drives, where cetaceans were driven ashore and killed, because they were believed to deplete Yellowtail Amberjack. Such systematic extermination is far from a usual threat to False Killer Whales, but some animals that rob fish lines may be shot. There is a small take of these whales for food in the Caribbean and off Japan but not along the west coast of the Americas. Bycatch in nets may be a problem in other areas.

Agreement	Conservation Status
U.S. ESA	Not listed
U.S. MMPA	No special status
COSEWIC	Not at risk
IUCN	Data deficient

Although False Killer Whales are largely an offshore species, high levels of mercury and organochlorines have been found in some individuals. Organic tin compounds, used in antifouling paint in boats, appear to accumulate in the livers of False Killer Whales more than in other species', and liver disease has been documented in stranded False Killer Whales. It is possible that any liver disease could make False Killer Whales more likely to strand. False Killer Whales that stranded on shorelines on the Straits of Magellan had their digestive tracks blocked with plastics, likely ingested when mistaken for squid. Plastics circulate and concentrate in seas around the globe, no matter how remote.

POPULATION: No official estimate exists for the number of False Killer Whales off western Canada, the United States, or Baja.

COMMENTS: "False Killer Whale"—what a wonderful name, each word carrying a dramatic undertone. This species might be thought of as the dominant high-order mammalian predator of the tropical and warm-temperate waters of the Earth, because it sometimes feeds on other whales, themselves at the top of their food chains. In colder, more poleward seas, its cousin the Killer Whale is far more prevalent. Yet neither species excludes other marine mammals from the oceans. How do the cetaceans divide the water planet? By type of prey? By latitude? By depth? By bathymetry? By learning and culture? As much as we think we know about this group of sea mammals, humans still have a great deal to study and learn.

REFERENCES: Alonso et al. 1999, Chivers et al. 2007, Nachtigall and Supin 2008, Odell and McClure 1999, Stacey et al. 1994.

MELON-HEADED WHALE   *Peponocephala electra*
Pl. 10; Fig. 76.

Adult male average length is 2.5 to 2.8 m (8 to 9 ft). Adult female average length is 2.4 m (8 ft). Newborn average length is 1 m (3 ft).

SCIENTIFIC AND COMMON NAMES: The American naturalist Titian Peale first described the species from a specimen collected in Hawaii in 1848, calling it *Phocaena pectoralis* (fig. 76). From then on, the species has had a confusing naming history, beginning with British naturalist John Gray who first placed the species in the genus *Lagenorhynchus,* naming it *Lagenorhynchus electra,*

Figure 76. Melon-headed Whale; note white lips.

but then later, in 1868, assigned it the name *Electra electra.* *Electra* is likely from the Greek name for the oceanic nymph Elektra. In 1966, however, mammalogists Masaharu Nishiwaki and Ken Norris described recent specimens of this rare species and concurred that it belonged to a new genus. Initially, they reaffirmed the genus *Electra,* but a genus of bryozoan already held that name, so they eventually named the genus and species *Peponocephala electra. Pepo* is Latin for "pumpkin," apparently an assumption of the authors that *pepo* meant melon in Latin. *Cephala* is Latin for "head"—and so the name, melon-headed nymph. A more apt name would be "pumpkin-headed." Only one species is currently in the genus of *Peponocephala* of the family of Delphinidae. Common names are Little Blackfish, Small Blackfish, Many-toothed Blackfish, Hawaiian Porpoise, Hawaiian Blackfish, and Electra Dolphin. Spanish names include *calderon pequeno* and *orca enana,* and a French name is *peponocephale.*

DESCRIPTION: Compared to other dolphins, the Melon-headed Whale is only moderately robust and appears long and slim with a longer tail stock. There is not much difference between the sexes, but males are slightly larger, and their appendages (flippers, tail, and dorsal fin) are proportionately longer. Nevertheless, the longest animal measured was a female at 2.8 m (9 ft). The heaviest was a male at 275 kg (737 lbs).

The head is rounded and beakless, and a slight overbite is apparent when the whale surfaces in calm water, ruffling the water in front. The species has no distinctive melon as seen in other blackfish toothed whales, but instead the head is curved from the tip of the snout to the blowhole. From above, however, the tip of the rostrum looks slightly pointed. The mouth is full of 20 to 26 sets of small teeth in upper and lower jaws, an unusually large number. The dorsal fin (25 cm; 10 in.) is midback, is somewhat falcate, and leans backward, ending in a pointed tip. The flippers are narrow, curved, and tapered with slightly rounded tips, some 40 to 50 cm (16 to 20 in.) long. The tail is broad [42 to 65 cm (17 to 26 in.)], curved, and notched. The tail stock is quite narrow when viewed from above, from the perspective of boat or plane.

Body color is mostly charcoal gray to black, overlaid with a dorsal cape that is darker black than the rest of the body. The cape is draped over the back from behind the blowhole, widens at the dorsal fin, and then narrows behind. The face is also darker gray or black and looks like a "mask." Other coloration that often occurs around the face includes a darker stripe from the blowhole to the tip of the nose and white lips that appear as a line along the mouth. On the chest, there can be an anchor-shaped white or light gray patch, and at the belly button and genital openings, small round white patches. When observed closely, there is much scarring about the body, which provides marks for studying individuals, but the species is encountered too infrequently to follow consistently.

FIELD MARKS AND SIMILAR SPECIES: Melon-headed Whales resemble other "blackfish," at a distance, but they most resemble Pygmy Killer and False Killer Whales. The Pygmy Killer Whale does not occur, except rarely, in the California Current, and the False Killer Whale is about twice the size of the Melon-headed Whale. Good key marks for the Melon-headed Whale are the triangular-shaped head, sharper snout, slim tapered flippers, and narrow dorsal fin. The Melon-headed Whale is the only species with narrow and pointed flippers, and the False Killer Whale is the only species with elbowed flippers. The Melon-headed Whale has a slightly falcate and not lobed dorsal fin, whereas the False Killer Whale has a lobed dorsal fin midback. Melon-headed Whales also might be confused with the Short-finned Pilot Whale, which has narrow pointed flippers, but the latter's dorsal fins are broad based and lobed, easily seen from afar. Also, the pilot whale is more than twice as large. Melon-headed Whales are very different from Killer Whales, which possess a taller dorsal fin and large, paddlelike flippers. If examined closely, the immediate difference between Melon-headed Whales and all other blackfish is the large number of teeth (more than 19 per row) compared to the other species, which have less than 15 per row.

NATURAL HISTORY: The Melon-headed Whale is truly an oceanic nomad that has been little studied. Some knowledge of its life cycle comes from examining the carcasses of mass stranded animals and those drowned in fisheries. The maximum estimated age of members from strandings was 47, but most whales likely live around 30 years, and females appear to live longer than males.

They are very gregarious, gathering in schools of 15 to 2,000; in the eastern Pacific, they are generally found in herds averaging

200 whales. Large and small groups form and break up in what is described as a "fission–fusion" social organization. The forming up and breaking down of groups also may follow a daily pattern linked to feeding and resting. In one group studied near oceanic islands, Melon-headed Whales rested in the early morning after a night of feeding, and then in the midafternoon, they became more active socially. "Logging" whales are in deep rest when they lie motionless at the surface. After periods of rest, they engage in vigorous activity, including jumping completely out of the water, spy hopping, tail slapping, and diving deep with flukes exposed at the surface before they descend.

While socializing, they are very physical, engaging in activities that include gentle flipper rubbing, head rubbing, and resting a flipper on the side or back of an adjacent whale. They even swim side by side in a "holding flippers" pose with flippers touching. They are also very vocal when they are active, and they emit underwater sounds that include whistles and clicks. Whistles are always short and include up and down sweeps. Clicks can occur in bursts that include 1,200 clicks per second with sounds as high as 165 dB, equivalent to a shotgun blast.

Melon-headed Whales often swim in mixed herds with other mammal species such as Fraser's Dolphins, Spinner Dolphins, Rough-toothed Dolphins, and Common Bottlenose Dolphins. When in the company of Fraser's Dolphins, Melon-headed Whales are positioned along the edge of the herd and trailing behind.

Rarely do Melon-headed Whales approach boats in the eastern Pacific, although in other parts of their range, they can briefly bow-ride boats. One group, observed in the Caribbean, spy hopped to look at, and even approached, a research vessel. When the vessel started up its engine, however, or when other vessels came around, the whales startled and were vocally silent.

Unlike pilot whales and other small cetaceans, these whales do not do well in oceanariums because they are aggressive and unresponsive to human handlers. Even in the wild, they have been aggressive toward human divers.

REPRODUCTION: The reproduction of Melon-headed Whales is insufficiently studied, and much must be inferred. Almost all information is derived from the inspection of individuals in mass strandings. Based on an examination of tooth layers, females are sexually mature between 4 and 11 years of age, and males at around 7 years. Gestation is believed to be 1 year, and calves are likely born in summer with a June to August peak, but year-round reproduction is also possible. One newborn was seen in June in Hawaii, but none have been seen in the eastern Pacific. As with other species of its size, a 2- to 3-year calving cycle is likely.

FOOD: Not much is known, but much is assumed, about what Melon-headed Whales eat. Beached whales often have empty stomachs, and so examination of contents is rare. They are believed to primarily eat squid and small fish, but also invertebrates. They prey in the deep-scattering layer, on meso-pelagic species that live down to 1,500 m (4,900 ft) but migrate up to within 200 m (650 ft) of the surface at night. One of the most abundant species is lanternfish, a deepwater prey favored by many marine mammal species. Given their numerous teeth compared to the fewer teeth of other blackfish whales but similar in number to Fraser's Dolphin, with which they feed, they likely prey more on small fish than larger deepwater squid. When feeding during the day, they often drive prey to the surface, where seabirds such as the Parkinson's Petrel can also benefit, and they have been observed swimming in a "chorus line" formation.

MORTALITY: Natural mortality is virtually unknown in Melon-headed Whales, except for examples of mass strandings. Like pilot whales, they will strand in mass for reasons that remain mysterious. Parasites may be one reason; another may be because they have strong social bonds. In one group of stranded whales, all suffered from a trematode parasite, but infestation of parasites in air passages and the brain is not present in all strandings. Another theory is that strandings are related to lunar cycles. The theory goes that the whales move inshore around oceanic islands during the day and offshore at night to feed in deep water, and so during lunar cycles, they would be confused by changes in light. However, this theory was subsequently debunked after a thorough review of several strandings. Human causes are likely in some cases. Cookie-cutter shark bites have been documented on dead stranded whales, but they were not the cause of death. No doubt, Killer Whales prey on this smaller whale, but there are no records of such attacks.

DISTRIBUTION: The Melon-headed Whale is truly pantropical, occurring in the deep tropical oceans around all the major continents of the globe. They range offshore between latitudes

Range

MELON-HEADED WHALE

20° N and 20° S; although rare, sightings have extended as far north as 40° N and south to 35° S. Despite this wide occurrence, some regional groups appear very habitual in their movements. They are associated with deep oceanic waters that are 1,000 to 2,000 m (6,600 to 9,800 ft) and are rarely close to shore, except near deepwater marine canyons or islands. In the eastern Pacific, and elsewhere, they are attracted to upwelling zones, where food is more abundant, and around oceanic islands, where they often gather during the day while resting and socializing. As with many tropical species, they do not migrate north–south but may move inshore and offshore in response to movements of prey. Wherever they roam, they are fast swimmers, traveling at speeds of 20 kph (12 mph), and often porpoising as they travel.

CONSERVATION: Human inhabitants of oceanic islands such as Hawaii and Japan historically caught Melon-headed Whales that were resting inshore of the islands during the day, for food and oil. Called a drive fishery, people herded the whales that were resting around islands toward the shore, forcing them to beach.

In Japan, the whales are not popular for food now, however, because their meat does not taste good and because pollutant levels of PCB of a few specimens were some of the highest recorded in Japan. Melon-headed Whales are rarely drowned in tuna purse-seine fisheries in the eastern Pacific, but the effects of offshore driftnet and longline fisheries are unknown.

Agreement	Conservation Status
U.S. ESA	Not listed
U.S. MMPA	No special status
COSEWIC	Not listed
IUCN	Least concern

Naval active sonar was the primary reason for the near mass stranding of Melon-headed Whales in a cove of Kauai, Hawaii, in 2004, when around 200 whales entered the shallows of the bay shortly after the U.S. Navy began using active sonar nearby. They swam directly toward shore in a wavefront formation as though "fleeing from a stimulus." The whales were agitated and milled about in a tight group in extremely shallow water and would not leave the bay even with guidance and encouragement from humans in canoes. The whales finally left the cove only after the Navy ceased operations at the end of the day at the request of the National Marine Fisheries Service. The following day, a flotilla of canoes and kayaks was able to encourage and guide the whales out of the cove, leaving one dead calf floating in their wake.

Melon-headed Whales, along with other blackfish and beaked whales, may be exceptionally vulnerable to injury or mortality from naval sonar because of their strong social bonds. Mass strandings of whales long preceded the use of active military sonar, so other factors may also cause strandings, such as illness.

POPULATION: In 1997, the population estimate for the entire eastern tropical Pacific was 45,000, and no data are available for the west coast of North America. Recent estimates for the U.S. waters in the Pacific for the Hawaiian stock range from 154 to 2,900. No trends are possible to discern with such a paucity of information.

COMMENTS: Mass strandings of beaked whales and Melon-headed Whales have near-conclusively been linked to military sonar. Many strandings can be inferred based on the coincidental presence of military activities, and experimental studies into the effects of sonar on whales are complicated. The example from Kauai is insightful because of the detailed study of the reactions of the Melon-headed Whales during the entire event. Their extreme reaction was to seek refuge in protected bays, where sonar signals might have been attenuated. Perhaps they could only achieve true relief from the sonar signals when in shallow water or through head-raised spy hopping.

REFERENCES: Brownell et al. 2009, Bryden et al. 1977, Jefferson and Barros 1997, Nishiwaki and Norris 1966, Perryman 2002, Watkins et al. 1997.

ROUGH-TOOTHED DOLPHIN   *Steno bredanensis*
Pl. 11; Fig. 77.

Adult average length is 2.8 m (9 ft); average weight is 160 kg (360 lbs). Newborn average length is 1 m (3 ft).

SCIENTIFIC NAMES: *Steno* is Greek for "narrow," referring to the head of the original specimen. *Bredanensis* is Latin, meaning "of Breda." Jacob van Breda was a prominent Dutch biologist and geologist who sent in a sketch of the skull to Baron Cuvier. Cuvier did not recognize it as a new species, but the French naturalist René-Primevere Lesson did, naming it in honor of van Breda.

COMMON NAMES: The English name Rough-toothed Dolphin refers to the dolphin's unique faintly ridged teeth. Other English names include Slopehead, Steno, Rough-toothed Porpoise, and Black Porpoise. In French, its name is simply *sténo*. In Spanish, *esteno* may be heard, as well as *delfín de dientes rogosa,* the Rough-toothed Dolphin.

DESCRIPTION: Reptilian or primitive is how a Rough-toothed Dolphin appears to many experts (fig. 77). They are struck by this dolphin's smoothly sloped, cone-shaped head, which lacks a crease between the beak and melon. The Rough-toothed Dolphin's relatively large flippers are set farther back than those of most dolphins. The variably hooked dorsal fin rests at midback. Some larger males have a keellike ventral hump before

Figure 77. Rough-toothed Dolphins.

the flukes. The flukes are notched and slightly pointed. Females are only slightly smaller than males.

Rough-toothed Dolphins are countershaded, being darker above and lighter below. A dark cape runs along the back, which is pinched in over their flippers. Their sides are a lighter gray below the cape. Their "lips" are whitish or whitish pink, with pink lips being a diagnostic field mark. Whitish or light pink or sometimes light yellow blotches can mark their lower jaws, lower sides, and bellies. The unusual wrinkled teeth of this dolphin are formed by vertical ridges of enamel. From 20 to 27 of their conical teeth rest in each row.

FIELD MARKS AND SIMILAR SPECIES: Close up, the smooth, conical head shape and narrow cape readily distinguish the Rough-toothed Dolphin from all others. At the limit of vision, Rough-toothed Dolphins might be mistaken for or overlooked among bottlenose, Spinner, or Pantropical Spotted Dolphins, all of which do have a distinct crease between beak and melon that the rough-toothed lacks.

NATURAL HISTORY: Little was known about the natural history of Rough-toothed Dolphins until this century. Survey ship data show that they often occur in small groups of 10 to 20, and occasionally in larger groups. Sometimes they associate with bottlenose dolphins, Short-finned Pilot Whales, Pantropical Spotted Dolphins, and Spinner Dolphins, as well as with floating debris. Rough-toothed Dolphins are noted for their highly synchronized shoulder-to-shoulder swimming. They also sometimes travel rapidly with beak and chin out of the water, which has been called skimming or surfing.

It is hard to infer the social networks of offshore species from the flying bridge of a survey vessel. In contrast, island-based researchers have recently begun to reveal much more about the social lives of the Rough-toothed Dolphin in nearby waters. By developing photo-identification catalogs and making underwater observations, they have discovered that some Rough-toothed Dolphins have restricted ranges near islands, and that individuals may have long-term bonds.

Rough-toothed Dolphins often travel in such tight formations that they touch each other at times. The dolphins can also move individually in a loose group. When they move in a tight formation, typically only a single dolphin echolocates, but when in an independently swimming group, more individuals echolocate. One theory is that dolphins eavesdrop on each other in a synchronized formation and can orient by listening to the echoes returning to a nearby companion, thus saving energy and limiting echo confusion. There is some evidence from captive Common Bottlenose Dolphins to support this contention, with a listening dolphin putting its head near an echolocating dolphin to solve a problem.

Underwater observers witness a lot of touching in Rough-toothed Dolphins: the dolphins swim or rest with parts of the body touching, they rub one another with their flippers and sides, and they mouth one another. Full body rubs have been seen, where the dolphins moved the whole length of their bodies against one another. Rubbing seems to be an important element in Rough-toothed Dolphin life and may reinforce or express social bonds.

Social behaviors may include attempted rescues. Several dolphin species have been reported to support weak, dying, or dead group members. Rough-toothed Dolphins have been observed carrying dead calves and an adult female near the water surface.

In captivity, unusual cross-species social behavior is possible. For example, a male bottlenose mated with a female

Rough-toothed Dolphin, producing a female calf that lived for nearly 5 years. In another case, a stranded, weakened, Rough-toothed Dolphin calf that was a few months old was given a female bottlenose enclosure mate once it began to recover. The bottlenose immediately began to swim with the calf, and in 12 days began to lactate.

Rough-toothed Dolphins are curious. They may inspect and echolocate on human swimmers, hydrophones, propellers, and other underwater novelties. They play with objects, such as in the case of a pair that passed a piece of plastic back and forth as they swam, each offering the plastic to a juvenile by releasing it near its mouth.

In an early study at an oceanarium, two Rough-toothed Dolphins learned to invent tricks. Each had its own well-rehearsed repertoire to please audiences, but both were also trained to create novel behaviors on demand, such as air spinning, inverted tail slaps, backward flips, and upside-down porpoising. According to their trainer, on one occasion, their holding cages were accidentally reversed before a show, and the dolphins were called out to perform in the wrong order. Each then attempted, albeit imperfectly, to do the other's regular tricks. All of this implies that dolphins know what they and other dolphins are doing and how to do things differently—that is, that they have knowledge of themselves and of others, and of alternative courses of action. The Rough-toothed Dolphins seem especially competent in training, unusual since oceanic dolphins rarely adapt to captivity.

REPRODUCTION: Off Japan, males become sexually mature at about 14 years of age and 2.25 m (7.4 ft) in length, and females become mature at about 10 years of age and 2.1 to 2.2 m (7.25 ft) in length. Little, however, is known about this oceanic dolphin. A stranded female in Florida was 42 when she perished.

FOOD: Rough-toothed Dolphins take fish and squid. Locally abundant top smelt, jack smelt, and neon flying squid were in the stomachs of two specimens that stranded in winter in Oregon. In Hawaii, Rough-toothed Dolphins are known to capture mahi mahi that were a meter (3 ft) or more long, and they may specialize in feeding on these fish. Both species are often found beneath large flotsam. The dolphins' slightly ridged teeth may help them keep hold of this slippery prey. In captivity, one dolphin consumed some 60 kcal per kg per day.

MORTALITY: Mass strandings of Rough-toothed Dolphins are possible. For instance, in 1997, 62 Rough-toothed Dolphins stranded together at Cape San Blas in Florida. All were returned to the ocean, but later, half stranded again. This stranding may have been linked to kidney damage seen in necropsies.

Some of the scars seen on the flanks and bellies of Rough-toothed Dolphins are the work of cookie-cutter sharks, but large squid may also inflict injuries.

DISTRIBUTION: Rough-toothed Dolphins swim in the open oceans of the world and near oceanic islands in warm seas. For example, in surveys off Hawaii, they were seen most often in waters greater than 1,500 m (4,900 ft) deep. Rough-toothed Dolphins in the lee of the island of Hawaii were often resighted, suggesting a small population with site fidelity. Those in the lee of Kauai and nearby Niihau were also resighted, but less often, indicating a larger population with some fidelity.

Rough-toothed Dolphins are reported in the eastern tropical Pacific and off the coast of Baja, and in the Gulf of California.

Range

ROUGH-TOOTHED DOLPHIN

This species is seldom associated with the California Current, and only rarely do individuals venture further than 40° N, though one stranded as far north as Washington.

CONSERVATION: Rough-toothed Dolphins are sometimes taken in drive fisheries off Okinawa, Papua New Guinea, the Solomons, and West Africa, and by harpoon off Japan, West Africa, and St. Vincent in the Lesser Antilles. The extent to which these drive fisheries affect populations is not known.

Agreement	Conservation Status
U.S. ESA	Not listed
U.S. MMPA	No special status; not listed as a North American West Coast stock
COSEWIC	Not listed
IUCN	Least concern

Plastics may be swallowed or may wrap around the bodies of Rough-toothed Dolphins.

Rough-toothed Dolphins are sometimes caught in purse-seines, gillnets, and driftnets.

Like other toothed dolphins, Rough-toothed Dolphins bio-accumulate POPs (persistent organic pollutants) such as PCBs, DDTs, and chlordanes. Nursing females pass much of their burden of contaminants on to their firstborns, while males never shed these pollutants.

POPULATION: About 8,700 Rough-toothed Dolphins were estimated for the EEZ surrounding the Hawaiian Islands in a 2002 survey. It was the second most populous species encountered off Hawaii. There is no estimate for the United States west coast or Baja California.

COMMENTS: Rough-toothed Dolphins are one of many odontocetes that we have come to know more intimately over the last 10 years. Another decade of study, photo-identification, genetic sampling, electronic tracking, and other evolving techniques will reveal more of the hidden lives of these deep-ocean cetaceans.

REFERENCES: Gaspar et al. 2000, Götz et al. 2005, Götz et al. 2006, Jefferson 2002a, Kuczaj and Yeater 2007, Miyazaki and Perrin 1994, Pitman and Stinchcomb 2002.

PACIFIC WHITE-SIDED DOLPHIN

Pls. 14, 15; Fig. 78.

Lagenorhynchus
obliquidens

> Adult average length is 1.7 to 2.5 m (5.6 to 8.2 ft); average weight
> is 85 to 150 kg (175 to 330 lbs). Newborn average weight is 15 kg
> (33 lbs); average length is 97 cm (3.2 ft).

SCIENTIFIC NAMES: The genus name *Lagenorhynchus* derives
from the Greek *lagenos,* which means "bottle" or "flask," and the
Greek *rhynchos,* meaning "snout" or "beak"; the species name
obliquidens stems from the Latin *obliquus,* meaning "slanted,"
and *dens,* meaning "tooth" (fig. 78). It was given to the species
by W. P. Trowbridge in 1865, from specimens collected near
San Francisco. We might translate the entire scientific name as
slanted-toothed flask beak.

COMMON NAMES: The Pacific White-sided Dolphin has been
called by other English names: Pacific Striped Dolphin,
White-striped Dolphin, and Hook-finned Porpoise. The most
common English name on the west coast of North America is
the nickname Lag, short for its generic name *Lagenorhynchus.*
Few whale-watchers will take the time to say the phrase "Pacific
White-sided Dolphin" when the syllable "Lag" is quick and easy to
shout above the sound of the surf or an engine. *Lagénorhynque à
flanc blanc du pacifique* is the French name for the Pacific White-
sided Dolphin—an exact translation of the English. The Spanish
name *delfín lagenoringo* means the *Lagenorhynchus* Dolphin.

DESCRIPTION: The Pacific White-sided Dolphin is sturdy and
robust, and it evolved for speed. There is little sexual difference in
body length, but males can reach a maximum of 200 kg (440 lbs)
in weight and 2.5 m (8 ft) in length. The rounded beak is small
and thick, with relatively tiny teeth for a toothed cetacean—23
to 32 pairs in each jaw. The lag's dorsal fin is distinctive, large,
and usually hooked, although in older animals it may be lobe-
shaped. This dolphin's flippers are large for its body size and are
slightly rounded. Its flukes are notched at the center.

The body coloration of the lag is complex. The basic body
color is gray above and white below. Upon this background of
color are etched complex markings. The eyes are ringed in black,

Figure 78. Pacific White-sided Dolphin.

and the lips, black as well, are bordered in whitish below and light gray above. A black line extends backward from the lips to the front of the flipper, and then it continues from behind the flipper to the rear of the animal, sharply outlining the white belly. A pair of gray stripes reaches from the gray coloration above the black mouth all the way to the gray area around the base of the tail flukes: looked at from the bow of a boat, these stripes resemble suspenders. The side of the lag above the flipper is gray also, as is the rear of the tail.

FIELD MARKS AND SIMILAR SPECIES: In the southern part of the California Current, the Pacific White-sided Dolphin may be mistaken for a common dolphin, which has a smaller, less hooked dorsal fin, a longer beak, and a more complex color pattern, often featuring an hourglass pattern on the side. In the northern portion of the California Current, rapidly moving Dall's Porpoises at a distance might be confused with lags, but the Dall's Porpoises are distinguished by their low, triangular dorsal fin and striking black-and-white coloration. Although the small-headed Dall's Porpoise may also stir up spray in rapid travel, they do not leap out of the water—as the lag does.

Pacific White-sided Dolphins often associate with Risso's Dolphins and Northern Right Whale Dolphins, but these are easy to distinguish. The round-headed, stocky Risso's is whitish or light gray and is often scarred, and it completely lacks a beak. The slender Northern Right Whale Dolphin is all black, except for a white hourglass on its belly, and it lacks a dorsal fin.

NATURAL HISTORY: Exuberance is the Pacific White-sided Dolphin's hallmark. The lag is one of the liveliest dolphins in the northern Pacific, leaping clear of the water, belly flopping, somersaulting, and speeding in and out of water in a rooster tail of spray. The lag will often approach vessels, sometimes unexpectedly, and ride the bow waves.

The life of a lag is fast and dangerous. Everything is in the open: fish, squid, Killer Whales, rivals, allies, mates, and offspring. There are no hiding places in the open ocean. And life is also a social whirl, with massive schools forming and parting, traveling offshore and inshore, moving with great speed. Despite this challenging oceanic life, the maximum recorded age of a female is 46 years, and of a male, 42 years.

The Pacific White-sided Dolphin can join schools of a thousand or more. It is difficult to understand the social structure of high-seas cetaceans that join in such large schools, among the largest of any cetacean. Most schools, however, are smaller—50 animals or fewer.

As with other small-toothed and highly gregarious dolphins, lags have a complex vocal repertoire, and individuals have a distinct signature sound. Most of our understanding of their vocalizations has come from studies of captive animals, however, and little is known about communication in wild populations—except that lags are more vocal when feeding and at dawn and dusk.

Pacific White-sided Dolphins sometimes venture close to shore and can be easier to study there. In Monterey Bay, one researcher relied on unusually colored dolphins as "herd markers" and, by sighting these recognizable individuals, discovered that particular dolphins return to the bay seasonally and in different years. A study in British Columbia inland waters relied on photographic records to understand the dolphins' movements and social alliances. There, dolphins within larger schools, which were composed of males, formed tight groups of extensively scarred individuals with extremely hooked dorsal fins. Another study in British Columbia collects photographs of Pacific White-sided Dolphins: scars and nicks on the dorsal fins have proved sufficient to identify more than one-third of the lags.

Pacific White-sided Dolphins often associate with Risso's Dolphins and Northern Right Whale Dolphins. Foraging advantages or predator evasion and defense are the two usual explanations given for the formation of large single- or multiple-species groups. Dolphins may join others that have found schools of fish, thus spending less time hunting and more time feeding. Sharks or Killer Whales may hesitate to attack hundreds or thousands of dolphins and may instead seek isolated individuals or smaller groups. California Sea Lions also join Pacific White-sided Dolphins feeding in Monterey and Santa Monica Bays in California.

The Pacific White-sided Dolphin may dash energetically throughout the day and later rest at night. A study in a research facility, not in the open ocean, found that lags can spend most of the night moving slowly in a tight formation, perhaps a dolphin's width from their neighbors. The lags would stir and signal at intervals, but the echelon at rest was silent. The dolphins tended to keep an open eye on their schoolmates, which would maintain their position visually. The silence would not attract Killer Whales. Field data suggest that lags can be active at night, however, so they may not always rest then.

REPRODUCTION: Gestation of Pacific White-sided Dolphins takes about 1 year, and most are born in the summer. At birth, calves weigh 15 kg (33 lbs) and average about 97 cm (3.2 ft) in length, and by the time they are weaned, they have grown to an average of 138 cm (4.5 ft). Weaning does not occur for 1 year or more, and so females usually give birth every other year or every 3 years, after mating in summer or fall. Females are usually sexually mature when they reach approximately 185 cm (6.1 ft) in length and less than 10 years of age, the youngest being around 8 years. Males average about 10.5 years at sexual maturity and are 179 cm (5.9 ft) in length. They are born with brains about 67 percent of adult size, which, at sexual maturity, reach 81 to 85 percent of adult size.

FOOD: Pacific White-sided Dolphins eat a variety of small schooling fish and cephalopods, with more than 35 species identified, including herring, salmon, anchovies, pollock, hake, Capelin, rockfish, Pacific Sardine, shrimp, and Market Squid. The diet varies by region. Offshore dolphins pursue midwater fish that are found at depths of 500 to 1,000 m (1,640 to 3,280 ft), such as lanternfish, deep-sea smelt, and pearleyes. Coastal dolphins, in contrast, mostly eat surface schooling fish and squid such as anchovies, sardines, salmon, hake, and Market Squid.

With their many tiny teeth, the dolphins seize their prey and swallow them whole, underwater. In pursuit of prey, dolphins may swim at bursts of 28 kph (17 mph), but average speeds clocked for dolphins in Monterey Bay ranged from 5 to 9 kph (3 to 6 mph).

MORTALITY: Great White Sharks and Killer Whales are known to eat Pacific White-sided Dolphins. In British Columbia, in the 1980s, Killer Whales killed dolphins by driving them into small bays. This no longer happens, however, perhaps because the dolphins learned to avoid this trap. Lowly parasites, such as trematodes and nematodes, are better known causes of death for dead dolphins that wash ashore.

DISTRIBUTION: The Pacific White-sided Dolphin is endemic to the northern Pacific Ocean. The species ranges from the South China Sea in an arc to the Aleutians and down the west coast of North America to the tip of Baja California and into the

PACIFIC WHITE-SIDED DOLPHIN

Range

Sea of Cortez. The dolphins occur along the entire California Current and are generally regarded as an offshore species but will approach coasts in inshore passages, along the continental shelf, and over submarine canyons.

There are two genetically distinctive forms of Pacific White-sided Dolphin off the west coast of North America. One form is typical off Baja California, and the other off the area north of Point Conception in California. The two types vary in the size and shape of their skulls and probably also in the sound spectra of their echolocation pulses. The two forms overlap in the Southern California Bight, but because they do not differ in color pattern, they are impossible to distinguish in the field.

Seasonal movements of Pacific White-sided Dolphins have been noted throughout their range. For example, in British Columbia, they were most numerous in winter, and off southern California, their abundance was higher from November through April. Off Washington and Oregon, dolphins are more abundant in May. Generally, both north–south and offshore–onshore movements have been noted from summer to winter, with dolphins no doubt following the movement of abundant prey; the species can be observed year-round throughout its range, however.

The Pacific White-sided Dolphin is an oceanic species, but individuals are most common over the continental shelf and break, at mean depths of 1,000 m (3,300 ft). Lags also will come closer to shores where there are submarine canyons, such as in Monterey Bay and the Southern California Bight. Also, after a long absence, Pacific White-sided Dolphins have returned to the inland waters of British Columbia.

VIEWING: Because lags are attracted to boats, these "whales" often come to the whale-watcher. In fact, cetacean census takers have to use different techniques to extrapolate Pacific White-sided Dolphin numbers, because most other whales do not show up in the same way, to be counted.

CONSERVATION: Pacific White-sided Dolphins were taken or scavenged by prehistoric peoples in Japan, Canada, and the Californias, according to remains found in middens. Japanese whalers continued to harvest low numbers of lags until recently. But bycatch became a more serious problem in the central Pacific in the 1970s. Perhaps 100,000 Pacific White-sided Dolphins perished in now defunct squid driftnet fisheries, conducted by

Japan, Taiwan, and Korea, before the United Nations enacted a moratorium in 1993. This mortality from bycatch may have substantially depleted the population in the northern Pacific.

Agreement	Conservation Status
U.S. ESA	Not listed
U.S. MMPA	No special status
COSEWIC	Not at risk
IUCN	Least concern

After the introduction of "net pingers" attached to nets that alerted dolphins to the presence of nets and the launching of an education campaign, mortality in the California driftnet fishery dropped considerably. Perhaps half-a-dozen lags still die annually in this fishery. There has been a swordfish drift-gillnet fishery off the coast of Baja California that has some impact on lags. In British Columbia, pingers are also used at fish farms, which may cause cetaceans to avoid nearby areas. Pacific White-sided Dolphin catch numbers fell dramatically when deterrent devices were activated during the late 1990s near the Broughton Archipelago in British Columbia. Lags may also become entangled in farm gear, but this is uncertain because reports of drowning are not required in Canada.

Pacific White-sided Dolphins have been popular in oceanaria, and at least 128 Pacific White-sided Dolphins were taken for public display in oceanariums from the late 1950s to 1993. Now, ecotourism has become a major way that people prefer to see the species. Eager lag bow-riders are drawn to ecotourist vessels and serve as cetacean "good will" ambassadors.

POPULATION: The current United States estimate of the population of these dolphins is a little less than 25,000 within 555 km (300 mi) of the North American coast. A Canadian estimate is an additional 26,000 inshore of British Columbia. There is no estimate for Baja.

COMMENTS: Oceanic dolphins are brainy creatures. The 1.15 kg (2.5 lbs) brain of the Pacific White-sided Dolphin weighs more than that of any terrestrial species, with the exception of our own and those of elephants. The relative size of the lag's brain—the ratio of the brain weight to the total weight—also surpasses that of most land mammals. The white-sided dolphin's brain is 4.5 times the weight of what we might expect of a similarly heavy terrestrial mammal. Some specifications of the dolphin brain exceed our own: their cortex is more convoluted and actually has more surface area than a human's. Their brain is built for speed, for danger, and for darkness and light.

REFERENCES: Brownell et al. 1999, Goley 1999, Morton 2000, Soldevilla et al. 2008, Van Waerebeek and Wursig 2002.

RISSO'S DOLPHIN

Grampus griseus

Pl. 11; Figs. 1, 79.

Adult average length is 3 m (10 ft); average weight is 300 kg (650 lbs).

SCIENTIFIC NAMES: The genus name *Grampus* may be a contraction of the Latin *grandis,* meaning "grand" or "large," and *piscis,* "fish." It may also be that the name derives from the French *grand poisson,* a "great fish," or from the French *gras poisson,* meaning "fat fish." The species name *griseus* is from Medieval Latin, meaning "gray," and refers to the whitish gray skin of this dolphin (fig. 79).

COMMON NAMES: This dolphin is named after the Italian–French naturalist Antoine Risso, who first described the species to the zoologist Cuvier in 1812. Risso's Dolphin is sometimes called the Grampus in English. This can lead to confusion when reading older literature, however, because the Killer Whale was also called Grampus in the past. Other names include Gray dolphin, Gray Grampus, and Whitehead Grampus. *Grampus* is also a French common name, along with *dauphin de Risso* (Risso's Dolphin) and *le marsouin gris* (Gray Porpoise). Spanish names are *chato* (Snubnose), *fabo calderon* (Bean Pot), and *delfin de Risso* (Risso's Dolphin).

DESCRIPTION: Risso's Dolphin is the fifth largest member of the ocean dolphin family and is the largest cetacean that is called a dolphin. Males are only slightly larger than females, with the largest of either reaching 3.8 m (12.5 ft) in length and 500 kg (1,100 lbs) in weight. This robust dolphin is stockier in its

Figure 79. *Upper:* Risso's Dolphin breaching.
Lower: Risso's Dolphin from above, showing blunt head.

front half than in its tail end and thus is often combined with the blackfish group of dolphins. The Risso's is blunt headed and is scarred at maturity, looking almost white, with relatively large dark fins and flippers. The underside of a Risso's Dolphin is marked in white or gray patches. The dorsal fin is quite tall for this dolphin's length, is centered along the back, and may be falcate. The flippers are long and sickle-shaped. The head is indented by a furrow that runs from the upper jaw to the upper forehead, although this can be seen only at close range. This deep groove is unique among cetaceans.

When healing, the skin of the Risso's Dolphin loses its pigmentation. Scars accumulate over time and can make the dolphin look as if it were spattered by white paint. Newborn Risso's Dolphins are gray or brown with a whitish belly. Calves darken with age, until they are nearly black except for the white dorsal patch. The scarring that begins when they become subadults is mostly linear, although some marks are circular. It is thought that the teeth of other Risso's Dolphins during play or fighting may cause the linear scars.

The mouth of the Risso's Dolphin points upward at approximately a 45 degree angle and is slightly downturned at the corners. There are two to seven pairs of teeth on the lower jaw but rarely any on the upper jaw. The lower jaw juts very slightly beyond the upper jaw.

FIELD MARKS AND SIMILAR SPECIES: In Monterey Bay and the Gulf of the Farallones, California, Risso's Dolphins often associate with Northern Right Whale Dolphins and Pacific White-sided Dolphins. It is easy to distinguish these three species. The square-headed, stocky, scarred Risso's is whitish or light gray and completely lacks a bill. The slender Northern Right Whale Dolphin is all black, except for a white hourglass on its belly and whitish marks on the base of its flukes; it also lacks a dorsal fin. The body coloration of the Pacific White-sided Dolphin is complex in comparison to that of Risso's Dolphin.

At sea, Risso's Dolphins may travel with other species such as pilot whales. The larger, dark pilot whale has a shorter, more curved dorsal fin, and it floats on the surface like a log with its head out of the water; in contrast, Risso's Dolphin is whiter and remains more submerged.

NATURAL HISTORY: Risso's Dolphins are typically encountered in small groups of from three to 50 individuals. Sometimes just individuals or pairs are sighted, and occasionally "super pods" form of up to 4,000. Risso's Dolphins often travel side by side. Spy hopping, breach slamming, and tail and fluke slapping are common. Risso's Dolphins rarely bow-ride boats, except in the notable case of an individual named Pelorus Jack, who accompanied vessels in Cook Strait in New Zealand for 24 years at the turn of the 19th century. The first cetacean protective law in the world was enacted, in 1904, for this individual dolphin, after a passenger tried to shoot Jack from a steamer. Risso's Dolphins have been known to bow-ride Gray Whales.

Studies have begun to shed light on the Risso's Dolphin's social structure. A school of Risso's Dolphins harvested in a Japanese drive fishery at Taiji was mostly composed of females and young, with a single mature male, which might suggest a harem type of social organization. In contrast, a study of stranded Risso's Dolphins in southern California hinted that the groups may represent a fluid society. However, photo-identification research on Risso's Dolphins in the Mediterranean showed that there can be long-term associations between individuals. This finding has been confirmed in deepwater observations of Risso's Dolphin off Pico Island in the Azores. Stable relationships between males persisted during the study. Shorter bonds were seen between females and their calves.

Risso's Dolphins echolocate like other toothed whales. However, unlike the other odontocetes, the echolocation beam of the Risso's Dolphin seems to be pointed downward. Usually, echolocation pulses emerge from the center of a dolphin's forehead, and the head must be submerged for echolocation to work. A Risso's Dolphin can echolocate with most of its head out of the water, however. The echolocation beam appears to exit above and parallel to its downturned mouth. Risso's feed largely on cephalopods, which are probably more difficult to target than fish with gas-filled swim bladders, because gas more readily reflects echolocation pulses. The downward-focused echolocation pulses of the Risso's Dolphin may help fix the location of its faintly reflective quarry. The function of the mysterious head indentation is unknown, but Risso's Dolphins can echolocate with this groove out of the water.

The echolocation pulses of Risso's Dolphin differ from those of four other toothed whale species recorded in the California Bight. This means that Risso's Dolphin can be identified on traveling or stationary hydrophones, which can help researchers trace their movements and estimate their numbers at sea.

Risso's Dolphins also use clicks as social signals, making a variety of blat, buzz, creak, and raspberrylike sounds. Sometimes they are heard to whistle, and this sound may serve as an individual signature.

REPRODUCTION: There is scant information on the reproduction of this species. The age of sexual maturity is not known for either sex, but they reportedly live at least 30 years. Estimates of the body lengths of animals reaching maturity given from around the world are virtually identical for the sexes: 2.9 m (9.5 ft) for males and 2.8 m (9.2 ft) for females.

Calving is believed to peak in the fall and winter months, but no other information has been reported, which is surprising given the cosmopolitan distribution of the species.

FOOD: Risso's Dolphins rely largely on cephalopods but also eat small fish. Squid are the primary prey, and squid bites may account for some of the scarring seen on their bodies. They consume prey all up and down the water column, from bottom- to surface-dwelling prey. Off Santa Catalina Island, one researcher noted that Risso's Dolphins mostly fed at night. Pursuing prey, they may swim up to 32 kph (20 mph), but usually swim around 7 kph (4 mph).

MORTALITY: Wounds on Risso's Dolphins indicate that sharks and Killer Whales may attack this ghostly dolphin. Small oval scars suggest the bites of cookie-cutter sharks and lampreys. A large number of internal and external parasites have been found in beachcast animals, including pathogens that also occur in humans, such as *Vibrio* and *E. coli*.

DISTRIBUTION: The Risso's Dolphin is cosmopolitan, being found in waters 400 to 1,000 m (1,330 to 3,300 ft) deep in all the world's tropical and warm-temperate seas between latitudes 60° N to 60° S. They can be sighted far offshore of the west coast of North America, as far south as the latitude of northern Baja California. There seems to be a gap in their distribution to the south before they are encountered again in tropical waters south of Cabo San Luis and the Sea of Cortez. Risso's Dolphins often favor continental slopes, seamounts, and underwater escarpments. Sea temperature may limit their distribution, because they prefer waters between 15 and 20 degrees C (59 and 68 degrees F) and rarely occur in waters less than 10 degrees C (59 degrees F).

Range

RISSO'S DOLPHIN

This association may at least partly explain Risso's Dolphins' more common appearance in Monterey Bay since the 1960s, coincident with warming sea temperatures.

Seasonal movements have been noted in different parts of the Risso's Dolphin's range. For example, 10 times as many Risso's Dolphins were estimated to be off the California coast in winter than in summer.

VIEWING: Risso's Dolphins are most often seen in the Southern California Bight, in Monterey Bay and the Gulf of the Farallones in central California, and in British Columbia. The number of Risso's Dolphins in the Bight has increased over recent years, especially around Santa Catalina Island. Risso's Dolphin may come closer to the coast on the continental shelf in their pursuit of squid.

CONSERVATION: The cosmopolitan lifestyle of this dolphin exposes it to many different threats, ranging from traditional whaling to naval sonar. Some traditional whaling for Risso's Dolphins takes place off Sri Lanka, Taiwan, and Japan. Hundreds have been killed in Japanese drive, harpoon, and small-vessel hunting each year. Oceanic cetaceans are not as exposed to chemical pollutants as coastal species and ecotypes are. Yet there is plentiful evidence that the Risso's Dolphin is not exempt from industrial contaminants such as PCBs; as a consequence of whaling, it passes on this burden to consumers in the Asian marketplaces.

Agreement	Conservation Status
U.S. ESA	Not listed
U.S. MMPA	No special status
COSEWIC	Not at risk
IUCN	Least concern

Fisheries mortality off the U.S. west coast is estimated at 3.6 dolphins a year. Risso's Dolphins were sometimes shot in the squid purse-seine fishery off southern California. This has been illegal since 1994, under an addendum to the MMPA. However, two shootings in 2002 were probably the work of squid fishermen.

Gas-bubble damage to the liver was found in beached Risso's Dolphins in Britain and was thought to be related to regional naval sonar exercises. This is of concern, given that the U.S. Navy conducts exercises using active sonar on the west coast of North America. Risso's Dolphins, like other offshore squid-hunting cetaceans, may be especially vulnerable to this kind of damage.

Risso's Dolphins may be harassed by pleasure boaters when sighted in coastal waters. In an extreme case in the Mediterranean Sea, more than 100 pleasure craft drove and cornered a seasonally resident school of Risso's Dolphins. The Risso's were distressed and disoriented, swimming erratically at high speeds, colliding with one another, and breathing rapidly, until a path for their escape was cleared. On the west coast of North America, this would be unlikely to happen, given the sensitivity of whale-watching captains, crews and passengers, and federal regulation. However, whale-watchers should always be wary of subtle effects of their vessels on cetaceans.

POPULATION: Perhaps 12,000 Risso's Dolphins are present off the west coast of North America. Risso's Dolphin strandings, including several mass events, have been recorded on the British Columbia coast, but no offshore surveys have been conducted. There are no recent data for Baja California.

COMMENTS: Photographs are enlightening us about the social lives of many species of whales. From scars, nicks, and distinctive color marks, we can say who goes with whom and sometimes what they do together. Offshore species are the hardest to photograph, but from clear images and the patient work of the archivists and catalogers we are learning more about how these far-off cetaceans live, and biologists are beginning to unravel one of the great mysteries of the sea.

REFERENCES: Amano and Miyazaki 2004, Hartman et al. 2008, Kruse et al. 1999, Philips et al. 2003.

COMMON BOTTLENOSE DOLPHIN *Tursiops truncatus*
Pl. 11; Figs. 12, 50, 80.

Adult average length is 2 to 4 m (6 to 12 ft); average weight is 150 to 200 kg (330 to 530 lbs). Calf average length is 1 m (3.2 ft); weight may be as much as 20 kg (50 lbs).

SCIENTIFIC NAMES: The genus name *Tursiops* stems from the Latin root *tursio,* which means "porpoise." The Greek suffix *ops* means "face" (taxonomists do not hesitate to mix Latin and Greek in inventing words). The Latin *truncatus* can be defined as "cut off" or "maimed"; the term *truncatus* referred to the worn-down teeth of a possible 18th-century specimen.

COMMON NAMES: The short and stubby beak of this familiar dolphin extends beyond its "forehead," thus suggesting its English name, bottlenose (fig. 80). The word "dolphin" ultimately derives from the Greek word *delphis,* meaning "with a womb"— probably an ancient Greek's comment on what appeared to be a fish with a womb. Other common names include Cowfish and, simply, Porpoise. French words include *souffler,* or Blower, and *grand dauphin* (with *grand* meaning great and *dauphin* signifying a prince as well as a dolphin). Spanish terms include *delfín mular* (Mulelike Dolphin), *delfín de pico largo* (Big-beaked Dolphin), and *tursion,* after *Tursiops.* The species is

Figure 80. Common Bottlenose Dolphin.

so widely recognized that it is a kind of archetype in a dolphin creation story.

DESCRIPTION: Because they are often exhibited in marine parks and pictured in the media, Common Bottlenose Dolphins are perhaps the most recognizable of small cetaceans, but they are difficult to distinguish from other dolphins in the wild. On first sighting, this species is impressive, because it is fairly large for a dolphin and is powerfully built. Maximum weights have been reported up to 650 kg (1,700 lbs). Despite this imposing first appearance, the bottlenose is one of the dullest of dolphins in terms of coloration. Animals are countershaded: blackish, slate gray, brown, or bluish on top; typically lighter on the sides; and whitish or paler pinkish gray beneath. Sometimes, there are spots on the belly and a stripe between the eye and the flipper insertion. Their short [7.5 cm (3 in.)] beaks, or rostrums, seem to hold a Mona Lisa smile, and their well-demarked melons resemble a balding forehead, suggesting a kindly human countenance to some. Sounds are produced beneath a single blowhole on top of the head through "phonic lips." Their mouths hold a moderate number of teeth for a small cetacean, 18 to 26 pairs of conical teeth in each jaw, which likely reflects their generalist diet. The bottlenose's moderately hooked dorsal fin serves as a keel; up and down motions of their broad thin flukes propel them through the water, and rounded flippers help steer. The outer edge of the flukes curves somewhat at the edges and is slightly notched.

Like its northern relative, the beluga, the bottlenose has a relatively flexible neck, with five of the seven neck vertebrae not fused together as they are in other oceanic dolphins. The brain of the bottlenose weighs about 1.6 kg (3.5 lbs)—compared to the human's 1.4 kg (3.0 lbs)—and some have brains weighing 2.25 kg (5 lbs). Their brains are especially convoluted, and if spread out flat, one would have an area of more than 3 m^2 (97 ft^2)—greater than the surface area of any human cortex.

Males are a bit longer and more massive than females, and size also varies with ecotype and with latitude: warmwater and inshore bottlenoses may be smaller than coldwater and offshore races. In the eastern Pacific and some other areas, Common Bottlenose Dolphins are identified, therefore, as two ecotypes—coastal and offshore. The offshore ecotype in this region, however, is not well described, as is the case with pelagic Killer Whales in the California Current; this is mostly because the sightings of open-ocean dolphins are relatively infrequent. Although it is difficult to distinguish the two ecotypes in the water, biologists have found that, in the Atlantic, offshore dolphins have more hemoglobin and red blood cells—adaptations for deeper diving. Recent genetic evidence confirms that Pacific coastal and offshore bottlenoses are also distinctive, and (again similar to Killer Whales) there is debate about whether the two ecotypes might represent emerging species.

Bottlenose classification has been described as a muddle. One reason is that many populations of bottlenoses are local or regional, and are genetically different, and it is not clear why and what that means. Another reason is that Pacific populations seem to be more closely related to Atlantic offshore bottlenoses than to Atlantic coastal bottlenoses. One speculation is that wandering offshore dolphins "colonized" many inshore areas along the world's coasts, and these "colonies" subsequently grew genetically distinct through social isolation or local adaptation. All of this makes one hesitate to generalize from one bottlenose population to another.

FIELD MARKS AND SIMILAR SPECIES: Dolphins generally have long, sharp snouts, prominent hooked dorsal fins, and sharp conical teeth, which differ from the blunt snouts, triangular fins, and flattened teeth of porpoises. Common Bottlenose Dolphins might be mistaken for Rough-toothed Dolphins, Risso's Dolphins, and, off Baja, spotted dolphins. Rough-toothed

Dolphins have a conical head and body before the dorsal fin; their head lacks a crease between the beak and forehead. Risso's Dolphins have a blunt head that lacks a beak, white markings, and a taller dorsal fin. Spotted dolphins are spotted, although the intensity of the spots varies with age, and their beaks are more slender than that of the bottlenose.

NATURAL HISTORY: Common Bottlenose Dolphins are among the most well known and studied of all marine mammals, and research gleaned from observations of animals in captivity has yielded pioneering insights into the general biology, physiology, and sociology of marine mammals. Nevertheless, much of a dolphin's life in the wild remains poorly understood, because of the difficulty of studying wide-ranging marine mammals. We know that females are mature at 5 to 12 years, and males at 9 to 15 years, but we do not know how they navigate in a sea without boundaries. Based on looking at the growth rings of teeth from dolphins in the Atlantic, females may live longer than 50 years and males as long as 40 to 45 years, but the average lifespan is around 20 years. As with all long-lived mammals, learning, experience, and social activities guide much of their behavior.

Worldwide, Common Bottlenose Dolphins form many kinds of social groups—mother–calf pairs, bands of mothers and calves, pairs of allied males (called alliances), alliances between allied pairs, and large fission–fusion societies with variable memberships. Groups are often segregated by sex and age, and females often group together based on the ages of their calves or based on having no offspring. In the Southern California Bight, where North American west coast Common Bottlenose Dolphins are best described, they are usually encountered in schools of 10 to 20 dolphins. In Santa Monica Bay, these schools are mostly seen traveling and diving. More than 500 of these southern California dolphins are individually recognized from distinctive marks on their dorsal fins.

When in groups, they have a dominance hierarchy among individuals, with the larger males dominating; females have a looser hierarchy, with larger females being more dominant than smaller ones and juveniles. Dominance is maintained by biting, ramming, and tail slapping at subordinates.

Aristotle knew that dolphins could speak in the air: "The voice of the dolphin in air is like that of the human in that they can pronounce vowels and combinations of vowels, but have difficulties with the consonants" (quoted in Lilly 1962).

In the water, Common Bottlenose Dolphins make an incredible array of squeaks, grunts, grinds, and whines. These sounds fall into three categories: whistles, echolocation clicks, and pulse sounds. Bottlenoses "whistle," and each seems to have a special signature whistle it gives when alone, or when captured. Recently, researchers have suggested that this whistle acts like its "name." Signature whistles are difficult to study in nature because underwater human divers hear sound as coming from everywhere. But new technology helps divers localize the speakers, and eventually, it will let us understand more of the cacophony of calls we hear below the waves.

Dolphins also communicate nonvocally, and for a highly social animal that cannot easily convey information with facial expressions, touch can be a significant and complex nonverbal method to communicate. Although not facially expressive, they can communicate with a "gaze" expression while pointing with their beak.

Common Bottlenose Dolphins never fall completely asleep, because their breathing is under voluntary control. Thus, from birth to death, bottlenoses are always conscious. They do rest, and at these times, the brain waves from either the left or the right hemisphere tend to show the slow pattern associated with sleep in terrestrial mammals. When in this unihemispheric sleep, dolphins often close the eye opposite the sleeping hemisphere. This pattern seems to be universal in cetaceans, but it is best studied in bottlenoses and Pacific White-sided Dolphins. The quiet times are usually in the hours between midnight and dawn.

When not resting, dolphins are among the fastest swimmers of all marine mammals, achieving speeds of up to 35 kph (22 mph) in bursts of speed. When traveling, however, their average speed is around 8 kph (5 mph). The speed of bow riding dolphins can be clocked against the speed of the boat. Dolphins also bow-ride the pressure wave created by swimming Right Whales, Humpback Whales, and Gray Whales. A similar behavior is surfing in waves washing inshore. Common Bottlenose Dolphins have frequently surfed with human surfers.

REPRODUCTION: A newborn dolphin is 80 to 130 cm (3 to 4.25 ft) long and may weigh 15 to 30 kg (33 to 66 lbs), after a gestation of 11 to 12 months. When born, a second adult dolphin may assist in the delivery. These "auntie" whales can be either male or female and may continue to associate with the young calf, whereas the mother will avoid or repel other adults. During the first few days after birth, the mother vocalizes her signature whistle continuously to the calf, likely so that the calf learns its mother's name. Calves eventually develop their own signature whistle after 1 month or so.

During the first week, calves suckle several times per hour, but only for a brief 10 seconds. Young may suckle for more than a year (20 months), and the record is more than 3 years (38 months). Then calves may continue to stay with their mothers for a total of 3 to 6 years, with the record being 11 years. The time between calving is usually 2 to 3 years, but it can be much longer. Remarkably, females up to 48 years of age have given birth and raised young. Birthing can occur any time of the year, but in some areas of the world, seasonal peaks have been noted in spring, with a secondary peak in autumn. In the Southern California Bight, calving occurs mostly in the fall. Even in captivity, female dolphins retain a seasonal birthing cycle, but males have elevated testosterone year-round, which is one explanation for a polygynous mating system. However, females may also have many male mates, which would support the theory of a promiscuous mating system. Several researchers have noted that adult males banded together in groups and traveled from female group to female group, presumably seeking females in estrus.

FOOD: Common Bottlenose Dolphins are true generalists, eating a variety of small fish, as well as crabs, squid, shrimp, and similar prey. Species most sought after in the Pacific include schooling anchovies, sardines, and hake, but also bottom-dwelling toadfish and croaker fish. Diets vary regionally. For example, in a study of the coastal southern California stock, 62 percent of the diet consisted of croakers and perches, but the offshore population preferred mostly surface-dwelling fish such as mackerel and squid.

These dolphins' feeding strategies match their diverse diet. Often, individuals capture prey, particularly bottom fish, by first using sonar to detect fish buried in the sediment, and then digging small craters in the bottom to seize the prey. Individual dolphins were noted feeding in marsh channels by chasing fish completely out of the water onto mud banks, emerging on land themselves except for their tails, seizing the fish, and then sliding back into the water with fish in mouth. Dolphins may work as teams to herd in formation and concentrate shoals of fish, with individuals assuming different roles in synchrony while maneuvering fish. They may also forage in association with human fishermen, both eating discarded parts from fishing boats and stealing fish from nets and lines. Occasionally, as in Santa Monica Bay in southern California, bottlenoses feed alongside California Sea Lions, which may join them to find prey. The dolphins may "whack" fish to stun them, sometimes throwing the fish out of the water with their flukes. Usually, they dive less than 50 m (160 ft) in pursuit of prey, but their extreme range of depth while feeding extends from less than 1 m (3 ft) deep, when chasing prey into shallows, up to 700 m (2,300 ft), when eating oceanic prey. The most complex foraging strategy, however, is using sonar in shallow and turbid waters to capture prey. By swimming upside down, the dolphin projects a signal that is not distorted by the shallow bottom, and it can detect prey with its biosonar. In captivity, they eat as much as 6 to 7 kg (16 to 19 lbs) per day, equivalent to 4 to 6 percent of an average body weight.

MORTALITY: Worldwide, Great White Sharks, Tiger Sharks, Bull Sharks, Sixgill Sharks, and Dusky Sharks prey on bottlenoses, but in the California Current, only Great White Sharks and Sixgill Sharks are present and also large enough to attack a dolphin. Shark predation on dolphins is hardly ever witnessed, but healed wounds suggest that sharks attack frequently in some areas. Near Sarasota, Florida, nearly one-third of bottlenoses examined during capture bore shark scars. There was no direct way to determine how many others succumbed to sharks, so the scars underestimate predation. Offshore bottlenoses often show less scarring, but it could be that open-ocean predation is simply more deadly. Killer Whales may also prey on dolphins, but there are few confirmed sightings of this. When approached by predators, dolphins may flee or cluster in a defensive mode, but they rarely attack the shark.

Dolphins are more likely to strand on beaches because they are infested with parasites or disease than because of shark bites. Numerous diseases have been identified in Common Bottlenose Dolphins, but a large die-off in the Atlantic population in 1987 pointed to one virus, morbillivirus, which, either alone or in synergy with other factors, caused a nearly 50 percent decline in

Offshore Form
Coastal Form

BOTTLENOSE DOLPHIN

that population over a 1 year period. This virus has not, as yet, been detected on the Pacific coast.

DISTRIBUTION: The Common Bottlenose Dolphin swims in all the world's tropical and temperate seas, but the amount of residency varies tremendously by region. Some populations are strictly local; others migrate extensively. In Baja California, the coastal form can be found in the Sea of Cortez and along the western shore of the peninsula. In California, coastal Common Bottlenose Dolphins stay close to shore, most within half a kilometer of land. A map of the coastal form's range would look like a narrow ribbon tracing the shoreline from Cabo San Lucas to San Francisco. Offshore dolphins are found far beyond the coastal shelf of California and Baja California, over the deep oceanic plains. Little is known about the movements of these individuals or groups offshore.

Following the El Niño–Southern Oscillation (ENSO) of 1982–1983, southern California coastal bottlenose groups extended their range northward to Monterey Bay, and more recently to San Francisco, off Baker Beach just south of the Golden Gate, and into San Francisco Bay.

Prior to this range extension, the presence of Common Bottlenose Dolphins north of Monterey was considered rare. Remarkably, however, a fisherman snagged a skull with a fish hook in 1958, while fishing in San Francisco Bay. The skull was subsequently estimated to have settled in the mud of the bay from 50 to 100 years before the fisherman's hook brought it to the surface.

To the north of San Francisco Bay, sightings are rare. In Washington State, there is only one confirmed record of a Common Bottlenose Dolphin, although remains of five have been found in an Indian midden near Willapa Bay. In British Columbia there is a single unconfirmed record of a coastal bottlenose and only one unconfirmed record of a bottlenose offshore.

Southern California bottlenoses may travel over long lengths of the coast, some going as far north as Monterey Bay or as far south as San Quintin in Baja California. Such sightings may not represent mass migrations but simply individuals or small groups traveling within their long, narrow, linear home ranges. In the Southern California Bight, some groups of dolphins have home ranges associated with islands, but groups have also been identified moving between San Diego and Monterey Bay. The dolphins are seasonally present along the coast of California at various locations, but then will travel more than 500 km (310 mi) within a couple of weeks to another area. Some migration is related to calving, with females going to specific calving areas.

Specific habitats where dolphins are often seen include seagrass beds and tidal creeks, where they forage, and along the fringes of mangroves in lagoons, where they reside sometimes with young, perhaps to avoid predators or to rest.

VIEWING: Because the coastal ecotype clings so closely to shore, people often see Common Bottlenose Dolphins from headlands and beaches as well as from watercraft. Scientific surveys in the Southern California Bight have encountered Common Bottlenose Dolphins on more than half of the trips. Because they can be viewed from land, Common Bottlenose Dolphins and Gray Whales are perhaps the most commonly seen cetaceans along the southern west coast of North America.

CONSERVATION: The coastal form of Common Bottlenose Dolphins was hunted for food and other products (oil and leather) for centuries along both sides of the Pacific Ocean. Currently, large mesh gillnets for angel sharks, halibut, and other species may entangle California bottlenoses, but along the eastern Pacific, they are no longer the target of nets. They were one of the species killed in the tuna–porpoise controversy which fueled the boycott of tuna fish and eventually led to the passage of the Marine Mammal Protection Act, but they are now rarely trapped in nets set for tuna. Mexican gillnet fisheries probably take some coastal bottlenoses, but there are no data.

Agreement	Conservation Status
U.S. ESA	Not listed
U.S. MMPA	No special status
COSEWIC	Not at risk
IUCN	Least concern

By hugging the coast, coastal Common Bottlenose Dolphins accumulate pollutants concentrated there. California bottlenoses had record PCB and DDT contamination levels in the 1980s. The Southern California Bight had been a dumping ground for local industries in earlier decades. Dolphins are not able to detect refined oil products in the water, and so they are more vulnerable to these, which often are the most caustic types. Bottlenoses vanished from San Diego Bay after the 1960s, but once water quality improved, they returned, possibly reflecting the water's lower pollution levels.

Seven coastal and 27 offshore bottlenoses were captured for oceanaria off California over the last 40 years, but no permits for live captures are currently active.

Although ecotourism can take various forms with Common Bottlenose Dolphins including human swimmers' approaching and touching dolphins, swimming with dolphins is illegal along much of the eastern Pacific coast. In another part of the Pacific Ocean, Shark Bay, Australia, Common Bottlenose Dolphins became less abundant over 13 years of increasing ecotourism. Many behavior studies show short-term impacts of ecotourism on sea mammals; this long-term effect in Shark Bay may mean that an intense ecotourist business is not sustainable and that vessel tourism for bottlenoses may not be benign.

POPULATION: The California Common Bottlenose Dolphin coastal population is estimated to be a mere 323 dolphins. Numbers have been stable over the last 20 years. Comparisons of maternal genetic sequences show no overlap between coastal and offshore California Common Bottlenose Dolphins. The offshore bottlenose population in American waters within 555 km (300 mi) of the west coast of North America may comprise more than 3,000 dolphins. No sightings during surveys have been reported north of the Oregon border. To the south of the Mexican border, surveys have yielded an overall estimate of 336,000 bottlenoses in the eastern tropical Pacific, from the shores of Baja south to Peruvian waters and out to Hawaii.

COMMENTS: John Lilly ventured from conventional science with an orthodox anatomical and physiological study of the bottlenose brain to suggest that dolphins were highly evolved conscious life forms. Lilly noted that the killing of cetaceans would stop not with laws but when people understood that dolphins are "ancient, sentient earth residents, with tremendous intelligence." Much of his later research was unconventional, and the results questionable, but his delight in dolphin intelligence inspired the public to take action in the protection of dolphins.

REFERENCES: Connor et al. 2000, Dudzinski and Frohoff 2008, Lilly 1962, Lilly 1967, Montagu and Lilly 1963, Noren et al. 2008, Orr 1963, Reynolds et al. 2000, Wells and Scott 1999.

PANTROPICAL SPOTTED DOLPHIN *Stenella attenuata*
Pls. 12, 13; Fig. 81.

> Adult male average length is 1.7 to 2.4 m (5.6 to 7.9 ft). Adult female average length is 1.8 to 2.3 m (5.9 to 7.6 ft). Calf average length is 85 cm (33.5 in.).

SCIENTIFIC NAMES: *Stenella* comes from the Greek element *sten,* meaning "narrow," modified by the Latin suffix *ella,* meaning "small"—referring to this dolphin's beak (fig. 81). The species name *attenuata* derives from the Greek word for "thin" and also refers to the beak. British naturalist John Gray first named the species *Delphinus attenuata* in 1843, but eventually the genus

Figure 81. Pantropical Spotted Dolphin, offshore subspecies with few spots.

name of Stenella was assigned, applying to a group of slender long-beaked dolphins. Two such subspecies may occur in our area: *S. attenuata attenuata,* the offshore Pantropical Spotted Dolphin of the eastern Pacific, and *S. attenuata graffmani,* the coastal spotted dolphin that ranges from Baja to Peru.

COMMON NAMES: Common English names include Spotted Dolphin, Spotted Porpoise, Spotter, Bridled Dolphin, Slender-beaked Dolphin, and Slender Dolphin. The name Pantropical Spotted Dolphin became customary after the Atlantic Spotted Dolphin was recognized as a separate species, in 1987. In French, this dolphin has been called *le gamin* (The Child), *le dauphin a petites pectoralis* (Small-finned Dolphin), and *le dauphin veloce* (Swift Dolphin). In Spanish, the Pantropical Spotted Dolphin has been referred as *la delfin manchado* [Dappled (or Specked) Dolphin], *la estenala moteada* (Mottled Stenella), or *el moteado* (The Mottled One).

DESCRIPTION: Pantropical Spotted Dolphins are relatively small, slender, and streamlined, with pointed flippers and a strongly curved dorsal fin. Their long beaks are separated from their melon by a well-marked crease. Coloration changes with age and varies regionally, and it passes through stages with an individual's age. Newborn Pantropical Spotted Dolphins have no spots and are dark above and whitish on the belly. Later, a dark cape develops that runs from the forehead backward above the eyes and below and behind the dorsal fin. The tail stock remains dark above and light below, but a sweeping swatch of lighter gray appears on the sides with age. A dark stripe runs from the front of the flipper to the lower jaw. Around the eyes of older Pantropical Spotted Dolphins, a black patch develops, and this extends to form a line across the melon crease, joining the patches, as if the dolphin were wearing tiny Victorian *pince nez* sunglasses. In older juveniles, dark spots typically first appear on the belly, and later, lighter spots start to dot the cape, with both kinds growing in number and sometimes merging. The dark spots may finally fuse on the lighter belly to become a textured gray. The white spots may fuse, too, giving the dolphins a whitish cast at sea. In the eastern Pacific, coastal dolphins tend to be more spotted than offshore dolphins, while in the western and central Pacific, pantropical dolphins often virtually lack spots. The tip and lips of the beak may often be bright white in adults. In each jaw, there are two rows of 35 to 50 sharp conical teeth.

FIELD MARKS AND SIMILAR SPECIES: Seen closely, the Pantropical Spotted Dolphin's white lips and beak tip, dark cape, narrow dorsal fin, and white spots distinguish it from other narrow-beaked small cetaceans offshore in the eastern Pacific. Although spotting may not be obvious to far-off viewers, the two-tone coloration (dark cape and grayish belly) can serve as field marks at a distance. Their frequent companions, eastern tropical Pacific spinners, are usually a uniform steel gray, and Spinner Dolphin males may have forward-canted dorsal fins and humped tail stocks. Common Bottlenose Dolphins may have white-tipped beaks, but they have large heads and are stockier and thicker-beaked, with a varying cape blaze.

NATURAL HISTORY: Pantropical Spotted Dolphins are highly gregarious, often found in schools ranging, on average, from 28 to 83 and seen throughout the tropical eastern Pacific. These schools may consist of females with calves, juveniles, or adult males, or may contain mixed sexes and ages. Members of such groups dive, surface, travel, and rest synchronously. Small schools sometimes gather in much larger aggregations that can reach more than a thousand. Pantropical Spotted Dolphins are often joined by spinners and, occasionally, by other species of oceanic dolphins. For reasons that are unclear, such groups of dolphins frequently associate with Yellowfin Tuna. Several species of seabirds may flock to these lively high-seas assemblies as well.

The sounds made by a traveling school of Pantropical Spotted Dolphins are complex. Their staccato echolocation clicks snap against a background of descending and ascending whistles as the dolphins leap and dive, never colliding as they speed across the sea. The echoes and whistles may help orchestrate their delicate and deft maneuvers, but for humans, what could be harder to track, let alone analyze, than a thousand Pantropical Spotted Dolphins clicking and whistling in a feeding aggregation with seabirds screaming above.

REPRODUCTION: Females are sexually mature at 9 to 11 years, and males at 12 to 15 years, and coupled with slow maturity is long life, sometimes extending to nearly 50 years. The mating season peaks in spring and fall but can occur year-round, as is seen in many species of the tropics. The gestation period is around 11 to 12 months. Calves nurse for 1 to 2 years but can begin to eat solid food at a half-year of age. The calving interval is 2 to 3 years.

FOOD: Eastern Pacific spotted dolphins take as many as 56 species of fish and 36 species of cephalopods. In one study, lanternfish were the most common fish eaten, and flying squids were the most common cephalopod. Flying fish are favored prey in some areas. Pantropical spotteds also feed with Spinner Dolphins, in association with Yellowfin Tuna. Although not well understood, their association with tuna may help the spotted dolphins to find prey, and there is overlap in the species of fish and squid that they both consume. When in the company of tuna and pursuing fish, spotted dolphins must be swift, and have been clocked exceeding 35 kph (22 mph), although their average swimming speeds are 9 to 16 kph (6 to 10 mph).

MORTALITY: Sharks and Killer Whales attack and eat Pantropical Spotted Dolphins, as probably do blackfish, such as False Killer Whales, Pygmy Killer Whales, and Short-finned Pilot Whales. Spotted dolphins are particularly exposed to predation when released from tuna-porpoise purse-seine nets because they are confused and frightened when caught. Parasites also likely cause death, especially the nematode *Crassicauda* sp., which produces severe lesions in the skull.

DISTRIBUTION: True to its name, the Pantropical Spotted Dolphin is widely distributed in tropical, subtropical, and some warm-temperate waters. Although not usually associated with

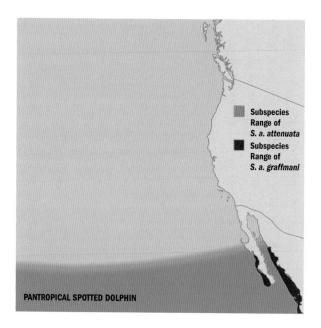

PANTROPICAL SPOTTED DOLPHIN

Subspecies Range of *S. a. attenuata*

Subspecies Range of *S. a. graffmani*

the California Current, abundance varies there and has not been well studied except in the tropical Pacific. Beginning about 36 km (16 mi) away from the coast of Baja California, the offshore northwestern subspecies of Pantropical Spotted Dolphin ranges out to about longitude 145° W. Waters with a surface temperature greater than 25 degrees C (77 degrees F), but with a shallow thermocline below, are favored by offshore Pantropical Spotted Dolphins.

The more heavily spotted coastal subspecies, *S. attenuata graffmani,* can be found from the tip of Baja California southward to Peru. There are at least four genetically distinctive coastal populations along the western shores of the Americas. Straying Pantropical Spotted Dolphins have reached as far north as Santa Cruz County in California and Cold Bay in Alaska. Tagging studies suggest that the eastern Pacific spotted dolphins tend to move toward shore in fall and winter and, later in spring and summer, away from the coast, likely following prey movements.

VIEWING: Ecotourists in Baja California may enjoy seeing coastal Pantropical Spotted Dolphins near Cabo San Luis or within the Gulf of California.

CONSERVATION: Originally, perhaps 7 million Pantropical Spotted Dolphins swam the warm seas of the Earth, with the species being most abundant in the eastern tropical Pacific. Some 4 million spotted dolphins were killed in the eastern Pacific tuna fishery, and that population never recovered. The Pantropical Spotted Dolphin today is the second most numerous delphinid worldwide, after the bottlenose, but it might once have been the most common cetacean of all. The collapse in pantropical numbers took place over 15 years, showing how quickly modern technology can cause devastation on an oceanic scale.

Agreement	Conservation Status
U.S. ESA	Not listed
U.S. MMPA	Northeastern offshore spotted dolphin stock depleted.
COSEWIC	Not listed
IUCN	Least concern

The fishery-related demise of Pantropical Spotted Dolphins began when mechanized purse-seine nets replaced traditional long poles in the eastern Pacific, around 1960. As before, herds of dolphins guided fishermen to schools of Yellowfin Tuna. But with the new technique, powerboats drew a long seine net around both dolphins and fish, because the tuna tended to stay below the dolphins. The speedboats pursued and drove dolphins to prevent escape. Dolphins were trapped and often drowned in folds of the fish net as it was drawn closed, or they were injured or killed as the nets were brought onto the processing ship. About 80 percent of the spotted dolphins in the eastern Pacific perished before public protest stopped the slaughter. As a result of advances in net technology and better management, the number of dolphin deaths dramatically fell from a high of some 300,000 a year. Currently, none are killed annually by U.S. boats, but about 3,000 a year are killed by other countries' fishing boats.

Several researchers have tried to understand the perplexing failure of this dolphin's numbers to come back to higher levels.

One factor may be the repeated stress of the chase, capture, and release. It has been estimated that large aggregations of dolphins may be encircled as much as once a week by tuna fishermen, with groups of 250 to 500 dolphins being captured two to eight times per year. The dolphins flee at high speed following release, revealing a strong fear response that might cause harmful physiological effects—a sort of cetacean post-traumatic stress disorder. The chaos of capture could also break up any social structure in groups, with unknown consequences. Indeed, there are data suggesting that some nursing dolphins become separated from their mothers, either in the confusion of netting or by being unable to keep up in the release flight. Suckling dolphins could starve if not reunited with their mothers, and older offspring may suffer from the lack of maternal guidance. There is evidence that Atlantic spotted dolphin mothers teach their calves to find prey. And predators such as blackfish have been reported to kill weakened dolphins when released from the seine nets.

The NMFS has created two programs to help restore eastern Pacific dolphin populations. The Dolphin Safe Program seeks to reduce dolphin deaths by developing fishing methods harmless to dolphins. The Dolphin Energetics Program studies how existing fishing techniques might lead to stress, separations, and other consequences limiting dolphin recovery.

POPULATION: The population of Pantropical Spotted Dolphins has been frequently surveyed since the 1970s and is currently estimated to be approximately 900,000 individuals, including around 150,000 of the coastal subspecies.

COMMENTS: Many populations of sea mammals have shown little sign of recovery following whaling and hunting at levels that began in 1750. Such species include Alaskan baleen whales, Peruvian Sperm Whales, and California Sea Otters. The spotted dolphins and Spinner Dolphins of the eastern Pacific were exceptional in that they were not hunted but were knowingly bycaught and thus unintentionally slaughtered. Despite years of research, scientists cannot say why these sea mammal populations have never regained their former numbers even though they are protected. It seems easier for us to upset the balance of ecosystems than to restore them.

REFERENCES: Archer et al. 2004, Chivers and Myrick 1993, Gerrodette and Forcada 2005, Noren and Edwards 2007, Perrin and Hohn 1994.

SPINNER DOLPHIN · *Stenella longirostris*
Pls. 12, 13; Figs. 6, 82.

> Adult average length is 1.65 to 1.8 m (5.5 to 6 ft); average weight
> is 61 kg (135 lbs). Calf average length is 80 cm (31.5 in.).

SCIENTIFIC AND COMMON NAMES: The Spinner Dolphin's generic
name, *Stenella,* which derives from the Greek root *sten,* meaning
"narrow," with the Latin suffix *ella,* meaning "small," refers to
the shape of this dolphin's beak. The species name *longirostris*
derives from the Latin roots *long,* meaning "long," and *rostr,*
meaning "beak" or "snout" (fig. 82).

British naturalist John Gray described this species in 1828
from a skull, but a Benedictine monk first published a detailed
account of live dolphins springing out of the waters of West
Africa in 1769. This dolphin species has borne many English
names, including spinner, Spinner Dolphin, Long-snouted Spinner Dolphin, Long-snouted Dolphin, Long-snout, Long-beaked
Dolphin, and Rollover. The element "long" in both the common
and scientific names distinguishes this species from *Stenella
clymene,* the Clymene Dolphin or Short-snouted Spinner

Figure 82. Spinner Dolphins, eastern form.

Dolphin of the Atlantic, first formally recognized in 1981. The French name *dauphin longirostre* repeats the Latin species name. The Spanish *tornillo* means screw, the related *delphin tornillon* means Great Screw Dolphin, and *estenela giradora* can be translated as the Spinning Stenella.

In the Pacific, four physically distinctive subspecies of *Stenella longirostris* dolphins have been identified: the Dwarf Spinner Dolphin, the Hawaiian (or Gray's) Spinner Dolphin, the Central American (or Costa Rican) Spinner Dolphin, and the Eastern Spinner Dolphin. The Eastern Spinner Dolphin *(S. l. occidentalis)* can be seen off the southern coast of Baja California and is the only subspecies we would expect to see along the California Current. The whitebelly form is recognized as a transition form between the eastern and Hawaiian subspecies and is typically found further offshore, and more to the south of Baja California.

DESCRIPTION: Spinner Dolphins are all small, streamlined, and slender, but they show strong regional variation in color and form. All Spinner Dolphins share long and thin beaks, distinct creases before a gently sloping melon, pointed flippers, and midback dorsal fins. Eastern Spinner Dolphins become sexually dimorphic as they mature: adult males are slightly larger than females, and there are differences in coloration as well. Immatures of both sexes are gray backed with a creamy throat and pale underbody. Adult females are gray except for varying white marks on the belly. Adult males usually look a uniform steel gray, which can appear bluish or purplish through the water. Males develop a distinct hump or bulge on the lower tailstock as they grow older. This "postanal" hump is thin and keellike. Over time, males' dorsal fins typically cant forward, and their fluke tips curve inward. One observer noted that the male's dorsal fin looks like it is stuck on backward. The male's distinctive features make it possible to sex this subspecies at sea.

Spinner Dolphins carry 40 to 65 rows of fine, slender teeth in each jaw. Spinner Dolphins have more teeth than any other cetacean except for the Franciscana River Dolphin of South America.

FIELD MARKS AND SIMILAR SPECIES: Far offshore of Baja California, the Eastern Spinner Dolphin may be confused with the whitebelly intergrade. The Whitebelly Spinner Dolphin shows off its namesake light underside as it leaps from the water, contrasting with the basic steel gray of the Eastern Spinner Dolphins. Also, Whitebelly Spinner Dolphins tend to have curved dorsal fins and straight flukes, in contrast to the Eastern Spinner Dolphins. Whitebelly Spinner Dolphins are not so obviously sexually dimorphic, and males lack conspicuous tailstock humps. The distribution of Pantropical Spotted Dolphins overlaps with that of Eastern Spinner Dolphins. Pantropical Spotted Dolphins all have a dark gray-brown cape. Younger pantropicals have a pale underside that darkens with age. Older pantropicals grow more spotted with time. Although, from a distance, Spinner Dolphins can be confused with other long-beaked dolphins, a close view can permit a positive identification.

NATURAL HISTORY: Eastern Spinner Dolphins are found in schools that average about 100 individuals, but they sometimes gather in groups of 1,000 or more. Group size tends to increase in the morning and then decrease later in the day. Eastern Spinner Dolphins may also join with Pantropical Spotted Dolphins and Whitebelly Spinner Dolphins in large feeding aggregations. Pure or mixed dolphin concentrations often hover over schools of Yellowfin Tuna, and sometimes yellow jack tuna. Flocks of pelagic seabirds can be drawn to the tuna as well. Mixed schools of Spinner Dolphins and Pantropical Spotted Dolphins are joined by the greatest variety of seabirds, but the birds seem more attracted to the tuna than the dolphins. The assemblies of pelagic-feeding birds, tuna, and dolphins tend to form in the mornings and break up later in the day.

Spinner Dolphins are acrobats: they can spin, somersault, spin and somersault together, breach, head slap, side slap, and tail slap. A school of Spinner Dolphins beats on the surface of the sea, and their chorus of splashes must indicate the group's mood and location to its members. Although such percussive signals in sea mammals are common, they can be difficult to describe and investigate. The spin of the spinner is an exception to this rule, and it has been much studied. Of all sea mammals, only the *longirostris* and *clymene* species of *Stenella* habitually spin out of the water. The cosmopolitan *longirostris* spinner spins more frequently and vigorously than the *clymene* spinner. The maximum spin ever recorded in a spinner was 7.5 turns, the longest time aloft was 1.5 seconds, and the maximum height was 3 m (10 ft). As many as 14 spins in a row have been recorded, but the number tends to diminish over the series. Spinner Dolphins begin their corkscrew course from below and, once freed of the viscosity of the water, they continue their pirouette into the air by momentum. Whale-watchers are often struck by the exuberance of these dolphin flights. Some researchers have theorized that the resulting side splash must mark positions of dolphins within a school. But the spinning can serve another purpose: sometimes remoras, "whale suckers," are dislodged with the centrifugal force of the leaps.

REPRODUCTION: Females are sexually mature at 4 to 7 years of age, and males at 7 to 10 years. The Spinner Dolphin has an annual and seasonal breeding cycle, but this varies regionally, ranging from late spring to fall. Gestation is around 10 to 11 months, and calves are very small at birth but nurse for 1 to 2 years. Consequently, the calving interval is 3 years. The female–calf relationship is close, and newborn dolphins usually swim in the "echelon position," behind their mother's dorsal fin, letting them keep up and even coast in her draft.

For many subspecies of Spinner Dolphin, and for the whitebelly form, males are moderately sexually dimorphic and have small testes, and some researchers have proposed that their mating system is polygynandry, whereby two or more males have an exclusive relationship with two or more females. However, the eastern subspecies is highly sexually dimorphic: along with the male's hump and canted dorsal fin, it has larger testes. Across many orders of other animals, big testes have been linked to a promiscuous lifestyle—many males mate with many females, and the male with the most or the best sperm attains fatherhood. In contrast, in species with pronounced sexual dimorphism, such as elephant seals, males tend to be polygynous, defending females from other males. Thus, we might expect male Eastern Spinner Dolphins both to be somewhat promiscuous and also to monopolize females in some fashion.

FOOD: Eastern Spinner Dolphins eat midwater fish, squid, and shrimp. Their prey are usually smaller than 20 cm (8 in.). The types of fish they consume include ones that range to depths of 200 to 300 m (600 to 1,000 ft), including deep-sea fish with descriptive names such as lanternfish and pearleye. They also eat deep-sea smelts and several species of deep-sea squid. Their habitual association with Yellowfin Tuna is not well understood, but possibly the tuna help the Spinner Dolphins find prey. Scientists have measured swim speeds averaging 16 kph

(10 mph), but likely they swim faster to keep pace with tuna, which attain speeds of 70 kph (43 mph).

MORTALITY: Sharks and Killer Whales take Spinner Dolphins, and blackfish such as False Killer Whales, Pygmy Killer Whales, and Short-finned Pilot Whales may also prey on Spinner Dolphins.

Remoras, suckerfish, attach themselves to a dolphin's head, fins, and back, irritating the dolphin's skin and adding to its drag in the water. They do not suck the dolphin's blood, but there is a debate about what they actually eat, with dolphin feces, food scraps, and skin bacteria being possible remora cuisine. One researcher attached a remora to his own back to see how it felt, and reported that it went into his skin and that it hurt.

DISTRIBUTION: Spinner Dolphins are cosmopolitan in tropical and subtropical seas. Numbers are highest in the tropics and mostly beyond the range of the California Current. The Eastern Spinner Dolphin subspecies ranges from latitude 5° N to 30° N and from longitude 145° W eastward toward the coast of North America, reaching about halfway up the Baja peninsula.

Range

EASTERN SPINNER DOLPHIN

These dolphins inhabit waters that are far from land and that have a shallow mixed layer, with either a shoaling or a sharp thermocline, where the water temperatures do not vary much during the year. Nevertheless, there are year-to-year differences in the oceanic features of their habitats, and so too in the distribution of the dolphins. For whatever reason, the Spinner Dolphin is seen more in association with a shoaling thermocline, and the whitebelly with a sharp thermocline. When they move, Spinner Dolphins have been documented to travel over 500 km (317 mi) in about 16 days.

CONSERVATION: The slaughter of Spinner Dolphins in the 1960s was a turning point in ocean conservation. The deaths of these charismatic cetaceans in the tuna nets of the eastern Pacific inspired widespread protests and helped launch the modern environmental movement.

Agreement	Conservation Status
U.S. ESA	Not listed
U.S. MMPA	Eastern Pacific stock depleted
COSEWIC	Not listed
IUCN	Data deficient

More than 6 million dolphins were killed in the eastern Pacific purse-seine fishery for tuna. In contrast, in all of 20th-century whaling, only 2 million large cetaceans were harpooned worldwide. The Pacific dolphins died as bycatch because they served as guides for fishermen. Fishing lookouts can locate Yellowfin Tuna by scanning the sea for lively concentrations of dolphins and seabirds. Boats can then surround the school of tuna with a long net, drawing the net closed with a power block. Because the dolphins mark the position of the Yellowfin Tuna below, they also are captured in the seine. When this tuna fishery first developed in the early 1960s, following the introduction of power gear and modern synthetic nets, trapped dolphins were often injured or drowned in folds of the nets or were hurt or killed as the nets were hauled aboard. The population of Eastern Spinner Dolphins, which lies within the target area for tuna, plummeted from more than 1.5 million to about one-third of that number, in just a couple of decades.

Public outcry at the deaths of these dolphins resulted in a series of legal protections. Nets were redesigned to spare the cetaceans' lives. By 1980, the number of Pacific dolphins dying each year dropped from 500,000 to 20,000, and the proportion of the population impacted fell to less than 0.1 percent by 2002. There were subsequent attempts to weaken dolphin-safe label legislation, but these efforts failed in an American court in 2004. Much of the international fleet abides by the rules of the dominant marketplace in the United States. Despite the impressive success of protective measures, however, the populations of both Spinner Dolphins and Pantropical Spotted Dolphins have never recovered.

There are several explanations for the lack of recovery, such as underreported kill, loss of the benefits from a tuna–dolphin relationship, and long-term sea changes. Furthermore, their reproductive rates have fallen in the eastern tropical Pacific. More research likely will help us understand the failure of these oceanic dolphin populations to recover.

POPULATION: The population of Eastern Spinner Dolphins has been frequently surveyed since the 1970s and is currently estimated to be approximately half a million individuals, although far, far fewer occur in the California Current.

COMMENTS: To understand why the Spinner Dolphin population is not recovering, American scientists have suggested that the nets may still kill dolphins—but more subtly and slowly than realized. One reason is that tuna fishing is so intense that, according to biologists' reckoning, most eastern Pacific dolphins are netted an average of 11 times in their lifetimes. The dolphins seem frightened by their capture. Repeated stress of such magnitude may result in an array of harmful physiological changes, as happens with other mammals. Netting can result in spontaneous abortions or the separation of newborn or young dolphins from their mothers, as they flee after release. Young dolphins need to breathe more frequently than their mothers do, and they cannot move as rapidly when swimming independently. Given the speeds involved in the chase, and in the flight after a release, many young dolphins likely fall behind their mothers. If not reunited, those that depend on their mother's milk starve. Several strands of evidence are consistent with the separation hypothesis. Accidentally killed nursing females often have been unaccompanied by their calves and the numbers of young Spinner Dolphins have dropped in years of heavy netting.

REFERENCES: Benoit-Bird and Au 2009, Fish et al. 2006, Gerrodette and Forcada 2005, Perrin 1998, Perrin 2002a, Perrin and Gilpatrick 1994, Wade et al. 2007, Weihs et al. 2007.

STRIPED DOLPHIN   *Stenella coeruleoalba*
Pls. 12, 13; Fig. 83.

> Adult male average length is 2.4 m (8 ft); average weight is 158 kg (422 lbs). Adult female average length is 2.2 m (7.2 ft); average weight is 136 kg (364 lbs). Calf average length is 1 m (3 ft); average weight is 10 kg (22 lbs).

SCIENTIFIC NAMES: A member of the Delphinidae family, the genus *Stenella* is named as a derivation from Greek and Latin meaning "narrow," *sten*, and small, *ella*, referring to this dolphin's beak (fig. 83). The species name *coeruleoalba* derives from the Latin words *caeruleus* for dark blue and *albus* for white, which allude to the species' distinctive blue and white stripes. The species was first discovered and named by Meyen in 1833 from a specimen collected in the South Atlantic off Argentina or Uruguay, but a specimen was not collected in the Pacific until 1848 by Peale. The genus name is applied to an assemblage of species that truly are not related, and Striped Dolphins are an example of a species that is more related to Common Bottlenose Dolphins than to other Stenella.

COMMON NAMES: Common descriptive names include the Striped Dolphin, Blue-white Dolphin, Long-snouted Dolphin, Harnessed Dolphin, Whitebelly Porpoise, Black-jawed Dolphin, and Streaker Dolphin. Gray's Dolphin and Meyen's Dolphin are common names associated with the zoologists who described the species. The French common name is *dauphin bleu et blanc*, and the Spanish is *delfín azul* or *delfín rayas*.

DESCRIPTION: The Striped Dolphin is fairly small and streamlined, with pointed flippers and a moderately curved dorsal fin. Its long beak abuts a distinct forehead, also called its melon, with a well-marked

Figure 83. Striped Dolphin.

crease, and its mouth holds many small teeth, some 37 to 59 pairs. The largest was 2.6 m (8.5 ft), and the heaviest was 156 kg (343 lbs), and males are only slightly larger than females. There is substantial geographic variation in size around the world, however, and those in the eastern Pacific are shorter than those in the western Pacific.

The Striped Dolphin's distinct color pattern is not seen in other oceanic dolphins. The stripes about the head are not unlike a harness, as one common name denotes. One long, very dark stripe stretches from the eye to the anus, and a second short stripe runs from the bottom of the eye to and along the insertion of the flipper. The long stripe often has a short spur of a stripe stretching down into the white above the flipper. All three stripes are dark blue or blue-black. The eye, too, is encircled in a dark blue color. The back is a lighter blue-black or gun metal gray, from the head to the lower back; this area, called a cape, is overlaid with a blaze of light gray. The blaze begins at the beak and splits into two arms extending around the head back toward the dorsal fin and from the flank back to the anus. The light belly has a typical countershading pattern, beginning at the chin and bordered by the dark stripe from the eye. The beak itself is dark, as are the flippers. The white belly may even appear pink, and other colors vary by individual and by light reflection, with some animals having more brown than blue tones.

FIELD MARKS AND SIMILAR SPECIES: The Striped Dolphin has a body similar to other dolphins, with a long, spindle-shaped but robust form and a long beak. Their distinctive striping, however, is unlike that of the other oceanic dolphins in the California Current. They might be confused with the common dolphin, particularly the Long-beaked Common Dolphin, which also has a line from the eye to the anus and the eye to the flipper, but the stripes of the Striped Dolphin clearly demark the darker cape from the lighter belly, whereas the common dolphin has a yellowish color above the stripe. Also, the common dolphin lacks the blaze between the head and the dorsal fin that is often present in the Striped Dolphin. The immature Pantropical Spotted Dolphin could be confused with the Striped Dolphin, but the former lacks any stripes along the flank. The Fraser's Dolphin also has an eye-to-anus stripe that is very thick, but these animals are more robust than the Striped Dolphin.

NATURAL HISTORY: Striped Dolphins are gregarious, but they do not regularly form extremely large schools, as is seen in some of the other oceanic dolphins. The average group size is around 100, and less than 15 percent are greater than 500; some estimates for average group size range from 28 to 83 in the eastern Pacific.

The composition of the schools in the western Pacific was described by researchers examining schools of hunted dolphins, who noted that the schools segregated into three types by age: adult, juvenile, and mixed. The schools were then further segregated based on the reproductive status of the members of the schools into breeding and nonbreeding. Breeding adult schools were composed of adult females and socially mature males. The males left the school after all the females were mated with, and the school then became a nonbreeding school of adult pregnant females. Once the females gave birth and attended their young, the schools became mixed, and once the young matured in 1 to 2 years and were weaned, they formed juvenile schools, separate from the adult schools. Determinations regarding whether this social structure exists in the eastern Pacific are also made in a different way than by examining dead dolphins. Researchers estimated the age and sex of members of schools by measuring lengths of members from photographs taken from helicopters and found similar segregation patterns. They also determined that juvenile schools were more likely to be associated with tuna schools. One theory for this association is that the dolphins benefit from associating with tuna because it takes them less energy to swim and find prey if they ride the drafts of water created by the large tuna schools. Lending some credence to this theory is the fact that juvenile Striped Dolphins are about the same size as the other dolphins most associated with tuna, the Pantropical Spotted Dolphin and Spinner Dolphin. Striped Dolphins will also school with common dolphins in the eastern Pacific. The relationship among the members of the group may be long lived because they are estimated to live longer than 55 years.

Whatever the group size, Striped Dolphins are very acrobatic, streaking across the wave tops—hence their common name, streaker. In their exuberance, they breach, slap their chins on the water, and execute an acrobatic maneuver called "roto-tailing," the envy of any gymnast. They are unique among dolphins in performing this feat, which involves leaping out of the water in a high arch while "performing several rotations with the tail" (Archer and Perrin 1999).

REPRODUCTION: Striped Dolphins are sexually different in size beginning as early as 2 to 3 years, but females are not sexually mature until 5 to 13 years, and males not until 7 to 15 years. The average age for females is 9 to 11 years in the eastern Pacific, and males may not be socially mature until 17. The younger maturity dates are likely related to intensive hunting pressures, which often result in animals breeding at younger ages.

If one just measured the weights of male testes, there would seem to be no apparent mating season; however, other studies support a spring and winter peak in number of calves. Based on the composition of schools (see above), most researchers presume that Striped Dolphins are polygynous. The gestation period is around 12 to 13 months, and calves are suckled for 1.5 years. Females then rest for about half a year before mating again, and consequently, the calving interval is 3 years.

Females are able to conceive up to around age 30, which suggests that an average female is capable of birthing five to seven young, and the oldest documented pregnant female was 48 years.

FOOD: Eastern Pacific Striped Dolphins eat a very diverse diet, a third of which is composed of midwater fish, a third of large and small squid, and a third of miscellaneous fish and invertebrates. Lanternfish are the most common fish eaten, and small squid the most common cephalopod. Prey preference and composition is widely different across the globe, however, and researchers have identified species from 20 fish families and 10 cephalopod families.

Based on the species of prey discovered in their stomach contents, Striped Dolphins likely hunt in pelagic waters but also near the bottom, in very deep water [200 to 700 m (656 to 2,300 ft)]. Deep-ocean prey include mostly lanternfish, but also ones with colorful names such as snipe eel and hatchet fish. Many of the species have organs that produce bioluminescence, for living in deep oceanic waters, but they also feed on midwater and surface-dwelling fish such as herring, cod, and hake.

MORTALITY: No doubt, sharks and Killer Whales, as well as False Killer Whales, Pygmy Killer Whales, and Short-finned Pilot Whales, attack and eat Striped Dolphins, but there is little information on predation. Instead, there is ample information that diseases such as cardiovascular pathologies, tooth infections, and parasites compromise their health.

DISTRIBUTION: Striped Dolphins range across the warm-temperate and tropical oceans of the world, including the Mediterranean,

Range

STRIPED DOLPHIN

and throughout the Pacific. Although they have stranded onshore north to latitude 43° N in Oregon and Washington, they typically occur only in offshore waters of California and Baja. Their southern limit is latitude 15° S in the eastern Pacific, but they also occur around New Zealand in the western Pacific at least as far as latitude 40° S. Their presence is likely, but they are not abundant in the midocean expanses, except around islands and convergence zones, where food is plentiful. Instead, Striped Dolphins are a creature of the continental slope and adjacent deep oceanic waters. Convergence and upwelling zones where there is hydrographic complexity with fronts and eddies are oceanic habitats that attract Striped Dolphins. In the eastern Pacific, this is usually where cold water converges with warm tropical water, forming a strong and shallow thermocline under the warm tropical surface water. Seasonal migration has not been observed in the eastern Pacific population of Striped Dolphins.

VIEWING: Ecotourists on offshore boats in southern California and Baja California may enjoy seeing coastal Striped Dolphins because they are one of the most abundant species in the subtropics, but not near shore.

CONSERVATION: Striped Dolphins were hunted intensively in the Japanese drive fishery dating back to the 15th century, and after World War II, catches were very high, exceeding 21,000 dolphins per year. In the 1990s, the numbers killed in this fishery, however, declined to 500 to 1,000 dolphins per year. Furthermore, there is likely concern for human health because of the high amount of mercury in the dolphins caught off Japan. Although not an issue in the eastern Pacific, in the Mediterranean, Striped Dolphins had some of the highest records for PCBs at 2,500 ppm of any mammal.

In the eastern Pacific, Striped Dolphins also were incidentally captured in the tuna-porpoise fishery in large numbers, and the

Agreement	Conservation Status
U.S. ESA	Not listed
U.S. MMPA	No special status
COSEWIC	Not at risk
IUCN	Least concern

effect on Striped Dolphins would have been disproportionately hard on the juveniles because that age group is more likely to associate with tuna. Striped Dolphins also are drowned in the drift gillnets of swordfish and shark fisheries in the Pacific off Baja and Mexico. These nets can extend up to 4 km (2.5 mi) long, fencing large expanses of ocean and ensnaring many non-targeted species. The placement of pingers to ward off dolphins can greatly reduce the number of entanglements. In 1990 to 1995, dolphins were drowned at a rate of 0.14 per net set, with an estimated number of net sets in 1992 of around 2,700. This would extrapolate to around 375 dolphins drowned per year.

POPULATION: In 1991, the population of Striped Dolphins off California was estimated to average 19,000, but to have a wide range of 8,000 to 46,000. More recently, in 2007, the population of eastern Pacific Striped Dolphins off California, Oregon, and Washington was estimated to be approximately 24,000 individuals. There is no estimate for Baja.

COMMENTS: Striped Dolphins are one species whose range may expand northward with the predicted warming sea temperatures of climate change because they already exhibit seasonal shifts and have been observed as far north as Washington. Because they are associated with convergence zones of warm and colder water, Striped Dolphins may be on the leading edge of this range shift of warmwater species north.

REFERENCES: Archer and Perrin 1999, Barlow and Cameron 2003, Perrin et al. 1994b, Perryman and Lynn 1994.

SHORT-BEAKED COMMON DOLPHIN   *Delphinus delphis*
Pls. 14, 15; Figs. 2, 39, 44, 47, 84.

> Adult male average length is 1.7 to 2.0 m (5.6 to 6.5 ft). Adult female average length is 1.6 to 1.9 m (5.2 to 6.2 ft). Average weight for both is 100 kg (268 lbs). Calf average length is 80 to 90 cm (31 to 35 in.); average weight is 7 kg (19 lbs).

SCIENTIFIC NAMES: The common dolphin is an iconic member of the Delphinidae family. Linnaeus named and described the species in 1758, providing the redundant derivation, from

Latin and Greek, of *Delphinus* and *delphis*—meaning "dolphin dolphin." In the eastern Pacific, Captain Charles Scammon first reported common dolphins in 1874. After much debate over regional differences, researchers in 1994 separated the common dolphin into two species based on analyses of genes and morphology; other new regional species likely will follow. The Short-beaked Common Dolphin was previously recognized as the offshore form of the common dolphin, and the Long-beaked Common Dolphin was the coastal form. Presently, the Short-beaked Common Dolphin retains the name of origin, *Delphinus delphis*, and the Long-beaked Common Dolphin is recognized as *Delphinus capensis*.

COMMON NAMES: Common descriptive names span the globe, including, in English, Short-beaked Common Dolphin, Saddleback Dolphin, Offshore Common Dolphin, Hourglass Dolphin, Crisscross Dolphin, and White-bellied (or Whitebelly) Dolphin. The French common names are *camus, dauphin vulgaire,* and *dauphin commun.* A Spanish common name is *delfín comun.*

DESCRIPTION: The Short-beaked Common Dolphin is the quintessential medium-sized dolphin, slender and spindle-shaped (fig. 84). The beak is distinct from the forehead, or melon, which rises abruptly and has a clear crease along the base separating it from the beak. The beak, at 23 to 27 cm (9 to 10 in.), is shorter than the long-beaked dolphin's. Of the

Figure 84. Short-beaked Common Dolphin.

dolphin family, common dolphins have the boldest and most clearly delineated color pattern. Seen in a flash, the colors are streaks of gray, white, and yellow, but with a closer look, the most distinguishing color feature is an hourglass pattern and a white belly patch. The hourglass pattern divides the side view into four sections: the front half of the body to the dorsal fin is pale yellow, cream or gray-green above a white belly, and the rear half of the body behind the dorsal fin is a light gray above to darker gray below. Draped over the back hour-glass pattern is a cape of dark black or brown-black that reaches from the melon to behind the dorsal fin.

Several other basic elements of the color pattern can be seen if dolphins are observed closely, as when they bow-ride. A narrow dark stripe extends from the flipper to under the chin, and a dark eye patch is linked to a dark line extending to the base of the melon. Seen from above, there is a blaze of lighter color from the tip of the beak to the blowhole, and if the dolphin rolls on its side, the lips of the beak often are black. The flippers and dorsal fin are slate gray but usually have a white patch in the middle of the fin, or can be entirely white. The dorsal fin is slightly falcate, pointed at the tip, and 40 cm (16 in.) tall. The tail is 50 cm (20 in.) wide and is scalloped in shape, with a median notch. The tail stock is laterally compressed so that viewed from above it appears narrow, but from the side broad and muscular.

All color marks are variable by region, and there are several regional groups in the California Current (see distribution section). There also is a color morph that completely lacks yellow color on the side patch; instead, it has an hourglass pattern that is dark to light gray. Other color phases include all-black and all-white. To add to the color confusion, juveniles are less distinctly patterned than adults; calves are dull-colored. Dead beachcast animals also lack the vibrant coloration of adults.

Generally males are only slightly larger than females, the largest recorded weighing 136 kg (300 lbs). The different sexes cannot be easily distinguished in the field, even from the bow of a vessel. Most adult males have a broad, dark blaze across the genital area that sometimes is linked to the dorsal cape with a stripe, whereas the female has a narrow dark band separating the countershading of black and gray. And too, some adult males have a hump behind the dorsal fin, seen also in other male dolphins.

If you had a common dolphin skull in hand, you would see 40 to 55 pairs of sharp pointed teeth. You also would see a difference between this species and all other members of the family by looking at the skull—two deep grooves along the palate can be seen in both fresh dead and older decomposed common dolphins.

FIELD MARKS AND SIMILAR SPECIES: Short-beak Common Dolphins and Long-beaked Common Dolphins cooccur in Southern California and are very difficult to distinguish at sea, even for trained observers. They do not cooccur in the same herds, though; instead, Short-beaked Common Dolphins cooccur with Pacific White-sided Dolphins, Striped Dolphins and Common Bottlenose Dolphins, with all of which they may be confused.

The Short-beaked Common Dolphin has a long, spindle-shaped body and a narrow beak, similar to other dolphins of the subtropical Pacific, but its bold and distinctive hourglass marks and white belly are not like those of the other oceanic dolphins in the California Current, except for the Long-beaked Common Dolphin. At sea, the two can be separated by the shorter beak and more robust body of the Short-beaked Common Dolphin. The Short-beaked Common Dolphin also lacks two dark stripes that are present in the long-beaked species: one from the eye to the anus and one from the eye to the flipper. The short-beaked species has a melon that is slightly larger and rounded than that of the long-beaked, and the dorsal fin and flippers often have white patches.

Both these common dolphins are often confused with the Striped Dolphin, which also has a line from the eye to the anus and the eye to the flipper, but the common dolphins have a yellowish color above the stripe whereas the stripes of the Striped Dolphin clearly demark the darker cape from the lighter belly. Also, the common dolphins lack the blaze between the head and the dorsal fin that is often present in the Striped Dolphin. The Fraser's Dolphin also has an eye-to-anus stripe that is very thick, but its body shape is distinctly robust compared to the common dolphin's. From a distance, common dolphins may be confused with Pacific White-sided Dolphins, but the head shape and falcate dorsal fin of the Pacific White-sided Dolphin are easy marks to detect and are different from the common's straight dorsal fin and long beak.

NATURAL HISTORY: This creature inhabits ancient myths and is woven throughout the history of humans. Common dolphins, along with Common Bottlenose Dolphins, are the most familiar dolphins to people around the world. Common dolphins are also the most gregarious and abundant of dolphins worldwide. In Southern California waters, they account for more than 50 percent of sightings by researchers and are the most likely species to be encountered by boaters.

They are very energetic, zipping from afar to bow-ride and then accelerating at speeds faster than the boat, leaving the impression that they are seeking faster vessels. For a dolphin, they are certainly fast, with bursts of speed of 40 kph (25 mph), and clocked passing ships traveling at 28 to 35 kph (17 to 21 mph). When swimming at speed in large schools, common dolphins can make the surface waters white with froth. Their normal travel speeds, though, are 10 kph (6 mph).

Many dolphin species are known to enjoy bow-riding both boats and whales, but worldwide, common dolphins are the most renowned and skilled for their delight in bow-riding. Large mixed groups will charge toward a boat and then jockey for position at the bow, layering themselves two to three or more deep, depending on the size and speed of the vessel. The smaller common dolphin often initiates the bow-riding but will be displaced by groups of the more robust Pacific White-sided Dolphin. Not all members of a given group of common dolphins bow-ride, though. A researcher tagged bow-riders in Southern California and, after a while, found that all bow-riders had tags, though not all of the dolphins of the school were tagged.

The leaps of common dolphins are remarkable, reaching heights of 6 to 7 m (20 to 23 ft). One distinctive aerial display is called "pitch poling," whereby a dolphin leaps straight out of the water and then falls back lengthwise, creating a substantial splash. Dolphins can have clean entries with no splash, or can create distinctive splashes, including by chin-slapping with a tail-lob splash, or landing sideways or on their back, creating a large splash.

Often members of a group swim in synchronization, surfacing and diving side by side; common dolphins do so even with other species such as Striped Dolphins. The side markings of striped and common dolphins are thought to aid them visually in swimming together, or in synchronization, perhaps to prevent them from running into each other when large schools swim at great speeds. Some schools of common dolphins in the Eastern Pacific are composed of 2,000 to 10,000 members during the day. The schools then separate into smaller feeding groups of 20 to 200 in the late afternoon and at night. This form of social organizing is called fission-fusion and is commonly seen in oceanic dolphins, likely in response to the patchy and ephemeral distribution of prey in offshore waters. At sunset, the large schools subdivide into small groups to feed in the deep-scattering layer; the small groups coalesce back into large groups in the morning and retain them through the day while they mostly socialize and rest.

Large schools are often composed of several species, with Short-beaked Common Dolphins mixing with Striped Dolphins and Spinner Dolphins. "Sympatric" is a term used to describe the cooccurrence of species in the same habitat, and several species of dolphins benefit from sympatric associations, especially in open ocean habitats, where the different strategies brought forth by different species can enhance the seeking, pursuing, and capturing of prey. Remarkably, Short-beaked Common Dolphins and Long-beaked Common Dolphins do not form schools together, even though they cooccur in the Southern California Bight. This is particularly interesting because they both feed on abundant anchovies: apparently there are enough differences in their diets that they do not feed together; they also seem to segregate in different habitats.

School size also changes seasonally. Smaller schools of 50 to 200 are seen in spring and summer, likely representing segregated groups of different sex and age. The smallest grouping, though, is composed of 20 to 30 individuals. This basic social unit of common dolphins was discovered by researchers who chased a large school of common dolphins until it began to break up: they continued this until the group no longer broke up but instead formed a tight, stable unit. The researchers surmised that when stressed, common dolphins will bunch close together. In the Eastern Pacific, researchers have subsequently determined that these small social units are genetically discrete breeding groups. Common dolphin social behavior includes aiding injured members and supporting them near the surface to breathe, a behavior also observed in other cetacean species. Mythological accounts chronicle common dolphins assisting humans in distress, as well.

Bow-riding, though a favored dolphin activity, does not give an accurate accounting of how many dolphins occur in an area.

Researchers increasingly rely not only on counting what is seen but also on acoustic sounds. They have done this by dragging a hydrophone behind a research vessel surveying across the Pacific. Signals of some species, including common dolphins, are distinct enough to use as the basis for identifying animals (although the whistles of Striped Dolphins are sometimes confused with those of common dolphins). The common dolphin repertoire of sounds is extensive and includes clicks, whistles, squeaks and creaks. Clicks are used for echolocating, and whistles are usually associated with communication, and are heard more during daylight hours. However, when dolphins are feeding at night, synchronous whistles are often detected, suggesting vocal coordination while they are pursing prey. Whistles are unique to individuals and are important in individual recognition and communication between dolphins. Learned from infancy, each individual produces a distinct whistle that others recognize and to which they respond.

REPRODUCTION: Information on mating and calving is inconsistent for Short-beaked Common Dolphins, in part because common dolphins were only recently divided into two species. Nevertheless, more is known about the reproduction of common dolphins than is known for most other small cetaceans. Common dolphins are sexually mature from age 7 to 12 years (average is 9) in males and 6 to 12 (average is 8) in females. The mating system has not been studied, but their modest sexual dimorphism suggests a promiscuous mating system. In the northern regional group of common dolphins, in California, there is a seasonal pulse in mating and calving, with a peak in births from June through September. Other studies, though, have documented two calving periods: March to May and August to October. In the tropical waters of Baja, the southern group of common dolphins gives birth year-round.

The gestation period is 11 to 12 months, and calves nurse for 6 months, though nursing for 16 to 17 months has also been noted. Regardless of weaning dates, calves can eat solid food after only 2 to 3 months. The birthing cycle is not particularly long compared to some other dolphins', occurring every 2 to 3 years. Parental care, though, is augmented by "auntie" females that assist with young calves.

FOOD: The Short-beaked Common Dolphin feeds mostly at night, in small groups, targeting prey in the deep-scattering layer. Their prey base associated with that layer is composed of midwater fish, squid, and crustaceans that all migrate to the surface at night, and that then retreat to deep water during the day. During the day, the deep-scattering layer is deep (365 m or 1200 ft), but at night this assemblage of prey rises to the surface (55 m or 180 ft), where they are accessible to dolphins. The dolphins can pursue prey down to 280 m (920 ft) but usually forage at 9 to 50 m (30 to 164 ft).

At night, dolphins feed mostly on lanternfish and deep-sea squid associated with the deep-scattering layer, but they also feed during the day on surface-schooling fish, such as herring, anchovies, mackerel, flying fish, Market Squid, and even pelagic red crabs.

Not surprisingly, what they eat varies with season and location, with over 30 species of fish and cephalopods identified in their diet in California; common dolphins are very flexible in switching prey to whatever is locally abundant. In fall and winter, prey consists mostly of anchovies, Market Squid, and hake, but in spring and summer, their diet is more diverse and includes some crustaceans but mostly deep-sea smelts, squid, and lanternfish. One study in the Southern California Bight determined that 62 percent of the diet was anchovies from September through January, when anchovies move into the Bight to feed. In the eastern tropical Pacific, common dolphins feed with spotted and Spinner Dolphins in association with Yellowfin Tuna. This tuna–porpoise association is not well understood, but possibly the tuna help the dolphins find prey.

MORTALITY: Sharks and Killer Whales, as well as False Killer Whales, Pygmy Killer Whales, and Short-finned Pilot Whales, likely attack and eat common dolphins, but bite marks from sharks are rarely observed on dolphins. The dolphins are just as likely to be felled by a lowly nematode that causes brain lesions, and in California and Oregon some 14 genera of parasites were identified from stranded dolphins.

Remoras also attach to dolphin's heads, fins and backs, irritating the dolphin's skin, and slowing their ability to swim, but not likely causing mortality.

DISTRIBUTION: Common dolphins have the widest distribution of all the whales and dolphins of the world, ranging across all the temperate and subtropical regions and throughout the Pacific, Atlantic, and Indian oceans. Overall their range is between

SHORT-BEAKED COMMON DOLPHIN

latitudes 40° N and 50° S, and west to 135°, but this distribution is not continuous. The core range is from Southern California south of Point Conception to tip of Baja and into the Sea of Cortez, and south to Central America. One dead dolphin was recorded at Victoria B.C., but live common dolphins only occur north of Santa Cruz occasionally. If new species are distinguished within this dolphin complex, though, the current range could change for the common dolphin.

In the Eastern Pacific, the National Marine Fisheries Service recognizes and manages three stocks, based on distinct breaks in their distribution ranging from British Columbia down to Chile: the three are northern common, central common, and southern common. A resident stock ranges around Baja California islands and the Channel Islands, called the central stock; and there may be two separate stocks off California.

Dolphins migrate north and farther offshore in the spring and summer, but these movements are strongly influenced by ocean conditions and change both seasonally and between years. They occur in a surprisingly wide range of oceanic waters, with sea surface temperatures of 10 to 28 degrees C (52 to 88 degrees F). Daily onshore–offshore movements may also be influenced by the sun. In the Eastern Pacific, common dolphins' seasonal movements are mostly north–south and are associated with prominent bottom features such as ridges and seamounts.

Generally, Short-beaked Common Dolphins inhabit both coastal and oceanic waters of the Pacific, and they occur along the continental slope at depths of 100 to 200 m (328 to 656 ft), further from shore, and in cooler water, than Long-beaked Common Dolphins. They particularly are attracted to seafloors with high relief where local upwelling of currents occurs, and they avoid sea bottoms that are barren and flat plains. When researchers followed radio-tagged dolphins in the Southern California Bight, they found that dolphins traveled along routes where there were bottom features such as seamounts, canyons and escarpments, particularly the Santa Rose–Cortes Ridge. Local upwelling of cooler waters occurs at these locations, in contrast to the surrounding tropical warmer water in the area, yielding higher biological productivity.

Between 1979 and 1991, there was a shift in the abundance of common dolphins from the eastern tropical Pacific north into California. There are also seasonal shifts in abundance of dolphins: for example, in 1991 to 1992 they were seen north of Point Arguello in the winter, especially in offshore areas, and were nearly absent there in summer. In Santa Monica Bay, though, common dolphins are present year-round far from shore, and they are present year-round in the Southern California Bight but are seasonally more abundant there in winter than in summer.

VIEWING: Short-beaked Common Dolphins can be seen inshore and further offshore from watercraft throughout the Southern California Bight and offshore of Baja, and commonly bow-ride.

CONSERVATION: Common dolphins have been kept in captivity in oceanariums, but they are not as easily trained as Common Bottlenose Dolphins. Instead, we know much of what we do about common dolphins because they were one of five species trapped and drowned in tuna purse seine nets in the Eastern Pacific. When this became known, the American public was faced with eating tuna fish that was caught in the same nets that drowned thousands of dolphins. The slaughter of common dolphins along with spotted and Spinner Dolphins in the 1960s

population have been estimated because of the large changes in number each year.

COMMENTS: New species of dolphins are likely to emerge from future genetic analyses of common dolphins because of their worldwide but disjunct distribution. Additionally, the range of common dolphins may shift with climate change, since in the past their presence off central California has been linked to warmwater years; this, in turn, may lead to new disjunct populations.

REFERENCES: Bearzi 2005, Danil and Chivers 2007, Evans 1994, Heyning and Perrin 1994, Hui 1979, Oswald et al. 2003, Perrin 2002b, Perryman and Lynn 1993.

LONG-BEAKED COMMON DOLPHIN · *Delphinus capensis*
Pls. 14, 15; Fig. 85.

Adult male average length is 2.0 to 2.5 m (6 to 8 ft). Adult female average length is 1.9 to 2.2 m (6 to 7.2 ft). Average weight for both is 135 kg (362 lbs). Calf average length is 80 cm (32 in.).

SCIENTIFIC NAMES: British naturalist John E. Gray first identified the Long-beaked Common Dolphin in 1828, from a specimen collected at the Cape of Good Hope, and he named the new species *Delphinus capensis*, meaning "dolphin belonging to the cape" in Latin. Although all common dolphins were subsequently lumped together as *Delphinus delphis*, of the Delphinidae family, some researchers continued to identify distinct forms of common dolphins, including one called the long-beaked form. In 1994, long-beaked dolphins were finally recognized as a species separate from the common dolphin when differences were conclusively demonstrated through genetic and morphological analyses. Presently, the Short-beaked Common Dolphin retains the name of origin, *Delphinus delphis*, and the Long-beaked Common Dolphin is recognized as *Delphinus capensis*.

COMMON NAMES: English common names include Long-beaked Common Dolphin, Saddleback Dolphin, Neritic Common Dolphin, Cape Dolphin, and Baird's Dolphin. Hourglass Dolphin and Crisscross Dolphin are also applied to the species in combination with the Short-beaked Common Dolphin.

DESCRIPTION: Long-beaked Common Dolphins have the shape of a prototypical dolphin of medium size and spindle shape (fig. 85). Their forehead, or melon, rises abruptly from the beak and is separated with an obvious crease along the base. The beak is long, up to 34 cm (13 in.), compared to the short-beaked dolphins. Consequently, they have more teeth, and the dolphins in California generally have 47 to 60 pairs of sharp, pointed, conical teeth.

Figure 85. *Upper:* Long-beaked Common Dolphin, immature. *Lower:* Regional geographic forms have different markings, as evident in this adult Long-beaked Common Dolphin.

Both long-beaked and Short-beaked Common Dolphins have the boldest color pattern of any members of the dolphin family. The hourglass pattern with white belly patch is not seen in any other dolphin genus. Seen from the side, the hourglass is composed of four sections: the front half of the body to the dorsal fin is pale yellow, cream, or gray-green above a white belly, and the rear half of the body behind the dorsal fin is a light gray above to darker gray below. Across the back is draped a cape of dark black or brown-black that reaches from the melon to just behind the dorsal fin.

Three very distinct stripes also grace the sides of the long-beaked dolphin: a wide stripe that extends from the base of the melon, under the eye, to the anus; a second wide dark stripe connecting the flipper to the beak; and a third stripe linking a dark eye patch to the base of the melon. The lips of the beak often are black. The dark cape, eye patch, and stripe to the flipper sometimes merge, giving the appearance of a dark mask to the face. The flippers and dorsal fin are slate gray but often have a white patch in the middle with an outline of dark gray. The flippers are slender with pointed tips. The dorsal fin is slightly falcate with a pointed tip. The tail is scalloped in shape, with a notch in the middle. Color patterns vary by region, and there is also an unusual color morph that lacks any yellow color on the side; instead, the hourglass pattern is dark to light gray. Other color phases include all-black and all-white. Calves are dull, and juveniles are less distinct compared to the adult coloration, both of which can increase the difficulty of positively identifying this species.

Males are slightly larger than females; the largest recorded weighed 235 kg (630 lbs). The different sexes are not easily distinguished in the field, even when viewed from the bow of a vessel. Adult males may have a broad dark blaze across the genital area that is sometimes linked to the dorsal cape with a stripe, and some have a hump behind the dorsal fin, also seen in other male dolphins.

FIELD MARKS AND SIMILAR SPECIES: Long-beaked Common Dolphins and Short-beaked Common Dolphins cooccur in southern California and are very difficult to tell apart, even for those trained in discerning their differences. They do not cooccur in mixed herds, though they mix with other species of dolphins.

The Long-beaked Common Dolphin has a long, spindle-shaped body with a narrow beak, much like the Short-beaked

Common Dolphin and other dolphins of the subtropical Pacific. Their bold and distinctive hourglass marks and white belly are unlike the markings of other oceanic dolphins in the California Current except for the Short-beaked Common Dolphin. At sea, the two can be separated by the longer beak and sleeker body of the Long-beaked Common Dolphin, which also has two distinctive dark stripes that are absent in the short-beaked species: one from the eye to the anus and one from the eye to the flipper. The melon is slightly smaller, and less rounded, than that of the short-beaked dolphin.

Both common dolphins are often confused with the Striped Dolphin, which also has a line from the eye to the anus and another from the eye to the flipper, but the common dolphin has a yellowish color above the stripe, whereas the stripes of the Striped Dolphin clearly demark the darker cape from the lighter belly. Also, the common dolphins lack the blaze between the head and the dorsal fin that is often present in the Striped Dolphin. The Fraser's Dolphin also has an eye-to-anus stripe that is very thick, but this species' beak is short compared to that of the Long-beaked Common Dolphin. From a distance, common dolphins may be confused with Pacific White-sided Dolphins, but the head shape and falcate dorsal fin of the Pacific White-sided Dolphin are easy marks with which to differentiate from a common dolphin's straight dorsal fin and long beak.

NATURAL HISTORY: Long-beaked Common Dolphins are very gregarious and abundant, with a worldwide but disjunct distribution, and with populations regionally isolated. They are also the most frequent dolphin encountered by boaters, because they are generally limited to coastal waters. Like their nearest relative the Short-beaked Common Dolphin, they are very energetic and frequently pursue vessels to bow-ride. They are fast swimmers, capable of bursts of speed of perhaps 40 kph (25 mph), although their normal swimming speed is 10 kph (6 mph). Common dolphins also excel at leaping, achieving heights of 7 m (23 ft). Acrobatic aerial displays include "pitch poling," whereby a dolphin leaps straight out of the water and then falls backward, creating a substantial splash. Their entries into the water fall into distinct styles ranging from a clean entry with no splash to a belly flop with an enormous splash. Perhaps many of these landings are performed merely for fun, but the distinctive chin slap with tail lob may communicate information to other dolphins. The chin slap involves slapping the chin on the surface of the water and creating a splash; similarly, a lob tailing uses the tail to create a splash.

When traveling or feeding, common dolphins engage in synchronous swimming that includes surfacing and diving side by side. Their distinctive side markings may aid them in swimming together, as has been demonstrated in large schools of fish, perhaps ensuring that neighbors do not collide while swimming at great speeds. Schools of common dolphins in the eastern Pacific are usually composed of 100 to 500 individuals, but they can assemble into larger schools of several thousand during the day. The schools separate into smaller feeding groups at dusk and throughout the night; these smaller groups feed in the deep-scattering layer that rises to the surface at night. At dawn, the small groups rejoin one another again into larger schools, mostly socializing and resting together during the day. Long-beaked Common Dolphins occur in smaller schools of 10 to 30, a fact perhaps related to family units or to segregation by sex or age groups. In the eastern Pacific, researchers discovered that some small social units are genetically related.

Long-beaked Common Dolphins and Short-beaked Common Dolphins cooccur in Santa Monica Bay but do not school together. This lack of mixing between the two common dolphin species is particularly interesting because they both feed on similar prey, such as abundant anchovies. Apparently enough differences exist between their diets and feeding strategies that they do not feed together. Long-beaked Common Dolphins do mix with other species such as Common Bottlenose Dolphins and Pacific White-sided Dolphins.

Bow-riding is play behavior enjoyed by many dolphin species, but common dolphins indulge in play more than most, stopping whatever else they are doing to dash over to meet a vessel, even from great distances. Large groups of bow-riders jockey for best position at the bow, layering two deep or more, depending on the size of the vessel. Not all individual common dolphins bow-ride, though; younger animals may be more playful than adults. Common dolphins vocalize while bow-riding, and their squeaks can be clearly heard by the human ear. The range of sounds produced by common dolphins includes clicks, whistles, squeaks and creaks. The clicks indicate that the dolphins echolocate to detect prey or objects, and the whistles are associated with communication. Whistles are unique to individual animals and apparently are used to communicate individual recognition between dolphins.

REPRODUCTION: Much of what is known about Long-beaked Common Dolphin reproduction is jumbled with information on Short-beaked Common Dolphins. Nevertheless, more is known about their reproduction than about most small cetaceans, because they occur nearshore and therefore are easier to study. Common dolphins of both species are sexually mature from age 7 to 12 (average is 9) in males and 6 to 12 (average is 8) in females. Males are only slightly larger than females, a very moderate sexual dimorphism that suggests a promiscuous mating system. In California, there is a seasonal period in mating, from spring to fall, followed by a gestation period of around 11 months, so that calving occurs during the following spring-to-summer period. In the more tropical waters of Baja, Long-beaked Common Dolphins likely give birth year-round. Calves are fairly small, but "auntie" females help rear and protect young. Calves nurse for 6 months but can eat fish and squid after only a couple months. The birthing cycle for an individual female occurs every 2 to 3 years, and in a lifetime an average female that lives to 40 years of age may rear perhaps nine calves, assuming that she calves regularly. However, oceanic conditions that affect food availability may have significant effects on female condition, resulting in longer periods between calving.

FOOD: The Long-beaked Common Dolphin has a diverse diet and is very flexible in switching to whatever food is locally abundant. This species may not be as dependent on prey in the deep-scattering layer as the Short-beaked Common Dolphin, because it occurs closer to shore, but what is known about its feeding habits has been combined in with the short-beaked dolphin and is therefore less than completely reliable. Long-beaked Common Dolphins can pursue prey down to greater than 200 m (656 ft) and likely feed on midwater prey such as lanternfish, hake, and deepwater smelt and squid. In the surface waters, they forage on schooling fish, such as herring, anchovies, mackerel, and sardines, and usually feed cooperatively, herding schooling prey into balls. Their diet also includes squid, such as Market Squid, and crustaceans, such as pelagic red crabs and krill. What they eat varies with season and location. In the Southern California Bight, more than 50 percent of their diet is anchovies in fall and winter, when anchovies are abundant. In the eastern tropical Pacific, some Long-beaked Common Dolphins feed with spotted and Spinner Dolphins in association with Yellowfin Tuna, though this association is less common than for other species of dolphins.

MORTALITY: The usual predators, including sharks, Killer Whales, False Killer Whales and Short-finned Pilot Whales, likely attack and may eat common dolphins, although bite marks from sharks are only rarely observed on dolphins. They are just as likely to be killed by a diminutive nematode that causes brain lesions; in California and Oregon, some 14 genera of parasites have been identified from stranded dolphins. Remoras also attach to dolphins' heads, fins and backs, irritating their skin and slowing their ability to swim; this is not likely, however, to cause mortality.

DISTRIBUTION: Long-beaked Common Dolphins are broadly distributed throughout the warm temperate and tropical coastal oceans of the world, but their distribution is disjointed into

■ Range

LONG-BEAKED COMMON DOLPHIN

regional concentrations with distinct breaks between. Central California to Baja is one such regional concentration, within a wider region that extends south to the eastern tropical Pacific, down to the Gulf of California and Mexico. A separate regional concentration occurs off South America, around Peru.

Whatever the region, Long-beaked Common Dolphins are associated with warmer and shallower water than Short-beaked Common Dolphins are, and are usually seen within 100 km (54 mi) of the coast. Generally, they inhabit coastal waters of the Pacific along the continental slope and shelf, within a depth range of 183 m (600 ft). They particularly are attracted to sea bottoms with high relief where local upwelling of currents occurs, and they avoid barren and flat plains. When researchers followed radio-tagged dolphins in the Southern California Bight, they found that dolphins traveled along routes where there were bottom features such as seamounts, canyons, and escarpments. Dolphin movements are strongly influenced by ocean conditions and change seasonally and between years. Daily movements are mostly associated with prominent bottom features such as ridges and seamounts.

Regionally, there may be shifts in abundance. In the Southern California Bight and Santa Monica Bay, while Long-beaked Common Dolphins are present year-round, they may be seasonally more abundant in winter than summer.

VIEWING: Long-beaked Common Dolphins cling closely to land, so people often see them from watercraft in inshore waters, especially around canyons and escarpments in the Southern California Bight.

CONSERVATION: Common dolphins, including Long-beaked Common Dolphins, have been kept in captivity in oceanariums and have even been hybridized with Common Bottlenose Dolphins. Much of what is known about Long-beaked and Short-beaked Common Dolphins has been learned by examining drowned dolphins in the tuna purse seine fishery in the eastern Pacific. Common dolphins—both species combined—were the third most frequently killed dolphin in the tuna fishery, with a total estimate of around 64,000 drowned between 1979 and 1993. The slaughter of common dolphins, along with Spotted Dolphins and Spinner Dolphins, put a dolphin face on the tuna consumed in the United States and elsewhere. The common dolphin, although not killed in as large a number as the other species, was more familiar to the public because of its mythology and worldwide distribution. Because of this, people changed the way they viewed marine conservation, which in turn eventually led to the Marine Mammal Protection Act of 1972, as well as other, international regulations. Today, this sector of fishery has reduced the number of dolphins drowned significantly to an average of around 100 per year in the period 2000 to 2004.

Agreement	Conservation Status
U.S. ESA	Not listed
U.S. MMPA	No special status
COSEWIC	Candidate
IUCN	Data deficient

Long-beaked Common Dolphins are also drowned in U.S. drift gillnets set for sea bass, halibut, shark, and swordfish, but the number thus drowned is low. Between 2002 and 2006, around 17 Long-beaked Common Dolphins stranded in California, with evidence of fisheries interactions with entangling nets, and hook and lines.

In coastal waters worldwide, indirect and direct hunting likely have reduced regional populations of Long-beaked Common Dolphins. Off Peru, common dolphins are occasionally still hunted for bait meat for use in shark fisheries.

Harmful algal blooms that produce domoic acid cause mortality in Long-beaked Common Dolphins, which succumb to this biotoxin when eating tainted schooling fish, particularly anchovies and sardines. Because they are a coastal species, they may be more susceptible to this biotoxin than more offshore species of dolphins.

POPULATION: For management purposes, the NMFS recognizes only one stock that inhabits the U.S. waters from California to Washington, and another stock south of California in the eastern tropical Pacific; members of the latter are also called the Baja Neritic Common Dolphins. The California stock is estimated at 15,335 (the average for two years, 2001 and 2005). Population estimates are based on an average number over multiple years of surveying because of high annual variability in numbers. No estimate is available for the stock in the eastern tropical Pacific south of California; rather, the species is lumped in with the northern common dolphin population estimate. Consequently, mortality estimates from fisheries interactions for Long-beaked Common Dolphins are not known. No trends in the population have been estimated because of the large changes each year; in 2001 the estimate for California was 20,076, but in 2005 it was only 11,714.

COMMENTS: The Long-beaked Common Dolphin's relatively new status as a species will stir the interest of many a researcher to make new discoveries, reconsider previous assumptions, and tease apart the subtle differences from its relative, the Short-beaked Common Dolphin.

REFERENCES: Bearzi 2005, Evans 1994, Heyning and Perrin 1994, Hui 1979, Oswald et al. 2003, Perrin 2002b, Perryman and Lynn 1993.

FRASER'S DOLPHIN
Lagenodelphis hosei
Pls. 12, 13; Fig. 86.

> Adult average length is 2.7 m (8.9 ft); average weight is 210 kg (460 lbs). Newborn average length is 1 m (3.3 ft).

SCIENTIFIC NAMES: In 1895, Ernest Hose collected a beachcast dolphin at the mouth of the Lutong River in Borneo. His naturalist brother Charles, an administrator for the Rajah of Sarawak, later donated its skeletal remains to the British Museum. In 1956, retired Indian Army surgeon and zoologist F.C. Fraser, who worked at the museum on both dragonflies and cetacea, identified this as a new species. Fraser thought the skull shared features of *Lagenorhynchus* and *Delphis,* and coined the new generic name *Lagenodelphis,* combining the names of the two older genera (fig. 86). The Greek roots of this name mean "flagon dolphin." Fraser named the species *hosei* to commemorate Charles Hose.

COMMON NAMES: Fraser's Dolphin is also called the Sarawak Dolphin or Borneo Dolphin, White Porpoise, Short-snouted Dolphin, and Short-snouted White Porpoise. In French, it is rendered the *dauphin de Fraser,* and in Spanish, *delfin de Fraser.*

Figure 86. Pale immature Fraser's Dolphin, lacking the distinctive stripe.

Fraser's Dolphin is unusual in having one person honored in its scientific name and a second in its common name.

DESCRIPTION: Fraser's Dolphin is stocky, with very small flippers, fin, and flukes, looking more like a miniature whale than a dolphin. It has a stubby but well-defined beak, 3 to 6 cm (1.2 to 2.4 in.) long, with 34 to 44 pairs of sharp, slender teeth in each jaw. Its dorsal fin may be falcate or triangular. The short fin becomes more erect as dolphins age, particularly in males. Most males develop a ventral hump before the anus when mature.

Fraser's Dolphins are often boldly patterned in alternating light and dark bands. A brownish or bluish gray cape drapes its back from head to halfway beyond the dorsal fin. The cape blends into a lighter grayish band from the melon to the tailstock. Below that, a striking dark stripe runs from the eye to the anus in many animals. The stripe is wider and more intense in older males. The stripe is variable in adult females. In some individuals, the black stripe joins with other dark facial markings and gives the impression of a raccoon's or bandit's mask. Another thinner dark stripe extends from the chin to the beginning of the flipper. The lower sides of the Fraser's Dolphins are creamy; the belly is white or pink. The characteristic dark bands are muted or absent in immatures and newborns and vary regionally. The flippers, dorsal fin, and flukes are all dark gray, though sometimes there is a pale spot in the center of the dorsal fin.

FIELD MARKS AND SIMILAR SPECIES: The Fraser's robust body and petite flippers, fins, and flukes should separate this species from other dolphins. The bold, dark eye-to-anus stripe is distinctive and is similar to that of the Striped Dolphins—but Striped Dolphins have longer beaks and larger appendages.

NATURAL HISTORY: A mostly oceanic dolphin, the Fraser's Dolphin has been seen in tight groups ranging from three to 1,000 or more. These schools, which are frequently large, may move quickly, with the dolphins splashing as they porpoise at low angles. Speeds of up to 28 kph (17 mph) are reached by those fleeing boats or pursuing prey. Sometimes Fraser's Dolphins form a broad "chorus" line as they move across the seas. These dolphins especially congregate with Melon-headed Whales in the eastern tropical Pacific and the Gulf of Mexico, and they sometimes mix with other blackfish such as False Killer Whales, Risso's Dolphins, and Short-finned Pilot Whales. Occasionally, they feed at the surface, where they feed alongside terns, swallowlike seabirds. These dolphins are not bow-riders, except briefly, perhaps because they have been chased by boats trying to harpoon them or encircle them in tuna–porpoise nets.

Seven opportunistic recordings of this infrequently heard dolphin reveal simple calls with few if any changes in inflection. The calls vary up to 0.93 seconds in length and fall mostly within the range of human hearing. Surprisingly, calls from the Pacific resemble those from the Atlantic, with no evidence of separate dialects.

REPRODUCTION: Not much is known about the reproduction of this dolphin. Analysis of 108 Fraser's Dolphins captured in a drive at Taiji in Japan indicated that males became sexually mature at 7 to 10 years and 220 to 230 cm (87 to 91 in.), and females at 5 to 8 years and 210 to 220 cm (83 to 87 in.). Based on like species, gestation is probably 11 to 12 months; the estimated calving interval is 2 years. In Japan, calving peaks in spring, and probably in fall, but in South Africa the peak may be in summer. No calving information is available for the eastern Pacific.

FOOD: Fraser's Dolphins eat a wide variety of fish, crustaceans, and cephalopods. In the southeast Caribbean, two schools of Fraser's Dolphins were observed cooperatively preying on fish near the surface. One school of 60 dolphins formed smaller groups that drove, circled, and caught rainbow runners, a common species of pelagic fish. Another school of 80 dolphins acted as a tightly organized group that chased near-surface fish. The number of short broadband clicks and whistles increased as hunting groups passed by other dolphins. Clicks were most often heard during hunting. In the eastern tropical Pacific, Fraser's Dolphins eat fish common at 250–500 m (800 to 1,600 ft), likely those associated with the deep-scattering layer. These midwater fish and squid are their primary prey, represented in over 30 families. To a lesser degree, they eat cuttlefish, isopods, and some bottom fish.

MORTALITY: Although no predation has been witnessed, Killer Whales, False Killer Whales and large sharks may prey on Fraser's Dolphins. Cookie-cutter sharks wound Fraser's Dolphins but do not cause mortality. Fraser's Dolphins also strand onshore, as individuals and in groups up to 17 individuals, for unknown reasons. Parasites such as trematodes that plug the middle ear and blow holes have been documented in some but not all incidents.

DISTRIBUTION: Uncommonly seen, Fraser's Dolphins have a wide but unevenly reported distribution in tropical and subtropical waters straddling the equator between latitudes 30° north and south. Strandings occur outside these limits, such as in Britain, and may reflect periodic oceanic variations, but along the west coast of Canada and the United States, there are no stranding records. Fraser's Dolphins usually favor the open ocean, and come near land when deep water approaches the coast. In the eastern Pacific, they gravitate to upwelling zones in the offshore eastern Pacific, where there are more opportunities for abundant prey, and around oceanic islands, where they often gather during the day while resting and socializing. They are rarely seen, however, in the waters of the California Current.

CONSERVATION: Fraser's Dolphins are taken in drive fisheries for food in Japan and Taiwan, and by harpoon and spear in Sri Lanka, the Philippines, Indonesia, and the Lesser Antilles. Whether this small-scale, largely traditional whaling has extensive effects on local or regional populations is unknown. Incidentally, Fraser's Dolphins have been caught in tuna purse seines, gillnets, driftnets, and trap nets in the Pacific, but numbers killed have been low. Some are even killed by antishark nets in South Africa. Like all

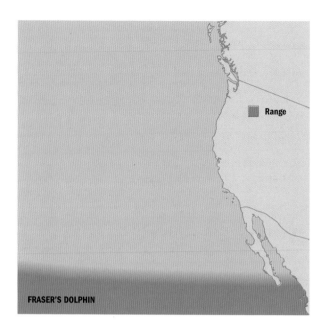

FRASER'S DOLPHIN

Range

cetaceans, Fraser's Dolphins bear a burden of organochlorines, such as DDT, PCBs, chlordane, and other pollutants.

Agreement	Conservation Status
U.S. ESA	Not listed
U.S. MMPA	No special status
COSEWIC	Not listed
IUCN	Least concern

POPULATION: The NMFS estimates the Hawaiian stock of Fraser's Dolphin to be between 8,000 and 19,000 dolphins. In the eastern tropical Pacific, there may be 100,000 to 289,000 Fraser's Dolphins, but there are no estimates for the U.S. waters of California.

COMMENTS: The sun may have set on the British Empire, but its legacy, and those of the Russian and French Empires, carries on. Many maritime naturalists from the 18th through the early 20th century accompanied research vessels or were otherwise supported by government positions aboard ships. Their individual interests were wide-ranging and lifelong. Many of us feel a bond with these adventurers and may want to join the expeditions of today's explorers as volunteers, to be part of the timeless search for knowledge of the sea.

REFERENCES: Dolar 2002, Jefferson and Leatherwood 1994, Perrin et al. 1994a.

NORTHERN RIGHT WHALE DOLPHIN
Pls. 14, 15; Fig. 87.

Lissodelphis borealis

> Adult male average length is 2.8 m (9.2 ft) long. Adult female average length is 2.3 m (7.5 ft). Average weight for both is 81 to 113 kg (217 to 303 lbs). Newborn average length is 0.8 to 1.4 m (2.6 to 4.6 ft).

SCIENTIFIC NAMES: The genus name *Lissodelphis* derives from the Greek *lissos,* which means "smooth"—referring to the finless back. The species name, *borealis,* is Latin for northern. Northern Right Whale Dolphins were first described by the American naturalist Titian Peale in 1848 from a specimen he collected off Astoria, Oregon, while on a discovery expedition. Right whale dolphins are members of the Delphinidae family of toothed whales, and the genus includes two species. In the southern ocean, the other species of the genus occurs, the Southern Right Whale Dolphin *(Lissodelphis peronii),* which is generally larger and which has bolder markings but is otherwise similar to its northern cousin.

COMMON NAMES: The namesake of the right whale dolphin is the right whale, which also lacks a dorsal fin. Another common English name is Pacific Right Whale Porpoise. A Spanish name is *delphin liso del norte,* and a French name is *dauphin a dos lisse boral.*

DESCRIPTION: Right whale dolphins are shaped more like eels than dolphins, but there the similarities end, for their finless, slender, streamlined form is sheathed in muscle and blubber,

Figure 87. Northern Right Whale Dolphins; note pattern variation of white.

creating a rigid body best designed for swift swimming (fig. 87). Their very narrow, tapered tail stock terminates with an equally small tail of a mere 33 cm (1 ft), with a deep notch. The forehead is smoothly sloped with no distinct demarcation of the melon, further accentuating the fusiform-shape. The small beak is short, filled with 45—sometimes up to as many as 54—pairs of conical teeth in upper and lower jaws. Small flippers are similarly slender and curved, with pointed tips. Right whale dolphins are small, as dolphins go, with no sexual dimorphism noticeable in the field. Males are slightly larger than females; the largest male was 3.1 m (10 ft) in length, and the heaviest weighed 113 kg (303 lbs).

Their coloration is as if their chest and lower lip had been dipped in white paint. The color pattern has been called primitive: mostly black, but with a distinctive broad white chest patch that narrows as a band to the tail and then spreads across the ventral surface of the tail. The belly band of white is wider in the genital region of females than males, perhaps to attract males. There is a white tip on the lower jaw. The black color can appear brownish black when observed from above, possibly due to reflection of light off the water. Flukes are dark gray in color on the top and white underneath. Individuals can have more white in their coloration, including a white chest extending up to lower jaw, and a white blaze on the side of the head and snout. One color variant is called "swirled." Calves are lighter than adults on their backsides, appearing gray or cream, but this color phase is lost by the first year.

FIELD MARKS AND SIMILAR SPECIES: The Northern Right Whale Dolphin is not likely to be mistaken for any other small cetacean in the California Current, because it is the only one that lacks a dorsal fin. In temperate waters, there are other boldly marked dolphins and porpoises such as Dall's Porpoise, but these are distinguished by their low triangular dorsal fin and robust body. From afar, however, porpoising California Sea Lions may look like Northern Right Whale Dolphins.

Northern Right Whale Dolphins often associate with Risso's Dolphins and Pacific White-sided Dolphins but are easy to separate from the other two. The round-headed, stocky Risso's Dolphin is whitish or light gray, is often scarred, and has a tall dorsal fin, in comparison with the slender, finless Northern Right Whale Dolphin, which is all-black except for a white hour-glass on its belly.

NATURAL HISTORY: A few to a few thousand right whale dolphins can rip the surface of the ocean open in their haste to flee approaching boats, but when in the company of Pacific White-sided Dolphin or the common dolphins, they are often tamed to bow-ride. They are a fast and synchronous swimmer but will create a great disturbance of the surface by their splashes. When evading boats, they swim quickly just below the surface, briefly surfacing to take a quick breath, and may appear similar to a herd of sea lions skimming across the surface, creating much disruption. They are not particularly acrobatic, and, when swimming fast, will make shallow leaps. They can leap up to 25 times in a row, sometimes belly-flopping in the process, creating a froth of commotion on the surface.

In the eastern Pacific, the size of herds averages 110 and, rarely, as large as 2,000. Many herds are composed of 10 to 20 members, but the definition of a herd is loosely based on how many dolphins are observed in a group. Actual herd size, however, can be much larger, based on acoustic communication with dolphins exchanging information over wider distances. As with many species of dolphin, Northern Right Whale Dolphins segregate by age, with juveniles forming separate groups from adults. Little is known, though, about social structure within the herds.

Researchers studying cetaceans from aircraft have noted herds of Northern Right Whale Dolphins in four patterns: a tight group with no subgroups, all moving in the same direction; a scattered group with separate subgroups, but with animals moving in unison in the same direction; a V-shaped line formation in single file, swimming in the same direction and keeping a constant distance from each other; and a "chorus line" formation traveling in the same direction in a line and maintaining a constant distance from each other. Each of these formations is considered to be related to different cooperative strategies for catching prey, but none has been studied closely enough to confirm.

Northern Right Whale Dolphins are very sociable with each other, as well as with other species. They have been observed with 14 other species of marine mammals ranging from California Sea Lions to Humpback Whales. Northern Right Whale Dolphins are often observed with Risso's Dolphins and common dolphins, but Pacific White-sided Dolphins are their most common companion, an association described when the species was first discovered by Peale. Foraging advantages likely explain the multiple species groups, with dolphins joining others in the vast ocean to find food. California Sea Lions join Northern Right Whale Dolphins and Pacific White-sided Dolphins feeding, which may lead to confusion between the species. When seen with large whales, Northern Right Whale Dolphins are often riding the pressure waves created in front of the heads of the swimming whales.

As with other small-toothed, highly gregarious whales, the vocalizations of Northern Right Whale Dolphins are complex. Their vocalizations often consist of "click trains," likely related to echolocation, that are similar to those of the common dolphin. They also produce moans and, rarely, whistles, favoring instead a complex burst-pulse sound that is thought to function the same as the whistle in Common Bottlenose Dolphins, aiding in the identification of individuals or groups.

REPRODUCTION: There is a paucity of information about Northern Right Whale Dolphins' reproduction, although one study has estimated the average age of sexual maturity at 9 to 10 years, regardless of sex. Females ovulate annually and gestation is slightly more than 1 year, but females skip years between calving. As a consequence, they give birth about every 2 to 3 years.

Seasonally, calving peaks in July and August in the North Pacific. Newborns have been recorded off the coast of Big Sur, California, and lactating females have been noted in southern California in September and October.

FOOD: Lanternfish and Market Squid are important prey in southern California, but Northern Right Whale Dolphins eat a variety of midwater and surface-schooling fish and squid. In association with Pacific White-sided Dolphins, they likely prey on herring, anchovies, hake, Pacific Saury, and sardines. From examining the stomachs of four dead beached dolphins in California, some 17 prey species from both offshore and nearshore habitats were identified, but the most common prey were several species of lanternfish (75 percent), followed by California Smooth-tongue (*Leuroglossus stilbius*), a deep-sea smelt that is a midwater fish down to a depth of 690 m (2,300 ft).

Northern Right Whale Dolphins feed in evening and early morning, following the rise and fall of the deep-scattering layer, an assemblage of fish and invertebrates that migrate up and down the water column in response to daylight. They likely hunt midwater and surface prey associated with this oceanic phenomenon. When in pursuit of prey or evading boats, dolphins may swim in bursts of 33 to 45 kph (20 to 30 mph), but average traveling speeds are 28 kph (17 mph).

MORTALITY: Parasitism of dolphins may contribute to their stranding on beaches where scientists can examine them. Parasites damage the air sinuses, the brain, and the intestines. One trematode, *Nasitrema,* infects the brain and causes death, although the effect of this parasite, among others, on the population is not known. No doubt Great White Sharks and Killer Whales prey on right whale dolphins, but there are no documented cases of such attacks in the eastern Pacific.

DISTRIBUTION: Northern Right Whale Dolphins live throughout, and entirely within, the North Pacific Ocean, ranging between latitudes 30° and 50° N, although incidental sightings have occurred as far north as the Kenai Peninsula, in Alaska. The eastern and western populations overlap off the Alaskan Aleutian Islands but otherwise appear separate. In the eastern Pacific, their range is very similar to that of the Pacific White-sided Dolphin, occurring along most of the California Current, though rarely down to lower Baja, at latitude 29° N. In the eastern Pacific, the core of the population appears to be centered on central

Range

NORTHERN RIGHT WHALE DOLPHIN

California between Points Sur and Conception, including the northern Channel Islands. There are fewer sightings north or south of this core area, except seasonally in winter, when they are seen more often in the Southern California Bight.

Seasonal movements are tied to seasonal changes in sea temperature, but there are great differences between years. Most sightings are linked with temperatures 8 to 19 degrees C (46 to 66 degrees F), and as ocean temperatures change with the seasons, the Northern Right Whale Dolphins move north or south. They concentrate off California during late winter and spring and then range north along Oregon, Washington, and British Columbia during summer, beginning in May. In winter, they generally congregate in southern California from October to May, with the highest numbers in January; they are absent from southern California the rest of the year. During the winter, they may even travel as far as Baja when temperatures drop below 19 degrees C (66 degrees F), and then return north along central and northern California in spring and early summer. Northern

Right Whale Dolphins move inshore and offshore, too, following the movements of Market Squid that migrate inshore in winter to spawn, and offshore in spring.

Northern Right Whale Dolphins are oceanic but most common over the continental shelf and break, at depths of 1,000 m (3,300 ft). They will venture closer to shore where there are submarine canyons, such as in Monterey Bay, and around islands such as those in the Southern Californian Bight. Much less is known about their affiliations midocean, where they disperse widely.

VIEWING: Northern Right Whale Dolphins are attracted to boats when with Pacific White-sided Dolphins and so often come to whale-watchers. They are commonly seen on whale-watching trips in the Gulf of the Farallones and Monterey Bay, and in the Southern California Bight, around the Channel Islands, in winter.

CONSERVATION: As with several species of small cetaceans, Northern Right Whale Dolphins were historically fished for food off the coast of Japan. They also were drowned as bycatch in the North Pacific squid driftnet fishery. This fishery likely devastated the population, with one estimate of up to 24,000 killed per year. The fishery was subsequently stopped in 1992 by the International Whaling Commission, but not before the population was depleted to as low as 24 percent of their original abundance. The fishery caused exceptionally high mortality, because the dolphins were caught in large groups while they were feeding on squid, and whole families were killed, either directly in the net or as abandoned young.

Agreement	Conservation Status
U.S. ESA	Not listed
U.S. MMPA	No special status
COSEWIC	Not at risk
IUCN	Least concern

Another fishery for swordfish and shark using drift gillnets also drowns Northern Right Whale Dolphins, but such incidents may have declined when pingers were installed on the nets in 1997. Nevertheless, the effect on the eastern Pacific population is not known, and the estimated mortality associated with fisheries may be significant, much as with the Pacific White-sided Dolphins with whom they often travel.

POPULATION: The recent estimated population averaged over 5 years at the turn of the 21st century was 16,400, and in 2007 was 15,300, but there is no indication of a trend in the population. For the entire North Pacific, and thus for the entire species, the population is estimated at 68,000.

COMMENTS: Northern Right Whale Dolphins are very social, associating with an exceptional number of other species, and yet their body form is so very different from that of the others that we may wonder what draws them together. Species intermingle usually because of what and where they eat; perhaps Northern Right Whale Dolphins take advantage of the food-finding abilities of other species as well as indulging in the pleasures of bow-riding with Pacific White-sided Dolphins.

REFERENCES: Ferrero and Walker 1993, Jefferson and Newcomer 1993, Leatherwood and Walker 1979, Lipsky 2002.

DALL'S PORPOISE
Phocoenoides dalli

Pl. 16; Figs. 49, 88.

> Adult average length is 1.8 m (6 ft), with a maximum of 2.3 m in males (7.5 ft) and 2.1 m (7 ft) in females. Average weight for both is 123 kg (270 lbs), with a maximum of approximately 200 kg (440 lbs). Newborn average length is 1 m (3 ft).

SCIENTIFIC NAMES: *Phocoenoides* is derived from the Greek *Phocoena* for "porpoise" and *oides* for "like." This species was named *dalli* to honor William H. Dall, the pioneering American naturalist who collected the first specimen, in Alaska, 1873. Dall described many new species, including the Dall's Sheep of western Canada and the United States.

COMMON NAMES: The English common name is Dall's Porpoise. A distinctively marked subspecies in the western Pacific was called True's Porpoise and was once proposed as a separate species, *Phocoenoides truei*. Older English synonyms for Dall's Porpoise include White-flanked Porpoise and Spray Porpoise.

Figure 88. Dall's Porpoise.

The French and Spanish names, *marsouin de Dall* and *marisopa de Dall*, respectively, are similar to the current English name.

DESCRIPTION: This lively toothed whale has a stocky body that is usually less than 2 m (6.5 ft) long (fig. 88). There are only slight differences between males and females, with the male larger and more robust. The back of the body, before the flukes, is humped above and below, with the ventral hump more prominent in males. The body is black or dark gray with white flank patches that join at the belly.

The flippers are small, well forward, and slightly rounded at the end. The flukes and dorsal fin are typically tipped with grayish white "frosting"; the dorsal fin's flecking is darker. The rounded head looks small relative to the powerful, deep body. The beak is not clearly distinct from the melon. From above, the head appears triangular. The short, triangular dorsal fin has a broad base and may be canted forward in large adult males. The tail stock is keeled above and below, and the notched flukes are small, both of which may have white tips.

The beak is short and narrow, with the lower jaw extending slightly beyond the upper. There are 19 to 23 pairs of teeth in the upper jaw, and 20 to 24 pairs in the lower. The spade-shaped teeth, the size of grains of rice, are the smallest of all cetaceans'. The tiny teeth are separated by horny, rigid protrusions called gum teeth, and a dental pad in the upper jaw, both of which may aid in grasping prey.

FIELD MARKS AND SIMILAR SPECIES: Dall's Porpoises might be mistaken for Harbor Porpoises at a distance. Close up, their striking black-and-white pattern distinguishes them from Harbor Porpoises, with their dark gray bodies and lighter gray bellies. In the inland waters around southern Vancouver Island, hybrids between Dall's Porpoises and Harbor Porpoises are regularly seen; sometimes they associate with Dall's Porpoises. These hybrids are a lighter gray than Harbor Porpoises but do not bear the Dall's Porpoises' white flanks or frost-tipped dorsal fins and flukes. Tissue studies indicate that all hybrids sampled were the offspring of Dall's Porpoise mothers. Perhaps Dall's Porpoise mothers cannot discriminate between suitors of the two species, or perhaps Harbor Porpoise males indiscriminately pursue Dall's Porpoise females. Mammalian hybrids are quite rare, but some hybrid Dall's Porpoises have been sighted on a majority of research cruises seeking to document them.

Both the Dall's Porpoise and the Pacific White-sided Dolphin generate spray around themselves as they swim quickly at the surface. The stocky body form and bold black-and-white coloration of Dall's Porpoises separate them from the more streamlined Pacific White-sided Dolphin, with its tall, hooked dorsal fin and complex pattern of gray stripes. It should be noted that three other rare color morphs of Dall's Porpoises apparently exist, including all-black, all-white, and gray forms; these have been very infrequently sighted along the west coast of North America.

NATURAL HISTORY: The Dall's Porpoise is one of the most common and easily identified cetaceans in the cool temperate waters of the west coast of North America. It also may be the fastest of all cetaceans, at least in short bursts, reaching speeds of 55 kph (34 mph). At times, Dall's Porpoises dash along the surface, pushing a rooster tail of spray around them, plowing a furrow in the sea. The pressure wave created in front of large, fast-moving whales such as Humpback Whales and Fin Whales may attract the Dall's Porpoise, with its inclination to bow-ride; the porpoises also pursue fast-moving freighters. They sometimes materialize unexpectedly ahead of fast motorboats and then bow-ride, zigzagging erratically in the vessel's path; they prefer boats that travel at least 26 kph (16 mph). Despite their penchant for speed, Dall's Porpoises are not acrobatic, and they rarely breach or porpoise at the surface. More commonly, Dall's Porpoises move quite slowly and roll quietly on the surface as they forage or rest in small schools of two to 20. Occasionally, larger groups form that may number a hundred or more, with aggregations into the thousands being reported. Dall's Porpoises also sometimes school with Pacific White-sided Dolphins and pilot whales.

Despite its abundance, not much of this porpoise's life is well understood, perhaps because the animals are hard for researchers to study or catch. Their maximum age is an estimated 22 years, based on tooth rings.

REPRODUCTION: The age of sexual maturity varies with subpopulation and has been reported at 5 to 8 years for males and 3.5 to 7 years for females. Gestation lasts 10 to 11 months, and it is likely that delayed implantation occurs, because calving takes place annually in spring and summer, with a peak in June and early July. Calving times, however, vary according to stock. Calving intervals also vary among different populations. In some areas, females may calve each year, unlike most cetacean species, but in others, females calve at 3-year intervals. Widely different lactation periods have been reported, but for Dall's Porpoises in the eastern Pacific it is around 4 months.

Dall's Porpoise males may guard females, a mating strategy that is uncommon in small cetaceans (especially when compared with land mammals). Males travel with, approach, stay near, and tend to dive synchronously with females, but they challenge other males.

FOOD: Dall's Porpoises catch a variety of prey, which vary regionally but are mostly smaller in length than 30 cm (12 in.). Both squid and schooling surface and midwater fish are taken, including juvenile rockfish, Market Squid, hake, Capelin, herring, anchovies, mackerel, and sardines along the west coast of North America. In the deep sea, they eat fish such as lanternfish, deep-sea smelt and, no doubt, deep-sea squid.

An opportunistic feeder, the Dall's Porpoise also feeds mostly at night, when the deep-scattering layer of fish and squid rises to the surface and becomes easier to access. How deep these porpoises dive is not known, but based on their body form and speed of swimming, they likely dive fairly deep, perhaps akin to depths reached by Pilot Whales with whom they often associate.

On rare occasions enormous groups of Dall's Porpoises have been documented swimming as a phalanx. One observation of this phenomenon occurred 1,200 km (750 mi) off Medford, Oregon, and involved an estimated 5,320 porpoises. The school formed several columns, with porpoises 3 m (10 ft) apart and each column separated by 9 m (30 ft). The school moved in "perfect sequence" at a rapid clip of 46 kph (29 mph), pursuing something unapparent to observers.

MORTALITY: Killer Whales take Dall's Porpoises, although the extent of this predation is unknown. The Dall's Porpoise's high-speed dash may be a successful defense against Killer Whales, if it can outrun a pod of mammal-hunting transient Killer Whales. In Monterey Bay, California, transient Killer Whales seem to take Dall's Porpoises by surprise, perhaps because Killer Whales are not resident there. Lingering and hidden sharks, too, may surprise and kill Dall's Porpoises, although evidence of this is rare.

Parasites may be more important in the mortality of Dall's Porpoise. The number of species and the burden of parasites in dead beach-cast porpoise are remarkable.

DISTRIBUTION: Dall's Porpoises are endemic to the North Pacific, ranging from the waters off Baja California in the east to the Sea of Japan in the west, and as far as the Bering Sea in the North. Along the west coast of North America, they are sighted from shelf waters to the deep ocean.

Migratory movements, both north–south and onshore–offshore, are documented in Dall's Porpoises, and the stimulus for such movements is likely linked to prey movements. Magnetic material has been identified in the brains of Dall's Porpoises, which implies that they may use geomagnetic anomalies for navigation. Since they often move offshore, this ability would have value. Movements may also be affected by El Niño–Southern Oscillation (ENSO) events, in association with prey. For example, commercial landings of anchovies were higher in the Gulf of the Farallones in 1986 than in 1987, and coincidentally, the concentration of Dall's Porpoise sightings on the continental shelf was also higher in 1986. Until more thorough studies occur, however, much of our understanding of the relationship between Dall's Porpoise and their prey remains inferential and speculative.

Dall's Porpoises usually occur in waters deeper than 91 m (300 ft). Sea temperature also affects their movement, for their

Range

DALL'S PORPOISE

occurrence in Baja is usually associated with unusually cool temperatures, and generally the species does not range south of latitude 32° N.

VIEWING: Dall's Porpoises may be seen in offshore and inshore waters along the coast from Baja to British Columbia. They frequent deeper inland waters between southern Vancouver Island and the San Juan Islands and may be easily observed there. In Monterey Bay, they appear in any month but favor seasonally cooler waters and the deeper waters around the Monterey Canyon. In the Gulf of the Farallones, they are the most abundant and commonly sighted cetacean, accounting for up to 30 percent of all marine mammal sightings there.

CONSERVATION: Dall's Porpoises are the most hunted cetacean in the world. Porpoises are harpooned as they bow-ride catcher boats off the Sanriku coast of Japan; a traditional harpoon "fishery" for these dolphins in winter takes more than 15,000 a year. Perhaps as many as half a million Dall's Porpoises have perished in this fishery since 1970. More than a third of counted porpoises in one landed sample were mature females, many presumably with calves. If abandoned calves perish, the toll of Dall's Porpoises is far higher than official estimates. The International Whaling Commission has recommended that this fishery be ended, but the Japanese position has been that dolphins are not whales, and are thus not subject to IWC rules. Dall's Porpoises are also bycaught in drift nets, particularly in Japanese high-seas salmon fisheries, so thousands may drown annually.

Agreement	Conservation Status
U.S. ESA	Not listed
U.S. MMPA	No special status
COSEWIC	Not at risk
IUCN	Least concern

Dall's Porpoises are also contaminated by a broad range of industrial chemicals, most recently including organotins used in antifouling paints, and flame retardants found in plastics and textiles. Dall's Porpoises do not escape persisting pollutants despite their preference for open-water habitats often far from land.

POPULATION: The IWC has identified eight breeding stocks of Dall's Porpoises. Except the distinctive True's Dolphin subspecies, which typically ranges from the Pacific coast of Japan to the Okhotsk Sea, the stocks are almost all indistinguishable at sea. Genetic studies confirm the regional nature of Dall's Porpoise populations, and the species' partition into several stocks. The National Marine Fisheries Service estimates a population of 57,500 off the west coast of the United States. Surveys of the waters of the Inland Passage in British Columbia yielded an estimate of 4,900 Dall's Porpoises. Numbers vary seasonally and over years; the stock structure of the population is still unknown.

COMMENTS: There is now open concern over pollutants in "whale" meat in Japan, since diners take in a cocktail of toxins with their meals. Toothed whales tend to bioaccumulate higher levels of pollutants than baleen whales, since fish- and squid-eating whales eat higher on the marine food chain. Consuming odontocetes such as Dall's Porpoises may be especially injurious for people anywhere on the Pacific Rim who traditionally eat marine mammals.

REFERENCES: Houck and Jefferson 1999, Jefferson 1988, Jefferson 2002b, Morejohn 1979, Willis and Dill 2006, Willis et al. 2005.

HARBOR PORPOISE

Phocoena phocoena

Pl. 16; Figs. 15, 89.

Adult male average length is 1.4 m (4.6 ft) long. Adult female average length is 1.6 m (5 ft). Average weight for both is 45 to 60 kg (100 to 132 lbs). Newborn average length is 68 to 99 cm (27 to 35 in.); average weight is 5 to 10 kg (12 to 22 lbs).

SCIENTIFIC NAMES: Harbor Porpoises were first described by Linnaeus in 1758 but were not assigned to the genus Phocoena until 1816, by the French naturalist Georges Cuvier. The genus and species name *Phocoena* comes from the Greek *phocaena*, which simply means "porpoise." Current research favors the traditional subspecies designation *vomerina* for eastern Pacific Harbor Porpoises, deriving from *vomer*, Latin for "plowshare"; this reflects the plowlike look of this animal's narrow jaws and dished skull. Thus the Pacific Harbor Porpoise's proper zoological name is *Phocoena phocoena vomerina* (fig. 89).

The English name Harbor Porpoise for *Phocoena phocoena* is now in conventional use on the Pacific coast, with the alternative name Common Porpoise sometimes used in British literature.

COMMON NAMES: The English word *porpoise* can refer to any one of six species of small, short-beaked, toothed whales of the family Phocoenidae, including the Harbor Porpoise and Dall's Porpoise of the west coast of North America. The modern word descends from the old French *porpais*, a contraction meaning "pork fish." *Porpais* may be a translation into French of a medieval Germanic term meaning "sea swine," such as the old Dutch *mereswijn*. The contemporary French term *marsouin*, as well as the Spanish term *marsopa*, seem to both derive from such Germanic roots. The full French common name for Harbor

Figure 89. Harbor Porpoise.

Porpoise is *marsouin commun,* and the Spanish name is similar: *marsopa común.* Historic English sailors' terms included Sea Pig, Sea Hog, Puffing Pig, and Herring Hog—all in a similar porcine vein. Makah Native Americans use the name *Tseelkh-koo* to identify the Harbor Porpoise.

DESCRIPTION: The Harbor Porpoise is the smallest cetacean on the west coast of North America, most under 1.9 m (5 ft) long and weighing 90 kg (130 lbs). To keep such a small whale warm, the blubber is relatively thick (10 to 25 mm; 0.4 to 1 in.), accounting for 40 percent of body weight. These smallest of cetaceans can further maintain their body temperature because their blubber contains a capillary countercurrent system to prevent heat loss. Females are slightly larger than males; whether male or female, the Harbor Porpoise is a little smaller than the average human.

PACIFIC AND ATLANTIC HARBOR PORPOISES

Pacific Harbor Porpoises differ genetically and behaviorally from Atlantic Harbor Porpoises, and the two populations probably separated before the Ice Ages, perhaps as long as 5 million years ago. Whether the Harbor Porpoise originated in the Pacific or the Atlantic Ocean is an open question.

Harbor Porpoises are dark gray or brown above, shading to creamy below. Their small oval-shaped flippers, low dorsal fins, and short, tapered tails are also dark. Often a darker stripe connects the corner of the mouth with the insertion of the flippers. Their bodies are chunky and their heads blunt. Their mouths are short, with 22 to 28 small, spade-shaped teeth on each side in the upper jaw and 22 to 26 in the lower. The mouthline tilts slightly upward. The conical head of the Harbor Porpoise has reminded many seafarers of a pig, which accounts for its common names over the centuries. Seen from a vessel in the water, these small porpoises look dark, sit low in the water, and show only an inconspicuous blow; they are hard to spot compared with bow-riding Pacific White-sided Dolphins or sounding Gray Whales. In calm, quiet waters, however, a kayaker can hear Harbor Porpoises puff loudly as they breathe.

FIELD MARKS AND SIMILAR SPECIES: Harbor Porpoises may be confused with two other small cetaceans in the region—Dall's Porpoises or Pacific White-sided Dolphins. The Dall's Porpoise is stockier, and its blackish body typically has a striking white patch on either lower flank. Dall's Porpoises often swim fast, sometimes creating a characteristic rooster-tail spray, and also bow-ride, unlike the timid Harbor Porpoise. Curiously, hybrids between Harbor Porpoise and Dall's Porpoise are regularly seen in Boundary Pass or Haro Strait in British Columbia. The hybrids are a lighter gray, bow-ride, and often associate with Dall's Porpoises. Genetic analysis indicate that fathers of these hybrids were Harbor Porpoises, and the mothers Dall's Porpoises, perhaps pursued by relentless Harbor Porpoise bulls. At the tip of Vancouver Island, some 1 to 2 percent of all porpoises are hybrids, a remarkable proportion for cetacea.

The Pacific White-sided Dolphin is larger than the Harbor Porpoise and had a complex set of stripes, but has a similar dark gray back, light gray sides, a bicolored dorsal fin, and a white belly. However, the Pacific White-sided Dolphin is far livelier, often leaping entirely out of the water, in contrast to the slow and sedate swimming of the Harbor Porpoise (which hardly even splashes). Harbor Porpoises may be confused with the smaller Vaquita, *Phocoena sinus*, but this species is restricted to the upper Gulf of California.

NATURAL HISTORY: Compared to what is known about the coastal Common Bottlenose Dolphins and Killer Whales, very little is known about the life of the Harbor Porpoise. Harbor Porpoises are shy and tend to stay away from vessels, making them relatively inconspicuous, and difficult to study intensively. They are also hard to see, because they occur mostly alone or in pairs. They are recorded in small groups, though, of eight to 20 in Santa Monica Bay, three to five in Washington State, and up to nine in British Columbia. Occasionally they form much larger groups of 100 or more, perhaps drawn by some local super abundance of prey. Their social bond appears stronger than many other cetaceans', with individuals known to support injured or young of their kind at the surface.

The structure of larger social groups of Harbor Porpoise is unclear. There is limited cataloguing of photographs of individuals, compared to what has been obtained for Bottlenose Dolphins and Killer Whales, and thus there exists no means of knowing the nature of long-term associations, if any, between individuals. Harbor Porpoises are short-lived, with only a few surviving longer than 25 years; 10 years is the age of the oldest known porpoise in Canada.

The echolocation sounds of Harbor Porpoises are ultrasonic, narrowly focused, close-range, low-intensity, and not always easy to detect in the field. (Ultrasonic clicks in the porpoise family in general may have evolved because they are not easily heard by Killer Whales.) Special hydrophones that record high frequencies must be used to hear this porpoise's sounds (the Harbor Porpoise hearing system has 10 times more cochlear parts than the human ear). Pioneering human observers used "bat ears," or electronic bat detectors, to slow down the hydrophone signals to the human range of hearing. Herring and some other fishes upon which these little porpoises prey are deaf to their specialized ultrasonic clicks, which could help explain the evolution of the sounds. Because the Harbor Porpoise's clicks are so distinctive, engineers have been able to devise automated detectors particularly for this species, helping in their conservation.

Whether they are fleeing predators or chasing fish, Harbor Porpoise are moderately fast for a cetacean, attaining maximum speeds of around 25 kph (16 mph) and average speeds of 9 kph (6 mph). Because they mostly inhabit nearshore waters and are small-sized animals, deep dives are rare for Harbor Porpoise: their dives likely do not exceed 100 m (328 ft). The small size of Harbor Porpoises may also explain why they often bask at the surface on calm, sunny days.

REPRODUCTION: Males and females are sexually mature as early as 3 years of age, but the average appears to vary with population, up to 6 years of age. Calves may be born from late spring through late summer, with peak time depending on region. They have been documented as early as April, but most are born in June and July. After birth, calves may nurse for as long as 8 to 12 months but can take prey well before weaning. Mothers start to wean their calves after 8 months but maintain close contact; calves, in turn, may stay with their mothers another 9 months after weaning. In some populations, females may calve every year—and be both pregnant and nursing. In other populations, they may be likelier to calve in alternate years. Gestation in this species lasts 11.5 months.

Harbor Porpoises have a strongly seasonal mating pattern. Most females in a given area become receptive for only a few weeks out of a year. Males in the area meanwhile will have developed "megatestes" that increase in size to as much as 4 percent of their body mass. This testicular increase is spectacular: a 50 kg (110 lb) male may have 2 kg (4.4 lb) testes. In contrast, a Harbor Porpoise's brain only weighs about 1 percent of their mass. The periodic testes growth has been taken to imply the males are "sperm competitors" whose role is to produce masses of sperm and then to intensively pursue or monopolize females during their brief period of receptivity.

FOOD: Harbor Porpoises eat a variety of small, smooth-skinned fishes such as herring, pollock, anchovies, and sardines, as well as squid. They will also eat hake and mackerel. The diet varies by region and seasonally. Herring, Capelin, and shad were identified in the stomachs of porpoises in Washington, and anchovy, juvenile rockfish, hake, and Tomcod were the primary food in north central California. Their stomach has three chambers: a forestomach for storing and mechanical grinding of fish and, for the chemical breakdown of the food, the glandular second chamber and the pyloric stomach.

Though not known to cooperate when feeding, Harbor Porpoise are often seen feeding along with seabirds, seals, and whales on large concentrations of prey.

Because of their small size, Harbor Porpoises feed daily at an average rate of around 10 percent (range 5 to 14 percent) of their body weight (larger dolphins consume an average of 4 to 6 percent).

MORTALITY: Transient Killer Whales kill Harbor Porpoises, sometimes stunning their victims with head butts and tossing them aloft. However, it is difficult to estimate the number of

Harbor Porpoises taken by Killer Whales, since most killings are not seen. Sharks, including the Great White Shark, also eat Harbor Porpoises. Shark predation on porpoises and other cetaceans has not been systematically documented, although single incidents have been described, and numerous carcasses have washed ashore in central California with characteristic bites. The scars from the suckers of the giant Pacific squid were identified on one dead Harbor Porpoise in central California. In Scotland, Atlantic Harbor Porpoises have been killed by Common Bottlenose Dolphins, and recently similar killing was reported in Monterey Bay, where scientists observed several Common Bottlenose Dolphins ram Harbor Porpoises with their heads. Predation apparently was not the reason for the killings; instead, Common Bottlenose Dolphins may be competing with the diminutive Harbor Porpoise for similar prey or habitat.

DISTRIBUTION: The range of the Harbor Porpoise extends throughout the northern latitudes of the Atlantic and Pacific Oceans in the boreal and temperate waters. In the Pacific, they

Range

HARBOR PORPOISE

have been observed as far north as the Bering Sea, Alaska, and as far south as Point Conception, California. Within this range, their distribution forms a continuous ribbon hugging the coastal zone.

Despite this broad distribution along the California Current, Harbor Porpoises are highly provincial in their movements and rarely migrate long distances. In the Atlantic, Harbor Porpoises from the Bay of Fundy and Gulf of Maine migrate along the eastern coast to visit herring spawning grounds off South Carolina. In contrast, there is no evidence for long-distance movements in Harbor Porpoises along the west coast of North America. Instead, genetics and trace pollutant levels suggest that Pacific Harbor Porpoises are regional. Accordingly, U.S. wildlife management teams have provisionally divided the American west coast Harbor Porpoise population into six genetically distinctive geographical stocks: Morro Bay, Monterey Bay, San Francisco–Russian River, Northern California/Southern Oregon, Oregon/Washington, and Inshore Washington. Canadians draw a distinction between Harbor Porpoise in the busy inshore Haro Strait Victoria area and those further north in British Columbia. The inshore Washington and lower British Columbian population may represent the same stock. Even these stocks may be subdivided, depending on future research.

HABITAT: Harbor Porpoises prefer bays, estuaries, and nearshore waters. Along the California coast, most Harbor Porpoises are sighted over the continental shelf, usually in waters less than 200 m (600 ft) in depth, within 8 km (5 mi) of shore, and lower than 17 degrees C (63 degrees F) in sea temperature. Their numbers typically are higher closer to shore, bounded by the 70 m (886 ft) depth line, although there are exceptions. Off Oregon and within the inshore seas of Washington State and British Columbia, Harbor Porpoises may frequent deeper waters, such as those off the San Juan Islands near the Canadian border.

VIEWING: As their name suggests, Harbor Porpoises are creatures of nearshore waters and thus are among the most familiar porpoise species to whale-watchers. These small porpoises may be seen from land or from small watercraft or larger ecotourism vessels. There are areas (and times) where they are abundant, such as in the Hecate Reef off Oregon and in the Gulf of the Farallones. Formerly seen often in the protected waters of the Pacific Northwest, Harbor Porpoises recently have become uncommon there. In southern California, researchers conducting surveys averaged one encounter with Harbor Porpoises on each survey. A persistent whale-watcher in a good area will, without doubt, soon observe Harbor Porpoises.

CONSERVATION: Seagoing Native Americans hunted Harbor Porpoises from precontact times until the 20th century. Middens along the west coast of North America testify to occasional hunting or scavenging for thousands of years. At Bear Cove in northern Vancouver Island, specialized hunters perhaps 6,000 years ago left behind a midden composed overwhelmingly of dolphin and porpoise bones. The coastal Salish Native Americans of Washington State and British Columbia harpooned Harbor Porpoises from quietly moving canoes in protected bays by moonlight. However, this smallest of the porpoise species was not of interest to the western whalers, so it probably was relatively unaffected by modern humans until the development of gillnet fisheries in the 20th century.

Agreement	Conservation Status
U.S. ESA	Not listed
U.S. MMPA	No special status
COSEWIC	Special concern
IUCN	Vulnerable

In central California waters, gillnet fishing for small schooling fish caused a marked decline in Harbor Porpoise numbers in the 1980s. The porpoise could not see the monofilament netting and so became entangled and drowned in the set nets. After area and depth closures for gill nets in the 1990s reduced the porpoise bycatch, there was some recovery of this population, and with a 2002 gillnet closure in depths less than 110 m (360 feet), few California Harbor Porpoises should drown in commercial fisheries today. A few gillnet fisheries remain in Washington State and British Columbia. It was estimated that 80 Harbor Porpoises were drowned in the salmon gillnet fishery in southern British Columbia in 2001. Several efforts have been made to develop nets, or to attach "pingers" to the nets, so that the porpoises can detect and avoid them.

Since sources of marine pollution tend to be concentrated on coasts, inshore sea mammals such as Harbor Porpoises become more contaminated than oceanic species. In Europe, Harbor Porpoises have accumulated HCPBs, a newer form of flame retardant, to levels a thousand times that seen in humans. The European Union is considering banning HCPBs. In North America, efforts are beginning to eliminate PCDES, an older flame retardant already banned in the European Union.

Because of the shyness of Harbor Porpoises, overeager waterborne ecotourists may distress some populations along the west coast of North America. The species' reclusiveness also makes it difficult for scientists to study the effects of disturbance, as they themselves are often in the offending watercraft. Part of the apparent decline in Harbor Porpoise numbers in southern Puget Sound, the sheltered waters of the Strait of Juan de Fuca, the Haro Strait, and near Victoria, British Columbia, has been attributed to increases in noisy sea traffic and ecotourists. Pollution and the fisheries collapse near Victoria, Vancouver, and Seattle may play a greater role, however.

POPULATION: Estimating the Pacific Harbor Porpoise stocks off the American west coast is extremely difficult because of these animals' small size and avoidance of boats. To better determine the numbers of Harbor Porpoise in various regions, much research has been carried out by air and sea; to be able to see the little porpoise, aerial surveys are conducted only under ideal conditions. Current estimates by stock are Morro Bay, 1,700; Monterey Bay, 1,600; San Francisco–Russian River, 8,500; northern California/southern Oregon, 17,800; coastal Oregon and Washington, 37,800; and inland Washington, 10,700. A Canadian estimate was 9,100 Harbor Porpoises in British Columbia.

COMMENTS: Although the tales of many California seaside cultures have been forgotten, early folklorists did gather some stories about Harbor Porpoises from more northern peoples. The Kwakiutl bathed in the sea before their porpoise hunt, rubbing themselves with spruce boughs, in a purification rite. Hunters had to approach silently, since porpoises were believed to have extraordinary hearing. Indeed, in one folktale a Porpoise guard atop a totem pole hears an approaching party before the Eagle on a neighboring pole spies it, and boasts, "I can hear what you cannot see."

REFERENCES: Barlow 1988a, Barlow 1988b, Carretta et al. 2001, Gaskin et al. 1974, Read and Westgate 1997, Read 1999.

LIVING MARINE MAMMALS under Carnivora fall within five families. Pinnipeds are grouped in three separate families under the suborder Pinnipedia, otters in the Mustelidae family, and polar bears in the Ursidae family. Polar bears are classed as marine mammals because they have evolved special features to live in the ocean, but they range far north of the California Current (CC).

Pinnipeds (Suborder Pinnipedia)

Pinnipeds, seals and sea lions, have "winged feet," from the Latin for "feather" *(pinna)* and "foot" *(ped)* (fig. 90). They lead a dual existence: like whales for their life in the sea, and like wolves for their life on land. Their flippers evolved for both swimming and locomotion on land. "Sea wolf" is an appropriate name, for pinnipeds are carnivores, a group of mammals whose teeth are made for seizing and eating living animals. Although carnivores all have large canine teeth for catching and tearing prey, pinnipeds, which eat bony fish or boneless invertebrates, lack the distinctive teeth (carnassials) that wolves have for crushing the dense, weight-bearing bones of their prey.

Pinnipeds have fur with textures and purposes that vary widely, but the group's primary distinction among sea mammals

Figure 90. The bull Northern Elephant Seal, a carnivore.

Figure 91. Pinniped colonies with elephant seals, fur seals, and sea lions on San Miguel Island, California—a Serengeti for marine mammals.

is a tether to land, where they give birth. This forces pinnipeds to follow a predictable and synchronous annual cycle, including delayed implantation of the embryo for 1 to 2 months. Terrestrial sites [variously called colonies, rookeries (for birthing sites), and haul-outs] are centers of social activity—seal cities—where individuals gather to birth, nurse, breed, and rest together. As a group, they can more readily detect terrestrial predators, and several species may commingle where suitable habitat is limited (fig. 91).

Pinnipeds often segregate by species. On land, elephant seals may sprawl along sandy beaches, sea lions crawl up rocky slopes, and diminutive Harbor Seals cluster on intertidal offshore rocks. At sea, Harbor Seals forage in the nearshore environment, sea lions more often along the mid-to-outer continental shelf, and elephant seals and fur seals in deep oceanic waters, at different depths.

Worldwide, all pinnipeds are classified into three families, but only two occur in the CC: earless seals (*Phocidae,* two species) and eared seals (*Otariidae,* four species). Walruses *(Odobenidae)* are limited to arctic and subarctic waters, though their fossils attest to their previous presence here.

Eared seals of the Otaridae family are known as "walking pinnipeds," because they can rotate their pelvis and swing their hind flippers beneath them to walk on all four limbs on land (fig. 92). Of some 15 species worldwide, four occur in the CC—two fur

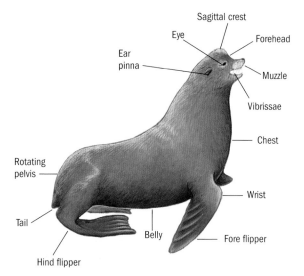

Figure 92. The California Sea Lion, an eared seal of the Otaridae family.

seals (Northern Fur Seals and Guadalupe Fur Seals) and two sea lions (California Sea Lions and Steller Sea Lions). Key distinguishing features of the family include external ear pinnae, smooth facial whiskers, rotating pelvis, and long front flippers. Swimming effort, called pectoral oscillation, is centered in the forebody with the "flapping" of the foreflippers. All flippers are mostly hairless, and the front ones have splayed digits with a hard leading edge. Sea lions and fur seals, which diverged around 3 mya, differ by fur type and blubber thickness. Sea lions have a single layer of coarse fur and a thick layer of blubber for insulation, whereas fur seals depend solely on their fur—dense, short underfur overlaid with coarse outer hairs. All four species are sexually dimorphic, with the females' size ranging from one-quarter to one-half the males'. Males have an external scrotum, perhaps to reduce overheating. Females have at least two pairs of retractable abdominal nipples, one pair anterior to the umbilicus and one posterior, and have a bicornuate uterus—two horns forming a single body of the uterus. Since females breed shortly after giving birth, this adaptation allows them to prepare one horn for the next fertilized egg while still holding a near-term fetus in the other.

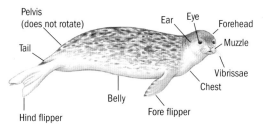

Figure 93. The Harbor Seal, an earless seal of the Phocidae family.

Earless seals of the Phocidae family lack the small but distinct ear flap of the Otariidae seals and are also called "true" and "crawling" seals (fig. 93). Only two of the 18 species in the family occur in the CC: the Northern Elephant Seal and the Harbor Seal. Movement on land is cumbersome, because the animals rely on the front flippers to heave their body forward, and then on their abdominal muscles to propel the body like an inchworm. This seeming inefficiency is offset by greater mobility in the water. Using short, wide rear flippers and pelvis, they swim with a back-and-forth sculling motion called pelvic oscillation. All flippers are covered with short, coarse hair, with no bare skin, and they rely on blubber for insulation. Facial whiskers of all earless seals are beaded with bumps, perhaps an important feature for detecting water movement of prey (Dehnhardt et al. 1998). Sexual dimorphism occurs in many species of earless seals, including elephant seals, but is not pronounced in Harbor Seals. All phocid males lack a scrotum, and females have a bipartite uterus—two horns share a small common area in the uterus. Mating takes place around 30 days after birthing, so females do not require another chamber to receive a fertilized egg. Females usually have one pair of mammary glands, though some elephant seals may have additional teats.

Otters (Suborder Caniformia)

Mammals classified within this suborder have doglike features and nonretractile claws. Both of these features are also shared by their relatives of the suborder Pinnipedia; sometimes pinnipeds are grouped under this suborder. Otters and polar bears are the only other marine mammals grouped under this suborder.

Mustelids have a weasellike shape and round head; the Sea Otter and River Otter fall within this family (fig. 94). Otters,

Figure 94. The Sea Otter, a member of the mustelid family, has the thickest fur of any mammal; note tag in rear flipper.

also called fissipeds because of their "paw-footed" feet, are one of the least evolved to oceanic life of the sea mammals. Whereas Sea Otters seek refuge in kelp forests to give birth to single pups, River Otters retreat to dens protected in terrestrial vegetation and raise litters. Sea Otter fur is the densest of any mammal's [130,000 hairs/cm² (19,300 hairs/in.²); Reynolds and Rommel 1999], and their specialized teeth, used for crushing shellfish, have no equivalent in mammals.

STELLER SEA LION   *Eumetopias jubatus*
Pls. 17, 19; Figs. 91, 95, 96.

Adult male average length is 2.7 to 3.7 m (9 to 12 ft); average weight is 545 to 1,136 kg (1,200 to 2,500 lbs). Adult female average length is 1.8 to 2.7 m (6 to 9 ft); average weight is 250 to 350 kg (550 to 770 lbs). Pup average length is 1 m (3 ft); average weight is 16 to 23 kg (35 to 50 lbs).

SCIENTIFIC AND COMMON NAMES: *Eumetopias jubatus* is Greek for the broad forehead, and Latin for the large mane of the male (fig. 95). The lionlike mane and the roar of the bulls accounts for the name. The species went through several name changes, starting with *Leo marinus*, meaning "sea lion," as described by the naturalist Georg Wilhem Steller in 1751. In 1828, Lesson named the species *Otaria stellerii* to honor Steller's discovery of

Figure 95. *Upper:* Steller Sea Lion bull.
Lower: Steller Sea Lion female, with pup.

the species on Bering Island, but eventually *Eumetopias jubatus* became the accepted name, in the 20th century. Common English names are Steller (or Steller's) Sea Lion and Northern Sea Lion. Common Spanish and French names are *lobo marino de Steller* and *lion de mer de Steller*, respectively.

DESCRIPTION: Steller Sea Lions are the largest member of the family of eared seals, and are highly sexually dimorphic in size and appearance. Bulls are two to three times the size of females and are distinguished by a very muscular neck accentuated by a mane of dense, longer fur. Otherwise, the body fur is short and coarse. Both sexes have a broad, curved, forehead that is more pronounced and bearlike in the bull. Coloration in adults is uniform blond or reddish, with dark brown flippers and underbelly. When wet, they remain pale tan in color. Juveniles are chocolate to cinnamon brown, which lightens with age after molting, at 4 to 6 months of age. Pups are born with a thick chocolate brown color. The long foreflippers are broad and mostly naked with black skin. The rear flippers have functional claws on the three middle digits, which are used in doglike scratching. The external ears are small. Whiskers are smooth and light in color. Females have two to six mammary glands, at least one pair posterior to the umbilicus and one anterior; four is the most common number.

FIELD MARKS AND SIMILAR SPECIES: Key field marks are very large size, light coloration, blunt head, and distinct, vocal roar. The bull Steller has a lionlike mane and lacks the sagittal crest of the bull California Sea Lion. The two species intermingle on land and in the water, making identification challenging, particularly of young. Young Steller Sea Lions, although similar in size and coloration to the California Sea Lion, have a broad, bearlike head that contrasts with the narrow, doglike head of California sea lions.

NATURAL HISTORY: Steller Sea Lions follow an annual cycle linked to reproductive and feeding requirements, and their presence onshore is closely tied to this cycle. Generally, they are nonmigratory, attending haul-out sites year-round. Abundance onshore, however, varies with season and food availability. The highest numbers occur between June and August, during the breeding season. Peak abundance on shore for bulls occurs in mid-June, for young males in late June, and for females between late June and early July. Most males depart rookeries in late July and August, and females (when waiting for pups to learn to swim) depart in late August to September.

Even within a single day, attendance at colonies may follow a pattern, with lower abundance during the night, when foraging, and higher during the day, when resting. Temperature and tide have some influence on this pattern, depending upon the site.

The annual molt varies with sex and age groups, females molting in spring and summer, and males from late summer into fall and early winter.

Steller Sea Lions are highly polygynous, but during the non-breeding season, groups onshore are mixed with all ages and both sexes. The breeding season begins in early May, when males arrive to establish territories prior to the arrival of females, and extends into late July.

The relationships among individuals are complex, and likely long-term. Females, for example, often arrive with a previous year's pup and are known to nurse newborns and yearlings simultaneously. Even more remarkable, researchers have documented females nursing a daughter that is in turn nursing a newborn pup.

Bull Steller Sea Lions roar and growl loudly, much like African lions, but neither males nor females bark. Females have a higher voice, and pups "bleat" much as sheep do, though their call is a deeper sound than California Sea Lions'. Underwater, the growls become a series of pulsed, underwater clicks that may be used socially or for foraging.

When foraging at sea, females and young form small groups of two to four; adult males usually are solitary. In Alaska and Canada, where the population is larger, females and young may form large groups of greater than 50 when foraging on schooling prey, and rafts of several hundred while floating on the surface, resting near colonies. Within a day, most sea lions come on shore from midday to the afternoon, when solar radiation is greatest; much of their behavior onshore is directed to maintaining a comfortable body temperature.

Because of their large size and weight, adults, young, and even pups have a lumbering gait on land and cannot gallop as they move on all fours. They also can overheat and enter the sea at air temperatures of greater than 21 degrees C (70 degrees F). Researchers speculate that males wet their rear flippers to cool exposed testes.

REPRODUCTION: Females are sexually mature at age 4 to 6, but males mature from age 3 to 8 and do not hold territories until they have the body mass to compete, at 9 to 10 years. The estimated lifespan is longer for females (20 to 25 years) than for males (15 to 17).

Pregnant females arrive at rookeries in late May, regardless of latitude; numbers peak in July and August. Females give birth to a single pup (rarely, two), with most born July 1 to 15. Mating occurs on land 10 to 14 days after birth, when the female comes into estrus. Gestation is typically 10 months, with delayed implantation in September to October. A female may nurse a yearling and newborn at the same time, but nursing usually lasts 32 to 44 weeks. Females will wean a pup at 1 to 2 years, just prior to giving birth again. In a lifetime, a female may produce a pup a year, skipping years only when food is scarce, nurturing some 15 or more pups. One study detected a diurnal pattern in birthing, with more births occurring in the morning (50 percent fewer at night).

Five to 15 days after giving birth, females will begin to go on feeding trips lasting from 9 to 40 hours, often leaving in the evening and returning the following morning. On average, mothers stay on shore to nurse pups for 21 hours before foraging for 36. The length of feeding time spent away from the pup progressively increases as the pup grows, so that by the sixth week, females may spend only 30 percent of their time ashore. When mothers are away feeding, pups form groups and begin to swim in nearshore pools. When they are 1 month old, pups begin to accompany females on short foraging trips.

Males arrive to breed in May and remain at the breeding colony for up to 2 months, loudly and aggressively defending territories. In contrast to females, territorial males mostly fast while holding territories. Males may defend up to 30 females, and territories may be no larger than 20 feet in diameter. Males defend aquatic and terrestrial territories, but are most successful defending semiaquatic ones. Site topography, such as rocky features, affects territory size and shape. Females are very social, and move across territories when accessing the water for foraging trips. Those males who are unsuccessful at defending territories are pushed to the periphery, where they form bachelor gatherings and engage in mock fighting. Steller Sea Lions give birth and breed on sloping, flat rocky areas and cobblestone or coarse sand beaches that are protected from high waves (fig. 96). Females appear to have favored spots for giving birth, and defend these areas from other females.

FOOD: Steller Sea Lions' food is primarily fish and squid, but they also prey on crustaceans and mammals. Seasonally abundant prey include flounder, Turbot, hake, sardines, herring, halibut, rockfish, mackerel, smelt, sand lance, sculpin, lamprey, skate, and salmon. Though they are thought to compete with commercial fishermen for salmon, in one study they fed mostly (87 percent) on lampreys (a major parasite of salmon) at the mouth of the Rogue River, Oregon. Invertebrates also may be a significant part of their diet, with squid and octopus accounting for 35 percent, and bivalves 20 percent, in one study. They feed occasionally on Harbor Seals, Northern Fur Seal pups, and Sea Otters. Fish are swallowed whole up to 2 kg (4.5 lbs) in size, but they break apart larger prey at the surface by holding one end of the fish and snapping it forcefully into pieces.

Figure 96. Steller Sea Lion rookery on rocky shelf at Año Nuevo Island.

Daily food intake models estimate that females require 16 to 20 kg (35 to 45 lbs/day) and bulls 30 to 34 kg (65 to 75 lbs), an estimated 2 to 6 percent of body mass per day.

MORTALITY: Pup mortality, predation and disease are the main natural causes for loss of life. Pup mortality can be high (10 to 15 percent) within the first month of life, a result of malnutrition, crushing (by fighting males), or drowning (when swept away in high surf), particularly in crowded island and reef rookeries. Though pups can swim, they are not adept swimmers when very young. El Niño–Southern Oscillation (ENSO) events affect food availability and severely reduce the ability of females to nurse and raise pups. Because of their large size, predators are limited to sharks (mostly Great White Sharks) and Killer Whales. Predation by transient Killer Whales may be significant in Canada, but is likely less important south of Washington because of the small numbers of Killer Whales. Hookworm parasites can infect young pups with high infestation levels, causing anemia and mortality. In California during the 1970s to 1980s, a virus (the San Miguel virus) that female sea lions contracted in southern and central California may have contributed to fetal miscarriages.

DISTRIBUTION: Steller Sea Lions are widely distributed around the Pacific Rim from Hokkaido, Japan, north to the Bering Sea, and south to San Miguel Island, in Southern California. The Aleutian Islands are the center of abundance during the breeding season. California is the southern edge of the range of Steller Sea Lions, but this range has shrunk over the past 30 years with the abandonment of the San Miguel Island colony in southern California. Año Nuevo Island is now the southernmost rookery and the largest one in California, accounting for 68 percent of the total number of breeding animals in California. Other breeding sites in California include the Farallon Islands, Fort Ross, Cape Mendocino, and Saint George Reef. In Oregon, breeding occurs at the Sea Lion Caves and at Orford and Rogue Reefs. There are no breeding sites in Washington. In British Columbia, Steller Sea Lions breed at offshore Islands such as the Scott Islands (70 percent of pup production in British Columbia), Forrester Island, Hazy Island, numerous small rookeries on offshore reefs and islands of Vancouver Island, and Cape St. James and Danger Rocks.

Although breeding sites south of Canada are few, there are over 200 haul-out sites on offshore rocks and islands distributed throughout Steller Sea Lions' range for resting during winter

STELLER SEA LION

feeding and dispersal. Examples in California include Cape Vizcaino, Point Reyes Headland, Fort Ross and Saint George Reef, Flatiron Rock near Trinidad Head, Sugarloaf at Cape Mendocino, the north and south Farallon Islands, and Lion Rock near Point Buchon. In Washington, colonies where 100 to 500 sea lions occur are on offshore rocks in the Split Rock Area, south of Tunnel Island in northern Washington, the Sea Lion Rocks, Carroll Island and Bodelteh Island, Umatilla Reef, and north of Cape Alava, Tatoosh Island, Skagway Rocks, and Fuca Pillar. Along the Straits of Juan de Fuca, Steller Sea Lions are found on offshore rocks west of Carmanah Point and on intertidal rocks and reef areas around Pachena Point.

Steller Sea Lions do not migrate beyond their breeding range, but most researchers agree that they disperse widely after breeding. In California, they tend to move northward along the coast, but those in Canada tend to move south; females and young may stay close to natal rookeries, such as Año Nuevo, all year. There is little evidence of females' shifting between rookeries to breed.

HABITAT: Generally, Steller Sea Lions feed nearshore on the continental shelf, but they have been observed at sea up to 483 km (300 mi) from land. With telemetry, more precise information is now gleaned about where they go at sea. In one study, juveniles made three types of forage trips that averaged 11 days: long-range, of greater than 16 km (10 mi); short-range, of less than 16 km (10 mi) offshore; and nearshore, between haul-out sites. Short-range foraging trips accounted for the majority of trips (87 percent), with an average distance of 4 km (2.5 mi) compared to long trips, with an average of 48 km (30 mi). Females with pups averaged 32 km (20 mi) from rookeries when foraging. During winter months, though, they travel 97 km (60 mi) within their home range during a feeding trip.

From satellite-tagged Steller Sea Lions in Washington, researchers documented that females and young of both sexes foraged in the Saint Georges Strait coincident with herring spawn, and that subadult males traveled to outer coastal locations on the west shore of Vancouver Island and northern Washington. In Alaska, Steller Sea Lions feed in persistent hot spots predictably filled with forage fish.

Sea lions use the intertidal areas of rocky shorelines within 18 m (60 ft) of the sea to rest onshore, but this distance is more likely a function of coastal topography. Researchers have identified four kinds of haul-out sites used by Steller Sea Lions, including rookeries, year-round haul-out sites, winter haul-out sites, and winter rafting sites. Rookeries in British Columbia tend to be sites farthest from land masses on islands that are exposed to oceanic swells. Year-round and winter haul-outs are located on islands or islets close to land masses exposed to oceanic swells. When no suitable haul-out space is available, Steller Sea Lions congregate in large rafts near shore, and near haul-out colonies.

VIEWING: In California, Steller Sea Lions can be seen year-round at Año Nuevo, the south and north Farallon Islands, Fort Ross, Sugarloaf Rock at Cape Mendocino, and southwest Seal Rock at Saint George Reef. Smaller numbers occur irregularly at Point Reyes Headland, Sea Lion Rock south of Cape Mendocino, Flatiron Rock and Little Flatiron Rock, Cone Rock Mainland, Turtle Rocks, and Castle Rock. Single males haul out seasonally in the fall at Pier 39, San Francisco. In Oregon, they cooccur with California Sea Lions at Oregon Caves and near the mouth of the Columbia River, at the tip of the South Jetty. In Washington, haul-out sites include offshore rocks in the Split Rock Area, south of Tunnel Island in northern Washington State, the Sea Lion Rocks, Carroll Island and Bodelteh Island Area, intertidal ledges of Umatilla Reef, and north of Cape Alava, Tatoosh Island Area, areas around Skagway Rocks and Fuca Pillar. Along Straits of Juan de Fuca, they occur on an offshore rock west of Carmanah Point and intertidal rocks and reef areas around Pachena Point. In British Columbia, they occur on offshore islands such as the Scott Islands, the north Danger Rocks, and Cape St. James.

CONSERVATION: Historically, Steller Sea Lions bred throughout the northern Pacific Rim from Japan to the southern Channel Islands of California. Native peoples from Alaska to California hunted them for food, fur, and hides, as did European sealers, who also commercially hunted them for blubber and dog food. The vibrissae were used by Chinese as opium pipe cleaners. In the past century, they were hunted to minimize potential competition with commercial fishermen; government-sponsored organized hunts occurred throughout their range to reduce populations. In Washington, the years 1944 to 1948 saw a bounty of $8 per sea lion. At one rookery in Canada, an estimated 20,000 were killed during 1923 to 1939, and by 2005, only a few nonbreeding sea lions used the site. In California, Steller Sea Lions recovered more quickly than other hunted seal species, and by the 1920s they were the most abundant pinniped in the southern Channel Islands.

Agreement	Conservation Status
U.S. ESA	Threatened (California Current); endangered (west of longitude 144° W; Alaskan population)
U.S. MMPA	Depleted
COSEWIC	Special concern
IUCN	Endangered

Since then though, their breeding range has contracted steadily in California, beginning in the 1930s and continuing, more sharply, since the 1960s. Counts in California between 1927 and 1947 ranged between 5,000 and 7,000, with no apparent trend, but have declined by over 50 percent from 1980 to 1998. In California, until the 1970s, Steller Sea Lions bred regularly in small groups on San Miguel Island, Año Nuevo, the

Farallon Islands, and the Point Reyes Headland, but since 1980 pups have been born only on the Farallon Islands and Año Nuevo. On the Farallon Islands, researchers discovered part of the reason for the sea lions' decline when they collected aborted fetuses and found elevated levels of PCBs, and the presence of the San Miguel virus. From northern California to Canada, the colonies have remained stable, or have increased slightly, for the past three decades, at an annual rate of 3.2 to 7.6 percent. However, in the western Gulf of Alaska, the population declined 85 percent during the period 1970 to 2000, with whole colonies nearly disappearing. A likely cause for decline in Alaska is commercial overfishing, although only limited direct proof of the relationship exists. The groundfish fishery, which includes pollock, cod, and mackerel, is the world's largest, taking annually 1.4 million metric tons of fish. Indirect effects are lack of prey to sustain pregnancy in females (consequently, the species has a low birth rate); low juvenile survival; and, possibly, traumatic death from drowning in nets or intentional shootings. Climate change, though, may also cause this species to shift its diet to focus on fish with less caloric content, reducing its chance of survival.

POPULATION: The eastern Pacific stock extends from southern Alaska at Cape Suckling to California, and was estimated to be 45,000 in 2007. The population of sea lions at haul-out sites that year was estimated at 2,400 in California; 5,300 in Oregon; 500 in Washington; and 15,400 in British Columbia. The numbers appear stable, except in California where they may still be declining.

COMMENTS: Scientists first proposed in the mid-1980s that overfishing in Alaska was affecting the amount of food available for Steller Sea Lions, as well as for other marine mammals. Not until around 2007, however, did researchers propose that juvenile sea lions were eating "junk food"—junk fish. In a captivity study, researchers fed juvenile sea lions pollock, a lean fish compared to the historically dominant and oily Capelin. The researchers found that the sea lions had to eat 20 percent of their body weight in pollock to sustain their weight, but that their stomachs could only accommodate 17 percent—so they lost weight.

REFERENCES: Angliss and Allen 2007, Bigg 1985, Bonnell et al. 1983, Fadley et al. 2005, Higgins et al. 1988, Jameson and Kenyon 1977, Kenyon and Rice 1961, Loughlin et al. 1987, Pitcher and

Calkins 1981, Pitcher et al. 2007, Scheffer 1958, Sydeman and Allen 1999, Whitfield 2008.

CALIFORNIA SEA LION   *Zalophus californianus*
Pls. 17, 19; Figs. 32, 92, 97.

> Adult male average length is 2.5 m (7 to 8 ft); average weight is 400 kg (700 to 880 lbs). Adult female average length is 2 m (6 to 6.5 ft); average weight is 120 kg (250 lbs). Pup average length is 0.8 m (2.6 ft); average weight is 9 kg (12 to 20 lbs).

SCIENTIFIC AND COMMON NAMES: The name *Zalophus* has a Greek derivation, from *za* for intensive and *lophus* for crest, referring to the crest on the top of the adult male's head. The primary common name is California Sea Lion in English, Spanish (*lobo marino de California*) and French (*lion de mer de Califonie*), but also simply Sea Lion or Sea Lion of California. There are three known subspecies of *Zalophus californianus* (the only members of the genus *Zalophus*): *Zalophus californianus californianus* occurs in the California Current; *Z. californianus wollebaeki,* the Galapagos Sea Lion, is geographically isolated from the California subspecies; and *Z. californianus japonicus,* the Japanese subspecies, is likely extinct, having disappeared in the 1950s.

DESCRIPTION: California Sea Lions are easily identified by their agility, which is accentuated by their long, wing-shaped front flippers: these transform their streamlined shape in water into a surprisingly nimble four-legged walk on land. A California Sea Lion is a midsized eared seal; an adult male is about the size of a female Steller Sea Lion. Female California Sea Lions can be more than half the length of males but weigh significantly less than half, because males are much bulkier, with well-developed necks and chests (fig. 97). A key feature of the mature bull is a crest on the top of the head (resembling a topknot), which is formed by muscles attached to the large "sagittal" flange along the top of the skull. The flange, similar to that of terrestrial bears, anchors strong jaw muscles used for seizing and dismembering prey. As a male matures, its crest grows in size and becomes more visible, with paler fur. The head of California Sea Lion is very doglike, with a long, narrow snout and

Figure 97. *Upper:* California Sea Lion bull.
Lower: California Sea Lion adult female with pup.

small external ear pinnae. California Sea Lions that breed in Baja are slightly smaller than those in California.

The body fur is short and coarse, lacking the dense insulating hairs of fur seals. The fur is uniform in color, not banded or spotted, and varies from dark chocolate brown to tan or silver (females tend toward tan and silver). Newborn pups are dark brown, changing to light brown and then tawny after two molts, during the first year. When wet, the fur color of all age classes darkens and, in males, can appear almost black. The long fore-flippers mostly lack fur, except to the wrist, and are black. The rear flippers have functional claws on the three middle digits, and the animals can be seen scratching, doglike, with these. The small external ears are narrow and pointed. Vibrissae, or whiskers, are smooth and light-colored.

FIELD MARKS AND SIMILAR SPECIES: The California Sea Lion is distinguished from the Steller Sea Lion and from fur seals by the distinct raised crest on the head of the bull, the doglike shape of the head, and the sleek body. Their porpoising behavior in the water can lead to confusion with porpoises, especially the Northern Right Whale Dolphin. At haul-out sites on land, their constant barking is distinctive from that of other pinnipeds.

NATURAL HISTORY: The annual cycle of this species revolves around the animals' reproductive and feeding requirements. During the breeding season, adults congregate on land; during the nonbreeding season, they migrate to areas with abundant prey. Year-round, sea lions come onshore at favored locations, but in numbers that vary through the year with season, time of day, cloud cover, temperature, sea state, tide, feeding opportunity, and human disturbance—and, between years, with changes in their population and distribution. From northern California to Washington, more sea lions occur onshore during the middle of the day, suggesting nocturnal feeding. On San Miguel Island, though, where temperatures are higher, more sea lions haul out in the morning, and retreat to the water from the midday heat. Because of their moderate size, sea lions are better able to tolerate temperatures onshore at subtropical latitudes than are the larger pinnipeds.

Each year, California Sea Lions molt their fur once; the time is dependent on sex and age. Immature and adult females molt on breeding rookeries in the summer to fall, but subadult and adult males may molt during migration.

California Sea Lions are highly social, both at sea and on land. Their mating system, like many pinnipeds, is polygynous, and males, when not breeding, form foraging groups at sea.

Compared to other seals and sea lions, California Sea Lions are very vocal, with barks consisting of several (two to five) repetitive staccato calls given in rapid succession. Other sounds include growls, grunts, and high-pitched calls, but these are generally obscured to the human ear by the incessant barking at a haul-out site. Barking can also be heard under water, when sea lions are in foraging groups.

Other distinct behaviors in water include porpoising, singly or in groups, when animals are swimming swiftly, as well as gathering (lollygagging) in large rafts while they rest and play. These rafts can include up to 200 animals near shoreline haul-out sites or at foraging areas. Sea lions also unload body heat when resting on the surface by raising a long foreflipper into the air like a sail. Because of their moderate weight compared to Steller Sea Lions, they can gallop short distances on land when chasing or fleeing competitors of their own kind or humans, for example. Other antics include body-surfing, bubble-blowing in the face of scuba divers, chasing of their own air bubbles, and play with inanimate objects.

REPRODUCTION: Females and males are sexually mature at 4 to 5 years, and live to around 20 years, though the longest recorded lifespan of a female in captivity was 34 years. Once mature, the sea lions' onshore cycle of pupping and breeding occurs between May and early August, with most pups born in June. Females give birth yearly to a single pup; twinning is known to occur, but rarely. Gestation is around 10 to 11 months, including the 2 to 3 months of delayed implantation, and full-term females usually arrive on rookeries a couple of days prior to giving birth. The bond between female and pup is very strong—through imprinting at birth, reinforced by vocalization, sight, and smell. To protect a very young pup from large sea swells or human disturbance, a female will carry it, doglike, in her mouth. About 1 week after birth, females begin feeding trips; the duration of these forays varies with food availability but tends to increase as the pup grows. A female's trip might range from 1 to 6.5 days (the average is 2.5 days), depending on her age and experience, as well as on food availability. Female foraging trips during the breeding season are limited by their need to nurse to an area within 150 miles of rookeries. Typically, females nurse pups from one of four nipples for around 6 months, but pups can nurse for up to 1 year. From 2 to 3 weeks of age, pups will form pods on the beach while females are at sea feeding; pod size varies from five pups to 200, and a pod will shift around the rookery. During years of extremely low food availability, such as during El Niño–Southern Oscillation (ENSO) events, females might abandon pups; early in the season, they can abort or reabsorb fetuses.

After females arrive on rookeries, males establish territories by ritualized fighting and maintain their territories through behaviors that include staring, head-shaking, incessant barking, and patrolling along territorial boundaries. Territorial males are evenly spaced along a beach at 25- to 45-foot intervals, defending tight clusters of females. A study on the San Nicolas Island rookery found an average ratio of males to females to be 1:14. On rocky shores, the boundaries between territories tend to follow natural features such as irregularities in the substrate. Bulls do not fast during the breeding season (this contrasts with the strategy of most territorial pinniped males), so their territories turn over, on average, every nine days, as males depart to forage. Nonterritorial males form smaller groups on the fringe of the rookeries and, from these areas, will attempt mating when a territorial bull is distracted. Researchers have observed males protecting pup pods from sharks foraging in nearshore waters—unusual behavior for male pinnipeds.

Females also defend spaces on a beach, but they regularly move across male territorial boundaries when seeking other females, their own pup, relief from heat, or food. Females are intolerant of aggressive males and will abandon territories of overbearing males that attempt to block their regular excursions. When they are in estrus, around 3 weeks after giving birth, females solicit males with whom to mate; most mating takes place on land. Aquatic copulations also occur, especially when males hold aquatic territories along the intertidal zone.

FOOD: California Sea Lions forage on what is seasonally abundant and feed cooperatively. In very large feeding groups of 500 individuals or more, sea lions behave like cetaceans—porpoising in long lines while herding big schools of fish. Researchers have determined by examining scat that sea lions feed primarily on Pacific Hake, sardines, Market Squid, rockfish, salmon, and Northern Anchovy. The proportion of these items in the diet varies annually and seasonally, with prey abundance, suggesting that sea lions are flexible in their diet. They do not dive particularly deep for their prey, averaging around 35 m (110 ft). Foraging trips of males average around 1 to 2 days, but during years when food is scarce can average more than 10 days.

In Washington, Oregon, and California, seasonal peak numbers coincide with spawning salmon at the mouths of rivers such as the Columbia, Klamath, and Smith rivers. In Oregon, on the Rogue River, though, sea lions feed mostly on abundant and slow-swimming lamprey and, opportunistically, on salmon and Steelhead. On the Columbia River, in contrast, one-third to one-half of sea lions feed on salmon. In Monterey Bay, abundance may be greatest during the squid migration, in spring. They concentrate when herring spawn in Tomales and San Francisco Bays from November through March. During ENSO years, California Sea Lions exploit the poleward movement of southern California species and switch from feeding on anchovies and squid to eating sardines and rockfish. Other species eaten include octopus, mackerel, sole, perch, Pacific Herring, midshipman fish, cutlassfish, bass, myctophids, and pelagic red crabs; the latter two species are consumed mostly during ENSO years.

Sea lions dive almost continuously during feeding trips, resting on the surface less than 5 percent of the time and for a few minutes between dives; peak diving sessions occur at sunrise and sunset. Near San Miguel Island, a maximum dive depth recorded for a female was 335 m (1,100 ft), but most females' dives are 15 to 40 m (50 to 130 ft). During ENSO years, sea lions modify their foraging behavior as they search harder for food, diving 9 to 27 percent deeper and 10 to 27 percent longer. The distance and length of foraging trips from the rookery for males increases from an average of 47 km (29 mi) to 124 km (77 mi; maximum = 800 km/500 mi), and from an average of 1 to 2.5 days (maximum = 13 days).

A daily intake of prey has been estimated at approximately 7 percent of body weight. One estimate for the amount of food eaten by California Sea Lions present in Monterey Bay in 1998 was 2,200 tons of sardines, 1,800 tons of rockfish, and 8,000 tons of salmon.

MORTALITY: The main causes of natural mortality in California Sea Lions are pup death, predation, and disease. Pups can drown and suffer injuries on rookeries that are crowded; they can also die of starvation if separated or abandoned by their mothers. Mortality within the first month of life may reach 15 percent.

This pinniped's primary predators are Blue Sharks, Hammerhead Sharks, White Sharks, and Killer Whales, but the level of predation is unknown. California Sea Lions may strand onshore, sometimes in large numbers, sick with the bacteria *Leptospira*. The bacteria are always present in the population, but sea lion die-offs are cyclic. Poisoning from a domoic acid biotoxin from harmful algal blooms (HAB) also causes cyclic mortality events such as occurred in 1978, 1986, 1988, 1992, and 2000. These diseases may act synergistically with pollutants in the environment, causing sea lion reproductive failure.

During the large ENSO events, northern and central California haul-out sites are inundated with thousands (an estimated 50,000 in July 1998) of young and female sea lions pursuing food not available in more southern waters. Emaciated individuals may strand on noncolony beaches, some begging for food from shoreline fishermen. During HAB events, they are also the species most likely to strand along California beaches from domoic acid poisoning.

DISTRIBUTION: California Sea Lions range from Baja California, Mexico, north to Vancouver Island off British Columbia. In recent years, males and some females have shown up on

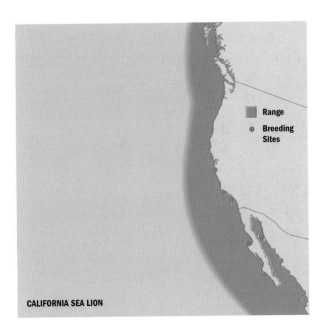

CALIFORNIA SEA LION

pinniped rookeries in Alaska, including on the Pribilof Islands, normally frequented by fur seals and Steller Sea Lions. The current breeding range for this species, however, is much smaller, extending from Año Nuevo Island, in central California, south to the tip of Baja California. The core breeding area centers primarily on the California Channel Islands, with San Miguel and San Nicolas Islands accounting for greater than 90 percent of the species' breeding in California. Only a few hundred pups are born to the north, on colonies such as the Año Nuevo, and a few have been born on the Farallon Islands. Small populations also breed in the Sea of Cortez, on islands off Baja California, and along the west coast of mainland Mexico to Nayarit.

After the breeding season, adult and young adult males migrate north from southern California and Mexican breeding sites in search of large concentrations of food. They range as far north as British Columbia, where they begin arriving in September. In northern California, numbers peak onshore twice, in fall and in spring, as the animals move away from and to breeding sites. Some males from Baja migrate north to the Channel Islands, arriving in December. The winter range of females and young is smaller and does not extend much past northern California. Males journey the furthest north and are known to rest onshore at more than 100 locations. As the population has grown, young sea lions are increasingly present during the summer on northern California haul-out sites such as Año Nuevo Island and the Farallon Islands.

HABITAT: California Sea Lions are coastal animals that feed in nearshore waters over the continental shelf. They are rarely seen further than 32 km (20 mi) from land, though satellite-tagged individuals have been reported as far as 1,127 km (700 mi) offshore during ENSO years. They travel up rivers such as the Klamath, Columbia, Rogue, Sacramento and Fraser; individuals have been reported as far as 235 km (146 mi) inland—and even on Highway Five, in Lodi, California.

The animals are most concentrated at sea near rookeries in the Channel Islands, California. Sea lions also are seen concentrated around marine features such as Cordell Bank and the Davidson Seamont, and near island and islet haul-out sites. Preferred haul-out sites are on remote island beaches such as the Channel Islands, but California Sea Lions also haul out on mainland promontories such as St. George Reef and the Point Reyes Headland. They use a variety of shoreline substrates, favor sandy beaches but will climb rocky bluffs higher than 50 feet, and often rest on human-made structures such as buoys, docks, floats, and breakwaters; the best known of these are Pier 39 in San Francisco Bay and the Monterey Bay breakwater. Pupping and breeding sites are primarily on sandy beach and flat, rocky areas on islands and along shores exposed to prevailing winds.

VIEWING: In California, this species can be observed by boat on the Channel Islands, the Farallon Islands, Año Nuevo, and Bodega Rock. Mainland viewing is limited to Piedras Blancas, Point Lobos, Pier 39 in San Francisco, the Point Reyes Headland, the Fort Ross area, Trinidad Head, Cape Mendocino, and St. George Reef. In Oregon, sea lions occur along the Columbia River on the South Jetty, on log booms anchored along Carroll Channel, and, at Astoria, on marina floats and jetties in winter and spring. In Washington and British Columbia, there are several resting areas, but the Ballard Locks in Seattle is the best known. In Baja, sea lions occur on most of the islands, especially Cedros and San Benitos.

CONSERVATION: Killing of California Sea Lions was banned in Mexico, Canada, and the United States in 1969, 1970, and 1972, respectively. Prior to these protections, California Sea Lions were hunted by Native Americans, and later by Europeans, for food, hides, glue stock, whiskers (for pipe cleaners), and blubber. Two sea lions yielded a single barrel (42 gallons) of oil. Later, in the early 20th century, they were hunted for their reproductive organs, to serve as pet food, to augment scientific collections, and to reduce fishery depredation. Females were the primary pinniped collected for live display in aquariums, zoos, and circuses. Many of their preferred prey (salmon, rockfish, herring, sardine, anchovy, and squid) are commercially important to the fishing industry, resulting in frequent interactions with commercial and sport fishermen.

California Sea Lions have the dubious distinction of having some of the highest concentrations of PCB and DDE of any pinniped

Agreement	Conservation Status
U.S. ESA	Not listed
U.S. MMPA	No special status
COSEWIC	Not at risk
IUCN	Low risk

in the world. The pollutants contributed to aborted fetuses in the 1970s, and scientists continue to document tumors in an alarming 18 percent of stranded sea lions in California. Sea lions are also particularly sensitive to changes in oceanic conditions, as seen during ENSO events, when they have difficulty in finding food.

POPULATION: The earliest population estimate, 7,000 sea lions, was gathered from aerial surveys in the 1930s. After protection, the population in southern California increased to 27,000 in 1975. The population has continued to increase at 5 to 6 percent per year through the 1990s to an estimated 110,000 in California and 92,000 in Mexico, with an overall population of 238,000 to 244,000 in 2005. California Sea Lions are now the most abundant pinniped in the California Current.

COMMENTS: State and federal agencies recently proposed selective removal of California Sea Lions that habitually forage at fish ladders, where they easily prey upon threatened and endangered salmon. The best known location is Ballard Locks in Seattle; a few sea lions were captured there and transported to California, with one individual, nicknamed "Hershel," notorious for returning, and subsequently being recaptured, several times.

REFERENCES: Ainley et al. 1982, Antonellis et al. 1983, Bigg 1985, Cass 1985, DeMaster et al. 1985, Feldkamp et al. 1989, Gilmartin et al. 1976, Heath 2002, Lowry and Forney 2005, Odell 1981, Peterson and Bartholomew 1967, Ripley et al. 1962, Stewart and Yochem 1984, Stewart and Yochem 1987, Weise et al. 2007.

NORTHERN FUR SEAL *Callorhinus ursinus*
Pls. 17, 19; Figs. 91, 98, 99.

Adult male average length is 1.5 to 2 m (5 to 7 ft); average weight is 181 to 272 kg (400 to 600 lbs). Adult female average length is 1.3 to 1.5 m (4.5 to 5 ft); average weight is 41 to 50 kg (90 to 110 lbs). Pup average length is 0.6 m (2 ft); average weight is 27 kg (60 lbs).

SCIENTIFIC AND COMMON NAMES: The Northern Fur Seal is the only species of the genus *Callorhinus* (figs. 98, 99). *Callorhinus* is Greek for "beautiful nose," and *ursinus* is Latin for "bearlike," though some researchers have taken the term to mean "beautiful fur." Initially, all Northern Fur Seals were known as "sea bears," based on Georg Wilhem Steller's description on the Commander Islands. The species was identified under several zoological names, including *Siren cynocephala,* for "dog-headed mermaid." One name, *Arctocephalus californianus,* which is Greek and Latin for "California bear-head," was based on a specimen collected in Monterey Bay. The common English name is Northern Fur Seal, and the French is *otarie des Pribilofs*; Spanish is *lobo fino del norte.*

DESCRIPTION: The head is very distinct, with a rounded convex shape, very short muzzle, and broad nose. The bull head looks proportionately small for the size of the body. Fur seals differ from other pinnipeds by relying on a dense underfur, rather than blubber, for insulation. One researcher estimated 9,000 hairs per square inch in a Northern Fur Seal, but no one has tackled the tedious task of comparing hair counts between sexes or age groups. The rear flippers are the longest of all eared seals', accounting for 25 to 33 percent of body length, and the long toes appear floppy. The digits are very long, and the middle three have nails. The front flippers are long and naked, with a distinct line where the fur stops at the wrist. The longer ear pinnae stand out because of the lighter-colored hair from the insertion to the tip.

Figure 98. Northern Fur Seal breeding colony on San Miguel Island.

Figure 99. Immature Northern Fur Seal.

The vibrissae are also longer in bulls than in other species and give the impression of a long mustache; one of the longest measured was 0.3 m (13 in.). In bulls, the vibrissae are black at birth but become white with age, and appear molted before adulthood.

The sexes are very different—even as pinnipeds go—in color, size, and shape. Bulls are black, dark brown, or cinnamon, with grizzled guard hairs on the mane, and females are light brown or gray with lighter silver, tan, or white hair on throat and chest and light markings around the muzzle. Pups generally are born black with light markings on the muzzle and chest, and shed to silver back with creamy belly. When wet, all appear dark brown or black; however, the lighter chest of females and young is distinguishable.

Bulls are four times larger than females, and develop a massive neck and shoulders with a longer mane. Females, in stark contrast, are petite, with the typical body form of eared seals. Sexual dimorphism is measurable even at birth.

FIELD MARKS AND SIMILAR SPECIES: Key features include dense fur, for insulation; rounded head, with very short muzzle; the longest rear flippers of any pinniped; and naked front flippers beginning at the wrist. Guadalupe Fur Seals are the only other fur seal in the northern hemisphere, and differ from Northern Fur Seals by their narrow, pointy muzzle and a hairline at the foreflipper, extending beyond the wrist. Although similar in size and shape, the head shape of the small California Sea Lion is doglike and quite different from that of fur seals.

NATURAL HISTORY: Northern Fur Seals are noteworthy as a pinniped, because they lead a mostly pelagic life (7 to 9 months) and come onshore only during the breeding season (2 to 5 months). This species is the most pelagic of all pinnipeds; young animals may stay at sea 95 percent of the year, even spending multiple years at sea before breeding. This pelagic period is only possible due to the extremely efficient insulating qualities of their fur, and their ability to molt at sea.

Despite their individual abbreviated time onshore, seals are present on rookeries for nearly half the year because of the different needs of each age and sex class. Males, for example, arrive early in the season to establish territorial dominance and leave after 2 to 3 months. Pregnant females arrive a few weeks after the males, give birth, and then, over the next 3 to 5 months, intermittently attend to pups onshore, in between foraging trips. Females are absent from the rookeries November through April.

Young and nonbreeding adult seals congregate briefly in large numbers late in the season, after most territorial males have departed. Although the species shows strong fidelity to specific rookeries, researchers have documented much exchange between some colonies. The rookery on the Farallon Islands, for example, was colonized in part by females from the Pribilof Islands and males from the Channel Islands. The annual cycle does not include molting onshore. The importance of fur for insulation explains why Northern Fur Seals do not have a distinct annual molt period, but rather shed slowly during their long annual pelagic period.

In addition to the seasonal cycle, they follow a daily cycle on rookeries and at sea. Attendance at colonies may fall during the night, when females often forage, and is higher during the day, when they rest and attend pups. Temperature and tide have some influence on this pattern, depending upon the site. Because of their thick fur, seals may retreat to the water for relief during warm days, particularly in colonies in California. When at sea, seals typically feed during the evening, night, and early morning, spending the rest of the day on the surface. Resting and preening at the surface can account for more than a quarter of time at sea.

Fur seals are very gregarious on land, and their social organization has been intensively studied longer than that of any other pinniped or marine mammal. Both Georg Wilhem Steller, a naturalist who was shipwrecked on Beaufort Island in 1749, and Gerasim Pribilof, who studied them on the Pribilof Islands in 1786, described the breeding season.

Their social structure is spatially typical for most polygynous pinnipeds: there is a core group of breeding males defending territories with females, idle males without females are on the fringe of the core area, and idle males and subadult males are on haul-outs outside the rookery areas.

When at sea, fur seals are mostly solitary but often gather in small feeding groups of 2 to 4. Large, loose gatherings of over 100 may occur where food is concentrated, but are rare.

The long rear flippers are highly vascular, and by waving them back and forth in the air, seals can offload heat when on the rookery. When at sea, they will hold the front flipper in the air for the same effect, or link front and rear flippers in a "jug handle" posture.

Male vocalizations are important for communicating social dominance. Males have a loud, deep, repetitive growl. Females can also be territorial during the breeding season, and will hiss to defend pups. The vocalization of pups is similar to the repetitive bleating of lambs. Females and pups communicate by calling, particularly when females return from foraging trips, and their voices are individually distinctive.

Fur seals are very playful among themselves as well as with other species, gleefully teasing the less agile pinnipeds. They will repeatedly body-surf into shore with the same determination as human surfers. Researchers who handle this species extol and curse their intelligence—and aggression: holding even a pup is submitting oneself to a buzz saw. During the breeding season, males and females are very aggressive and can intimidate even larger pinnipeds, such as male sea lions.

Females are sexually mature at 2 to 5 years, and males at age 3 to 4. Males are not fully mature until age 6, and rarely hold territories before age 8 to 9. The estimated lifespan is longer for females (20 to 25 yrs) than for males (18 to 20 yrs). The maximum age is estimated to be 35, but the average life expectancy is only 4 years.

REPRODUCTION: The breeding season begins in April and extends to November, with peak birthing from late June to early July. The annual peak breeding period varies by 2 weeks between Alaska and California rookeries and generally lasts 125 days (4.2 months), the length of time the pup nurses. In California, the breeding season begins with the arrival of males, as early as April, and extends into August, with peak abundance of all sexes onshore in June. The size of the male territory varies with rookery size and site characteristics, but usually males defend a group of 15 to 20 females. As colonies become large, the borders of territories are less distinct, and females move back and forth. In smaller colonies, though, males may seize departing females and toss them back into their territory. Females do not choose bulls, but gravitate to other females, and to good habitat for birthing.

Males defend actual boundaries of a territory rather than their position near females; some males will defend the same territories over multiple years. To defend their territory, males display a ritualized set of behaviors, including vocalizations, charging other males, and chest butting—but rarely overt fighting. Because males fast when on territory, fighting is minimized to preserve strength. The longest recorded fasts have been over 80 days (2.7 months), though the average is 40 days. Bulls can lose up to 20 percent of their weight while fasting.

Pregnant females start arriving in June in California, and give birth to a single pup within 2 days of arrival; twinning has rarely occurred. A female nurses her pup for 8 to 10 days, mates about 5 to 6 days after giving birth, and then goes on the first of a series of 12 to 14 feeding trips, returning to nurse the pup. The average length of a feeding trip on San Miguel Island is 4 days, and time onshore nursing is 1 to 2 days. As the breeding season progresses, the length of the feeding trip extends to 7 to 10 days. Round-trip distances of feeding trips in Alaska were 400 km (250 mi); one trip was 740 km (460 mi). Unattended pups form pods on the beach until their mothers return. Females continue this feeding/nursing cycle through September or October, for about 4 months, until the pup initiates weaning. The length of nursing may be closely linked to the female's success in finding food, and in lean years they will abandon pups. In contrast with sea lions, fur seal females do not take their pups to sea, but recognition between females and their older offspring have been documented on rookeries for up to 4 years. Pups remain at rookeries until November and then go to sea. Remarkably, many pups remain at sea for 2 or 3 years before returning to breeding islands, very often to haul-outs surrounding the rookery where they were born. Most females are in their reproductive prime at 8 to 13 years, and are pregnant every year, with the pregnancy rate gradually decreasing after 13 years of age.

FOOD: Fur seals are solitary and generalized feeders, with a list of nearly 70 species identified in their diet. The top 10 prey consumed by Pribilof seals were squids, Capelin, Pollock, mackerel, herring, deep-sea smelt, salmon, flatfishes, Sablefish, and sand lance. One study in British Columbia identified squid and herring as representing two-thirds of the diet. In California, food consists of at least 26 species of fish and nine of cephalopods, but the dominant prey are anchovy, Pacific Hake, herring, saury, lanternfish, squid, and rockfish. Fur seals occasionally also eat birds, such as oceanic petrels.

Seasonal and annual differences in diet have been documented from looking at the stomach contents of dead seals. From data collected between 1958 and 1974, researchers noted that female Northern Fur Seals wintering in California preyed primarily on Northern Anchovy from January to March, Pacific Hake from April to May, and Market Squid from January to June. One study estimated that Northern Fur Seals annually consumed 6 percent of the anchovy biomass off California, as well as 7.5 percent of the Pacific Hake and 20.5 percent of the Market Squid. Reproductive status may also affect the quality and quantity of food consumed, since nursing females spend, on average, 1.6 times longer feeding than nonnursing females do.

Although they favor waters off the continental shelf, fur seals do not dive very deep, but forage in shallow to midwater depths of deep oceanic waters. The maximum depth recorded during dives of nursing females in Alaska was around 160 m (525 ft); the average was 54 m (172 ft). Fur seals are mostly solitary at sea and individual foraging habits vary. Deep dives occur primarily during the day, and shallow dives at night, because at night seals feed on species that migrate to the surface then, such as lantern-fish and hake.

MORTALITY: In California, El Niño–Southern Oscillation (ENSO) events cause severe mortality. From 67 to 80 percent of the pups born at the San Miguel colony died before weaning during the 1982 and 1998 ENSO events, and the population did not recover for several years after the 1982 event. Heat prostration caused mortality of very young pups; mothers abandoned pups when unable to find food. Although not well documented outside of Alaska, predation is likely caused by sharks and Killer Whales. In Alaska, male Steller Sea Lions also prey on fur seals.

DISTRIBUTION: Northern Fur Seals are a wide-ranging offshore species found throughout the North Pacific Ocean, stretching

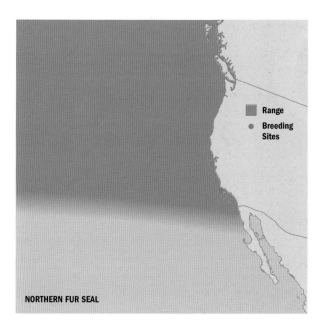

■ Range

● Breeding Sites

NORTHERN FUR SEAL

from the Bering Sea down to southern California in the east and central Japan in the west. Almost three-quarters of the global population breed on the Pribilof Islands of St. George and St. Paul in the southern Bering Sea. South of Alaska, they breed only on San Miguel Island and Southeast Farallon Island, in California.

Most seals are on or near the northern breeding grounds from May through October, but after the breeding season, seals migrate widely. Migratory trips can extend 6,200 miles in a season, the longest distance of any pinniped species except the Northern Elephant Seal. One young fur seal even traveled from the Commander Islands in Russia to the Farallon Islands in California. Some bulls start their migration in August; others stay onshore as late as November. Bulls from the Pribilof Islands generally migrate as far south as the Gulf of Alaska, but those from California remain in offshore waters of the state. Females and young from the Alaskan rookeries migrate near to the continental shelf from Canada down to California. They are the most abundant pinniped in the central and northern California offshore waters during the winter months, numbering perhaps 200,000. Despite their long travels, recorded swim speeds have ranged up to only 24 kph (15 mph).

HABITAT: At sea, fur seals occur mostly off the continental shelf in deep waters, 32 to 112 km (20 to 70 mi) offshore, concentrating in upwelling areas in water temperatures of 4 to 10 degrees C (40 to 50 degrees F). Underwater features such as seamounts and ridges and offshore islands are areas of concentration. Average densities in good foraging habitat can be $41/km^2$ ($16/mi^2$). Examples of areas of particularly high densities are offshore between Cape Vizcaino and Point Reyes, outside Monterey Bay south to the Sur Canyon, over the Mendocino Ridge west of Points Arguello and Conception, the Santa Rosa–Cortes Ridge, and the San Juan Seamont.

Because fur seals can walk on all four limbs, they can access habitat high up on a beach or rock, far above the surf zone. Generally, rookeries are near the continental slope on remote islands with rocky shoreline. On San Miguel Island, fur seals rest and breed on a flat sandy beach (Adams Cove) and on flat volcanic bedrock (Castle Rock).

VIEWING: Viewing of fur seals at rookeries is limited to only two locations south of Alaska, San Miguel Island and South Farallon

Island. People who venture on oceanic field trips, however, can land on San Miguel Island, Channel Islands National Park. In open waters, one can easily observe fur seals with their typical "jug handle" pose off the Channel Islands, Monterey Bay, and along the continental shelf break, from the Farallon Islands north to Cape Mendocino.

CONSERVATION: The prehistoric range of Northern Fur Seals extended from southern California north to the Bering and Okhotsk Seas and the Sea of Japan. In California and Washington, fossil records documented Northern Fur Seals breeding on mainland sites including the southern Channel Islands, the Farallon Islands, and along the mainland from as far south as southern California.

Agreement	Conservation Status
U.S. ESA	Not listed
U.S. MMPA	Depleted
COSEWIC	Threatened; subsistence hunting allowed
IUCN	Vulnerable

Northern Fur Seals' subsequent demise by European sealers is a textbook example of commercial hunting's taking a species to the brink of extinction. First described on the Commander Islands by the shipwrecked naturalist Georg Steller in 1742, the rapid depletion of the species was complete within 150 years. There were 2.5 million Northern Fur Seals discovered by Gerasim Pribilof between 1786 and 1787, but by 1909, only an estimated 300,000 remained. In California, fur seals were extirpated by the late 1800s; the logbook of one sealing vessel documented more than 150,000 slaughtered on the Farallon Islands in 1834.

In an effort to preserve the industry, fur seals were protected by the North Pacific Fur Seal Convention, signed in 1911 between Japan, the United Kingdom (for Canada), Russia, and United States, regulating the number hunted to include immature males only, and banning the hunting of seals at sea. Japan terminated the agreement in 1941 during World War II, but the agreement was reissued in 1957, and expired in 1984. With the protection provided by these conventions, the population rebounded to an estimated 1.5 million in 1957 despite annual commercial harvests of 66,000 young males. Fur seals recolonized San Miguel Island in the 1960s and the Farallon Islands in 1996. On San Miguel Island, the annual rate of increase in pups between 1969 and 1978 was 46 percent from a total of 28 to 635 pups. This increase was due, in part, to colonization of females from the Pribilof Islands. A similar recovery began on the Farallon Islands after an absence of 170 years, with 80 pups born within 10 years of the first birth.

Initially, this remarkable recovery was touted as a model for successful management. But an experimental hunt of females 1956 to 1968 (> 300,000 total) in addition to the annual harvest of young males (52,000) caused another population crash. Between 1975 and 1982, the population declined 6 to 8 percent per year; in Alaska, it dropped by more than half. By 1985, the United States ceased the annual harvest. After 2000, the population in Alaska continued to decline, and estimates of pup production in 2004 were 50 percent lower than 1977. Many causes for the decline have been proposed, including predation, fishery interactions, disease, contaminants, and climate change.

In the past, entanglement in high-seas driftnet fisheries have drowned many fur seals, but presently the pollack fishery in Alaska may be eliminating the fur seals' preferred prey, as has been proposed to explain the decline of Steller Sea Lions. The drastic reduction in the Alaskan population will be reflected in the number of females wintering in California waters.

POPULATION: The estimated breeding population of fur seals on San Miguel Island was 9,500 in 2004, and 200 on the Farallones in 2006. The total world population is estimated at 1,345,000 to 1,365,000.

COMMENTS: Before Native American and European hunting, fur seals were likely the most dominant pinniped in the eastern Pacific, with very large breeding colonies on the Farallon Islands and the Channel Islands, and in mainland rookeries such as those of Monterey Bay. Their generalist diet and domineering, aggressive behavior toward other pinnipeds may explain their previous, and perhaps future, success.

REFERENCES: Antonellis and Perez 1984, Gentry 2002, Gentry and Kooyman 1986, Irving et al. 1962, Loughlin et al. 1986, Newsome et al. 2007, Ream et al. 2005, Scheffner 1958, Towell et al. 2006.

GUADALUPE FUR SEAL

Arctocephalus townsendi

Pls. 17, 19; Figs. 100, 101.

> Adult male average length is 1.8 to 2.4 m (6 to 8 ft); average weight is 170 kg (375 lbs). Adult female average length is 1.2 to 1.5 m (4 to 5 ft); average weight is 50 kg (110 lbs). Pup average length is 0.6 m (2 ft); average weight is 4 kg (9 lbs).

SCIENTIFIC AND COMMON NAMES: The name *Arctocephalus* is derived from the Greek words for bear *(arktos)* and head *(kephale)*. Not until the 1970s did scientists confirm *Arctocephalus townsendi* as being distinctly different species from other fur seals. In 1892, Charles Townsend of the American Museum of Natural History observed a mere seven seals and collected old, worn skulls of the species on Guadalupe Island; shortly thereafter, Clinton Hart Merriam described the new "sea bear," based on the skulls, and named the species after Townsend (figs. 100, 101). Now commonly called the Guadalupe Fur Seal, because it breeds exclusively on Guadalupe Island, the species has also been called the California Fur Seal. Common names include the English Guadalupe Fur Seal, Spanish *lobo fino de Guadalupe,* and French *otarie de Guadalupe.*

DESCRIPTION: Guadalupe Fur Seals and Northern Fur Seals, are the only fur seals present in the northern hemisphere. Their thick layer of fine underfur, rather than blubber, prevents heat loss, also giving them buoyancy by trapping air. The Guadalupe Fur Seal, the smallest of eared seals in North America, has a large head with a long, pointed snout. The head shape is narrow, and flat above the eyes. The muzzle is narrow and one-third the length of the head, and the shape is accentuated by a slightly bulbous nose with downward-facing nostrils. The long front flippers are one-fourth the body length and mostly naked, except for some hair extending past the wrist. The middle three digits of the small rear flippers have nails, as do all five of the front flippers. As with all eared seals, bulls have distinguishing secondary sexual characteristics. Bulls are two to three times larger than females, have larger and more muscular necks and shoulders, and have distinct manes with light-tipped, coarse guard hairs.

Figure 100. Guadalupe Fur Seal bull.

Colors range from dusky black to chestnut, with the head and shoulders appearing paler and gray. The overall look is grizzled when dry, particularly in bulls, because the guard hairs are tipped gray to white. The back ranges in color from gray-brown to yellow. Brown to red markings are present around the eyes. For both sexes, coloration of the chest and neck is lighter—pale gray. Pups are black at birth, with tan hairs at the back of the neck, but acquire the grizzled look, when dry, by age 6 months. In adults, the whiskers are light to cream-colored.

FIELD MARKS AND SIMILAR SPECIES: Key features consist of dense fur with grizzled guard hairs; a low, sloped forehead; and a long, narrow, pointed muzzle. Guadalupe Fur Seals can be confused with Northern Fur Seals and California Sea Lions. Northern Fur Seals are the only other fur seal in the northern

Figure 101. Guadalupe Fur Seal immatures.

hemisphere and are only slightly larger, but the two species are very different. The Northern Fur Seal has a high forehead with a facial angle greater than 125° (i.e., more vertical) and a short snout. Also, on the Guadalupe Fur Seal's foreflipper, the fur extends beyond the wrist, whereas on the Northern Fur Seal's, it stops at the wrist. Guadalupe Fur Seals are more grizzled in color than Northern Fur Seals, and the exterior ear is the same color as the body (it is paler in the Northern Fur Seal). Guadalupe Fur Seals can be distinguished from California Sea Lions by their dense fur with its halo of coarse guard hairs, as well as by their flat (rather than concave) forehead and pointed snout. However, when young Guadalupe Fur Seals, Northern Fur Seals, and California Sea Lions commingle at haul-out sites, they are difficult to distinguish. When in the water, the Guadalupe Fur Seal and Northern Fur Seal can be distinguished by their head shape.

NATURAL HISTORY: Until recently, much of Guadalupe Fur Seal natural history was unknown—and inferred from the behavior of other species—because their population was small and geographically isolated. Through the use of new technologies, we now know that female and young Guadalupe Fur Seals are present year-round on and near Guadalupe Island, and that males (in contrast to females) follow a seasonal migratory cycle away from the island after the breeding season, ranging mostly north to the Southern California Bight, California. The species went through two severe genetic bottlenecks over 100 years, between the 19th and 20th centuries, and was assumed extinct until a small colony was discovered on Guadalupe Island, first in 1928 and, again, in 1954. Nevertheless, compared with other seal species that also experienced such bottlenecks, the Guadalupe Fur Seal's genetic diversity is considered relatively high. Comparison of the genes of "pre-bottleneck" samples (extracted from bone) with extant samples have showed that there was significant loss of ancient genotypes (24 of 25 identified), but the same genetic comparisons suggest to researchers that the species' preexploitation population was large and increasing. It is not known how long Guadalupe Fur Seals can live, but the lifespan is thought to be 13 years for males and up to 23 years for females.

Guadalupe Fur Seals are gregarious at their island colonies, but from the little information available, we suspect that they are solitary at sea.

Temperature, especially heat, has a major influence on how Guadalupe Fur Seals behave. When average temperatures are 20 degrees C (70 degrees F), fur seals on Guadalupe Island seek shade and windy areas, avoiding physical contact with other seals. They spend most of their time resting or grooming onshore, and frequently move back and forth to the wet intertidal zone. When resting in water, they float in a characteristic posture with head hanging down, and often expose the rear flippers, slowly waving them in the air—no doubt to cool off—or assume a "jug handle" posture, with front and rear flippers linked. When moving swiftly through the water, Guadalupe Fur Seals "porpoise" in a fashion similar to that of other eared seals. Play behavior occurs among juveniles, which have been observed chasing each other in the tide pools of Guadalupe Island.

Bulls bark, roar, or cough; some have likened the sound to that of a steamboat whistle. Spectrograms of males' calls display a distinct ascending scale, from hoarse growl to rasping tenor. Females produce a bawling sound, directed at pups.

REPRODUCTION: Guadalupe Fur Seals breed from mid-June to July, with most pups born in June. This species' mating system is polygynous, but harem size, at around six females (with a range of two to 10), is smaller than in other eared seals in North America. Mating and birthing occur in or around entrances of sea caves, so harem size may be limited both by cave size and by overall small population. Males can more successfully defend cave entrances than open habitat, and both sexes will defend territories from other seal species. Once on a territory, a male's tenure lasts from 35 to 122 days, and he will regularly bark at nearby males. Researchers have noted that Guadalupe Fur Seals on Guadalupe and San Miguel Islands are faithful to particular sites over a number of years.

Females start arriving at the breeding sites in June; each gives birth to a pup within a few days of her arrival. Pups are born from early June to early August, but late June is the median pupping date. After 7 to 9 days with the newborn, the mother mates, and then leaves on her first trip to forage at sea. Implantation likely occurs 4 months later, around the fall equinox. The female lactation–foraging cycle is very long compared to that of other eared seals, lasting 9 to 11 months. The mother spends an average of 9 to 13 days at sea before returning to land, for an average of 5 to 6 days, to nurse her pup. She fasts while on land, for as long as 14 days. The fat content of female milk was noted in one study as 50 percent.

FOOD: Guadalupe Fur Seals feed on squid and boney fish such as myctophids (lanternfish) and mackerel. Squid beaks have been the main content found in the stomachs of dead animals stranded in central California. This seal also ingests several species of crustaceans and fish, including Pacific Sanddabs, but also a deep-sea pelagic fish, Scopelogadus, also called ridgeheads. Fur seals feed by day and night, mostly on species that migrate up and down the water column toward the surface at night. The bite of a cookie-cutter shark was observed on at least one bull, indicating that seals may feed at the minimum depth of 91 m (300 ft), where this shark occurs. Research using telemetry on a group of adult females in the California Current discovered that they fed south of Guadalupe Island, making remarkable round trips of 703 to 4,023 km (437 to 2,500 mi), and that they foraged at depths of less than 30 m (100 ft). Longer trips were associated with ENSO years, when food was less abundant. One young female, treated and released with a satellite tag, traveled from central California to Guadalupe Island in 2 weeks.

A malnourished female Guadalupe Fur Seal, when treated in captivity, converted an average diet of 6 pounds of fish per day into 9 ounces of weight gain.

MORTALITY: Individuals, especially juveniles, strand on beaches, particularly during El Niño–Southern Oscillation (ENSO) years. One-third of the pups died in 1992, due in part to an ENSO and Hurricane Darby. Information on natural predators is scarce, but Guadalupe Fur Seals likely face the same predators as other fur seals—sharks and Killer Whales.

DISTRIBUTION: The rarest fur seal, the Guadalupe Fur Seal breeds almost exclusively on Guadalupe Island, Mexico; however, in 1999, a new colony with nine pups was discovered on Isla Benito del Este, off the coast of Baja, and in 1997 one pup was born on San Miguel Island within the California Bight. Outside the breeding season, this species' core range is from Guadalupe Island north to the California Channel Islands, but individuals are occasionally sighted as far south as Tapachula near the Mexico–Guatemala border; as far north as Mendocino, California; and east, in the

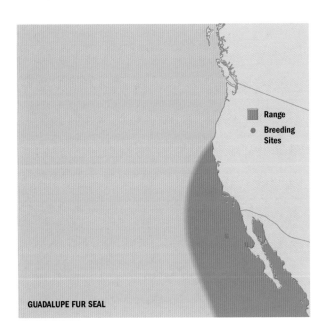

■ Range
● Breeding Sites

GUADALUPE FUR SEAL

Gulf of California. The northernmost record is a stranding north of the Columbia River, Washington State, in 1992. As they increase in number, Guadalupe Fur Seals are expanding their range and are regularly seen on San Miguel and San Nicolas Islands, and, occasionally, on the South Farallon Islands.

Very little is known of Guadalupe Fur Seals' migratory routes. Subadult and adult males are thought to move through southern California prior to the breeding season, and young may move into the Southern California Bight from February to April through May, after weaning. Individuals are seen regularly on the Channel Islands, but only rarely on the California mainland. With increasing frequency, however, young seals strand every few years, particularly during ENSO events. In 1977, an individual beached in Monterey Bay, and between 1988 and 2000, several individual seals stranded in central and northern California.

HABITAT: We can infer from their body size and diet that Guadalupe Fur Seals are mostly shallow divers, feeding in the midwaters near islands.

Their preferred habitats on islands are volcanic caves and rocky grottos where the substrate is solid rock, even when sandy beaches are available. Subtropical temperatures at the warm latitude of Guadalupe Island likely drive the species to seek out caves, crevices, and the shadows of large boulders for cover and a cooler environment. The sea caves are often at the bases of cliffs and are used for breeding. The use of caves may explain the survival of the species; seals have ventured up to 80 feet into caves on Guadalupe Island. On San Miguel Island, though, males have intermingled with California Sea Lions on sandy beaches. One female with pup on San Miguel Island was observed on a high rock shelf within a boulder-strewn rocky cove.

VIEWING: Rare sightings may occur at sea, on visits to San Miguel Island, or during ENSO years along the central California coast. Guadalupe Island is the only place to see this species regularly.

CONSERVATION: Prehistorically, Guadalupe Fur Seals likely ranged from the Pacific Northwest south to Revillagigedo Islands off Mexico, south of the California Current. The fossil record, based on bones identified in Native American middens, places the species near locations such as the Makah Ozette site at Cape Alava in northwestern Washington, the Emeryville shell mound in San Francisco Bay, and Chumash sites on San Miguel and San Nicolas Islands. Evidence for the core breeding range through history, though, suggests limits similar to today, from the Channel Islands south to the islands off Mexico.

Agreement	Conservation Status
U.S. ESA	Threatened
U.S. MMPA	Depleted
COSEWIC	Not listed
IUCN	Near threatened

As with Northern Fur Seals, Guadalupe Fur Seals were hunted intensively commercially for their fur. Records indicate that 73,402 skins were taken between 1810 and 1812 on the Farallon Islands alone, but researchers disagree about the identity of these animals. Sealers did not distinguish between Northern Fur Seals and Guadalupe Fur Seals, and the latter was not identified as a distinct species until after commercial harvesting died out. Townsend estimated that 52,000 Guadalupe Fur Seals were harvested, from Mexico to California, between the late 1700s and 1848. Sealing continued in Mexican waters through 1894, when the species was considered extinct. Townsend collected live specimens from the colony on Guadalupe Island in 1926. For nearly 30 years afterward, sightings were rare, and the Guadalupe Fur Seal again was thought extinct. George Bartholomew, a scientist from the University of California, discovered a Guadalupe Fur Seal on San Nicholas Island in 1949, and in 1954, a hopeful Carl Hubbs of the Scripps Institute led a party to Guadalupe Island to search for the island's namesake seal. On the last evening of the last day of the expedition, Hubbs finally heard an unfamiliar bellow near the mouth of a cave. He described this back-from-oblivion moment: "Full of expectation, we turned right into the surge, where in the early dusk I could plainly see the sharp, almost collielike snout, coarse guard hairs overlying silky under hairs, and other characteristics of the Guadalupe Fur Seal" (Hubbs 1956). Fourteen more animals were counted.

Smoothly polished lava rock faces on Guadalupe Island extend some 30 yards above sea level and back into sea caves, attesting to the possibility of large numbers of furs seals present there in the past. Estimates of the species' population size before depletion by commercial hunting ranged from 20,000 to 200,000, and the Guadalupe Fur Seal is the most abundant pinniped represented in Chumash middens on San Miguel Island.

POPULATION: The estimated population in 1977 was fewer than 2,000 seals, but by the late 1990s about 7,500 to 12,000 were counted, and the colony continues to increase exponentially, at about 10 percent per year.

COMMENTS: Guadalupe Fur Seals experienced a genetic bottleneck and consequently a loss of genetic variability in the 19th century, which suggests that they are more vulnerable to extinction; however, like Northern Elephant Seals, which also lost genetic diversity, their population is robust and growing. In fact, they may be more vulnerable to localized disasters such as oil spills, hurricanes, or climate change, because they breed only in a limited area. During the breeding season, these seals are sensitive to any form of human disturbance, and Guadalupe Island is a pinniped sanctuary.

REFERENCES: Antonellis and Fiscus 1980, Belcher and Lee 2002, Etnier 2002, Hanni et al. 2007, Hubbs 1956, Scheffner 1958, Weber et al. 2004.

HARBOR SEAL
Phoca vitulina

Pls. 18, 19; Figs. 16, 35, 93, 102.

Adult average length is 1.5 to 1.8 m (5 to 6 ft); average weight is 75 to 112 kg (200 to 300 lbs). Pup average length is 61 to 102 cm (25 to 40 in.); average weight is 6 to 9 kg (15 to 25 lbs).

SCIENTIFIC NAMES: *Phoca* means "seal" in Greek, and *vitulina* means "calflike" in Latin. Within the north Pacific there are two subspecies: *P.v. richardii* in the eastern Pacific, ranging from the Pribilof Islands in Alaska to Baja California and as far south as San Roque Island, Guadalupe; and *P.v. stejnegeri* in the western Pacific, ranging from Hokkaido, Japan, north to the Commander Islands in Russia. The naturalist John Edward Gray named the eastern subspecies *Halcyon richardii* after Captain G. Richards, the hydrographer on the voyage when the first specimen was collected on Vancouver Island in 1864 (although another account has the name as *Halcyon californica* in 1866). Later, the subspecies was renamed *P.v. richardsi.*

COMMON NAMES: There are historic names for the species from many cultures, such as Sea Calf and Sea Dog in England, and Selchies from Celtic mythology. Harbor Seal, Spotted Seal, and Common Seal are the usual English names for this worldwide species, in Spanish it is called *foca comun,* and in French *phoque veau marin* (marine seal calf) or *phoque commun.*

DESCRIPTION: Harbor Seals are the only pinniped species found in north temperate waters of the Atlantic and Pacific Oceans (fig. 102). They are the quintessential earless seal, having a spindle shape of medium size and short limbs. The head is doglike but broader, and the nostrils form a V. Their whiskers are light-colored and beaded, and they have well-developed claws on their front flippers, used in catching and holding prey, as well as less-functional claws on the rear flippers. The ear opening is large and easily visible behind the eye.

Onshore, they are cryptic, particularly on rocky substrates, because of their mottled coloration. Their color varies widely by latitude, as well as by individual. Two primary color phases follow a latitudinal cline in the eastern Pacific; a dark phase dominates in seals south of Point Conception (dark background

Figure 102. Harbor Seal. Note healed shark bite on lower flank.

with white rings around black spots), and a light phase (dark spots on a light background) from central California to the eastern Aleutian Islands. The light phase also tends to have the typical countershading of other marine mammals, with a denser spot pattern on the back fading to little or none on the belly. The fur is coarse, lacking insulating guard hairs; the seals rely on their blubber for insulation. The fur of newborn pups can, on occasion, include a downy fur layer colored silver to gray. Called *lanugo,* this natal fur is usually shed *in utero* in populations south of Alaska. The *lanugo* fur can trap air and insulate newborn pups born in the colder latitudes of Alaska. Bright red-haired seals are occasionally seen throughout their range, but in San Francisco Bay, nearly 40 percent of the seals sport this red fur. These "rusty seals" have iron oxide deposited on the hair shaft, for reasons that are still poorly understood. One theory is that seals acquire the iron oxide when seizing prey from the iron rich sediments of the Bay. This theory explains the more common red-headed seals, but not the full bodied red ones. Genetics, foraging location, and pollution may each contribute to the red seals of the Bay.

Even for a trained eye, identifying the sex of Harbor Seals is difficult, because they exhibit little sexual dimorphism. Males are only slightly larger, and have a genital opening on the abdomen, just below the belly button; large males may have a thickened and scarred neck from fighting. Otherwise, there are no obvious sexual differences.

FIELD MARKS AND SIMILAR SPECIES: Harbor Seals are distinguished from other pinnipeds in the region by being the smallest earless seal. They can be confused with young Northern Elephant Seals, but their spotted pattern of silver, black, and brown coats contrasts with the uniform gray to tan of Northern Elephant Seals. Other differences from elephant seals include smaller eyes and head, and claws in their rear flippers. Seen from afar while in the water, Harbor Seals differ from sea lions by rarely porpoising, and their lack of long front flippers. Often, Harbor Seals rest in kelp beds with only their heads exposed.

NATURAL HISTORY: Harbor Seals are the least pelagic of the pinnipeds, and rest onshore almost every day; their small size and reliance on blubber for insulation likely account for their terrestrial tether. The longest interval that Harbor Seals have been documented staying in the water was around 10 days.

Within a day, tide level, time of day, and season affect their presence onshore. Seals gather every day, during mostly daylight hours, and primarily when their preferred haul-out sites are exposed by receding tides.

Seals follow an annual calendar that has three seasons: breeding, molting, and fall/winter. The breeding season follows a latitudinal cline, with pups born earlier in the south (February in Baja) and later in the north (June in Canada—but, inexplicably, in May in the Hood Canal, Washington). One eplanation is genetic because the Hood Canal population is distinctly different from the other eastern Pacific populations. All age classes haul out at colonies during the breeding season, but females, both pregnant and with pup, segregate from males and juveniles. The molt season follows about 1 month after the breeding season, but precedes implantation of the embryo in females. Individual seals molt for some 3 weeks, but the season lasts from 2 to 3 months, since different age and sex groups molt at different times. Before molting, the fur looks very dull, and spot patterns are indistinct; when newly grown, the fur appears luminous. Exposure to sunlight is associated with enhanced cell division of hair follicles and decrease in the metabolic rate, allowing seals to spend less time feeding and more time onshore—an average of 10 to 12 hours. On a typical day, seals will haul out for an average of 4 to 7 hours between feeding bouts, and most seals forage at night from dusk to dawn, resting during the day.

Harbor Seals are highly gregarious at sites onshore and when feeding. Despite their gregarious nature, however, Harbor Seals rarely tolerate physical contact with nearby seals on land and will defend their half-foot space of beach with head-thrusts and snarls. Nose touching in greeting is common, though, and females with pups continuously touch.

Harbor Seals are the quiet pinniped, and their infrequent repertoire consists of grunts, growls, and snorts. The underwater calls made by maritorial male seals are difficult for the human ear to discern, because the sounds are indistinguishable from the shushing and gurgling sounds, made by breaking waves. Their relative silence and the obscure underwater call of males may help seals avoid predators. Pups are the only ones that are regularly heard vocalizing. They emit a "kroo" sound when begging to nurse, when interacting with the mother, or when lost.

On land, Harbor Seals are wary of people and terrestrial predators, stampeding into the water in response to perceived threats.

Once in the water, seals are often curious and will follow boats or divers. They will slap the water surface to alarm other seals, and when engaged in male jousting. While resting in the water, they assume a perpendicular position with head raised above the surface ("bottling"), and often will do so in protected kelp beds. Sleeping underwater for long periods is also possible, and researchers have identified aquatic "haul-down" sites where numerous seals rest on the bottom together, briefly rising to the surface for a breath and sinking back down to their place on the seabed.

REPRODUCTION: Harbor Seals are the only pinniped species in the region that does not have a polygynous mating system. For decades, the Harbor Seal mating system was described as promiscuous because males did not compete on land to defend territories (their lack of sexual dimorphism supported this conclusion). Underwater sound and video recordings, however, revealed that male Harbor Seals form leks (gathering places where competing males display to attract females) while in the water, performing displays such as slapping, rolling, and leaping out of the water. Very recently, researchers have identified male aquatic territories adjacent to colonies ("maritories"), having discovered the location and size of a maritory by playing back male underwater calls. Individual male seals consistently and vigorously attacked a hydrophone that produced male calls if it was within their maritory, and maritories were often adjacent to haul-outs where females with pups were dominant. In Monterey Bay, males defended maritories ranging in size from 0.01 to 0.08 km (3 to 20 ac) for multiple and successive years.

Harbor Seals breed throughout their geographic range with a latitudinal birthing cline, extending from February to June. Seals are born progressively later in the season as one moves north from Baja, where pups are born in February, to Canada, where they are born in June. In northern California, pups are born primarily in late May; in the Southern California Bight, in March. Latitudinal changes in daylight along with timing of abundant prey likely affect birthing dates.

Both sexes are sexually mature from 3 to 5 years (the oldest animal in captivity lived around 30 years). When mature, females give birth annually to a single pup and lactate from their two nipples for an average of 30 days. Gestation lasts around 9 to 10 months, and with 1 to 2 months of delayed implantation, birthing is annually synchronous. Unlike most seals, Harbor Seal pups can swim at birth and are attended by their mothers from birth till the onset of weaning. When very young, pups will cling to the neck of their mothers while swimming. For the first few days after birth, females may remain onshore for more than 20 hours at a time and, during the first week of life, spend an average of 10 hours per day resting and nursing. They do not fast at this time, but do not depart on long or distant foraging trips. Pups do not form pods on the beach as is done in sea lion rookeries; instead, socializing centers on the female–pup relationship. If pups and mothers become separated, they mutually identify each other by sound, sight, and smell. Rarely, to protect a very young pup from large sea swells or human disturbance, a female will carry it in her mouth, doglike.

Male sperm count is high as early as February, but females come into estrus around 30 days after birthing. Within a few days of mating, females may depart the colony, thereby weaning the pup after nursing for only 30 to 45 days. Pups can disperse widely from their natal areas, but individual behaviors vary substantially. In central California, some pups dispersed more than 161 km (100 mi) within weeks of weaning.

In a lifetime, a female may give birth to 10 or more pups, but will skip years following El Niño–Southern Oscillation (ENSO) events, when prey abundance is low during the preceding fall and winter.

FOOD: Harbor Seals feed locally and opportunistically on whatever is abundant, but in winter they may journey to areas where prey concentrate. Seals also forage cooperatively in small and large packs of more than 25. Their diet consists of small schooling fish, cephalopods, and crustaceans, the presence of which vary seasonally, regionally and annually. They mainly eat anchovy, sardines, herring, hake, flatfish, rockfish, sculpin, perch, sanddabs, and squid. Other prey of note that are commercially important are Tomcod, salmon, and Steelhead; noncommercially important prey are hagfish, lamprey, shrimp, octopus, and bivalves. An example of Harbor Seals' opportunistic nature is their preferential eating of the nonnative Yellow-fin Goby in San Francisco Bay after native fishes declined. The weaning of pups in summer in some areas coincides with swarming mysid shrimp and schooling anchovies, upon which the seals gorge.

In an effort to understand how seals forage, researchers attached dive recorders to seals' backs, and placed temperature sensors in their stomachs. When a seal ate a mouthful of fish, the temperature sensor would send a signal to the tag on the seal's back, which also recorded how deep and long the seal dived. When the seal molted its fur, the tag on the back of the seal detached, and the researcher collected it, downloading the dive data to a computer. Researchers discovered that Harbor Seals are not deep divers—which is not surprising, since their prey occur mostly nearshore. Most dives were less than 5 m (16 ft), but the deepest dives are impressive for such a small animal, ranging from 225 to 500 m (738 to 1,640 ft). When analyzing dive patterns, researchers noted a U-shaped pattern similar to that of elephant seals whereby the seal swam along the bottom (no doubt looking for prey) and confirmed feeding when the stomach sensor recorded a drop in temperature. They also discovered, however, that seals would swim to the bottom and, apparently, lie in wait for prey moving along the bottom or swimming overhead. This sit and wait tactic is common among predators, including Great White Sharks.

Shallow nearshore waters are often cloudy, but seals can detect prey with their whiskers, which sense water movement caused by swimming prey. One researcher blindfolded a captive seal and then watched it follow in the wake of a swimming fish. In another captive study, seals detected slight concentrations of dimethyl sulfide, a gas produced by plankton; presumably, seals know where to find prey by smelling where there are concentrations of plankton. Estimates of prey consumption by Harbor Seals range from 4 to 8 percent of body weight per day depending upon season.

MORTALITY: Great White Sharks are Harbor Seals' main marine predator, although Steller Sea Lions and Killer Whales likely are locally important predators. On at least one occasion, a Killer Whale beached at Trinidad, California, and ate a Harbor Seal resting onshore. Smaller terrestrial carnivores such as coyotes and bobcats are likely a greater threat, because their numbers are more numerous, and they prey mostly on young, naive seals. Turkey vultures, gulls, and ravens attack and sometimes kill newborn, unattended pups. In recent years, solitary male Northern Elephant Seals have killed many Harbor Seals at colonies such as at Jenner and Point Reyes, California. Canine and phocine distemper, and influenza virus, though not yet documented in Pacific Harbor Seals, have caused massive mortality of Harbor Seals on both sides of the Atlantic.

DISTRIBUTION: This smallest of earless seals is also the most widespread of all pinnipeds, occurring throughout the oceans of the Northern Hemisphere, extending over 16,000 km (10,000 mi) between latitudes 80 and 20 degrees north. In the eastern Pacific, Harbor Seals occur from Ascension Island, Baja California, northward along the Alaskan peninsula. But despite their widespread occurrence, there is no free exchange of seals. Individual seals infrequently travel more than 800 km (500 mi) from natal sites, and individuals rarely venture more than 80 km (50 mi). Instead, the population is comprised of regional stocks that have been recently identified by genetic analyses. Pups are seen at most Harbor Seal haul-out sites throughout their range, but the highest numbers of pups are found at colonies in estuaries and on inaccessible sandy beaches.

Until recently, Harbor Seals were not considered migratory, because haul-out sites were occupied year-round; however, with the aid of radio and satellite telemetry, researchers have tracked some seals moving great distances to and from breeding and

Range

HARBOR SEAL

winter feeding areas. The itinerary of a typical journey might include an 8 to 16 km (5 to 10 mi) trip from a breeding site to a nearby haul-out site. The seal might use this nearby site as a base from which it conducts daily foraging trips. The seal then might extend the journey north or southward, visiting haul-out sites along the way. The entire journey may last a few months, with the seal visiting several colony sites before returning to the "home" site the following breeding season. These journeys indicate that there is substantial mixing among Harbor Seal colonies during the nonbreeding season. Not all adult seals migrate, though, and some remain at the "home" site throughout the winter.

In winter, seals often gravitate to bays and estuaries from the outer coast, following seasonally abundant prey, such as herring. In central California, seals migrate both north and south from Point Reyes to San Francisco and Monterey bays, some reaching as far as the Klamath River. On the southern Channel Islands, seals travel between islands, and occasionally to the mainland. Seals from Oregon and Washington colonies converge on the Columbia River during the winter months, when salmon spawn.

HABITAT: Harbor Seals generally reside in estuaries or in nearshore waters with both rocky and soft-bottom substrates on the outer coast. Although they travel great distances between islands, they rarely venture more than 20 km (12 mi) from shore, and most reside within 8 km (5 mi) of a haul-out. They also travel far up freshwater rivers such as the Sacramento, American, Russian, Klamath, and Columbia rivers when pursuing spawning prey such as salmon.

Harbor Seals cannot dive as deep as their larger relatives, but they do visit nearshore marine canyons such Monterey Bay. Most seals feed in waters less than 35 m (110 ft) and dive for 3 to 5 min, but the deepest recorded dives were over 500 m (1,600 ft).

Seals congregate year-round at haul-outs where generations of seals have probably gathered for centuries. Colonies may be on intertidal rocks, tidal mudflats, or sandy beaches; they prefer gradual slopes or low rocks, because they cannot maneuver easily onto land. Also important are wide visibility (to aid in the detection of predators) and easy access to deep water (to allow for rapid retreat). Large breeding colonies of more than 1,000 seals exist at remote locations where females give birth and raise pups without undue disturbance from humans and terrestrial predators. Suboptimal haul-out sites such as piers, log jams, and breakwaters are used by seals for convenience while resting between foraging bouts.

VIEWING: Harbor Seals are the easiest pinniped to observe, because they haul out at numerous accessible, even urban sites throughout their range. The more famous urban locations include Children's Pool Beach in La Jolla and Castro Rocks in San Francisco Bay. While driving along Highway 1 in California, seals are easily seen at Point Lobos, Pescadero Rocks, Pillar Point, Bolinas Lagoon, and Jenner. In Washington and Oregon, large groups of seals occur along the Columbia River and in large estuaries such Willapa Bay and Gray's Harbor. In the San Juan Islands and Puget Sound, seals haul out on many of the scattered islets.

CONSERVATION: The current range of Harbor Seals has probably not changed much since prehistoric times, perhaps because they were not hunted with the same intensity as other marine mammals, and because of their shy nature. They were hunted first by Native Americans and European settlers for food and fur, and, later, commercially for scientific display and collections. A substantial number of Harbor Seals were also killed by bounty hunters in the early 1900s to reduce fishery depredation, and bounty hunting and bombing of terrestrial sites continued up to passage of the MMPA. Over a 3-year period in the 1950s, over 21,000 Harbor Seals were estimated killed by bombs in Alaska. In California, only a few hundred individuals survived in isolated areas, particularly on the Channel Islands and along the north coast.

Agreement	Conservation Status
U.S. ESA	Not listed
U.S. MMPA	No special status
COSEWIC	Not at risk
IUCN	Least concern

Modern-day concerns include drowning in fishing gear, and loss of habitat through human development and disturbance. Human disturbance, including a variety of sources from dogs barking to kayaks' approaching too close, can cause seal flight. Seals may habituate over time to predictable nonthreatening activities; nursing females barely lift a head when large oil tankers dock within 400 m (1,300 ft) but will take flight when a kayak comes within 150 m (500 ft). If colony sites are frequently disturbed, however, seals do not habituate. In San Francisco Bay, for example, Harbor Seals adjusted initially to disturbances by shifting haul-out patterns to night hours, but eventually abandoned sites entirely when disturbances became chronic. Incessant human-generated noise or nearby activity that startles seals several times an hour will drive seals to quieter places for their daily rest. The type of disturbance is different for each area, but the more common ones include people on foot, dogs, boats (especially small boats that can maneuver closely), and low-flying aircraft such as helicopters. An emerging issue is the growing mariculture industry in estuaries where disturbance by operations may displace seals from haul-out sites. Provisions for protecting seal habitat (which do not currently exist under the MMPA) would diminish this potential threat.

More insidious harm may come from exposure to pollutants. Because seals live nearshore and often in highly developed areas, they are invariably exposed to waterborne pollutants, as both toxic chemicals and pathogens. San Francisco Bay and Puget Sound seals carry heavy loads of PCBs and DDE, in some cases at a level that may affect their reproduction and survival. Industrial pollutants such as flame retardants may interfere with immune responses.

POPULATION: NMFS assessments of the status of Harbor Seals currently recognize three stocks along the west coast of North America: California; Oregon and Washington outer coast waters; and inland waters of Washington. The San Francisco Bay population may also qualify as a separate stock, pending further genetic study. Although the need for stock boundaries for management is supported by biological information, the exact placement of a boundary between California and Oregon is largely a political/jurisdictional convenience. Federal and state wildlife agencies have conducted aerial surveys of Harbor Seals for more than 30 years. Since passage of the MMPA, seal numbers have rebounded, increasing annually by as much as 15 percent per year but stabilizing in the past 10 years.

Recently, the estimate for the California population was around 43,000 using 400 to 600 haul-out sites. The population in Oregon and Washington together was estimated at 45,700, increasing at an annual rate of around 4 percent per year until stabilizing in the 1990s. A small number of Harbor Seals also occur along the west coast of Baja California, including on San Martin Island.

COMMENTS: In contrast to the California Current Harbor Seal populations, those in the Central Gulf of Alaska have declined by nearly 50 percent in the past two decades, the causes for which are unclear and a point of controversy. Many researchers believe that Harbor Seals, like fur seals and Steller Sea Lions, are declining because there is insufficient food to sustain them (the "junk food" hypothesis). But there is some evidence, too, that the Killer Whales that once intensively hunted whales have shifted to pinnipeds and Sea Otters. As in many stories of ecology, interactions are complex, and causal relationships are rarely linear.

REFERENCES: Allen et al. 1985, Allen et al. 1993, Bigg 1981, Boulva and McLaren 1979, Brown et al. 2005, Burns 2002, Carretta et al. 2009, Gray 1864, Grigg et al. 2002, Hayes et al. 2004, Jeffries et al. 2005, Johnson and Acevedo-Gutierrez 2007, Scheffer 1958, Stewart et al. 1988, Suryan and Harvey 1999, Thompson et al. 2001, Yochem et al. 1987.

NORTHERN ELEPHANT SEAL *Mirounga angustirostris*
Pls. 18, 19; Figs. 25, 34, 90, 91, 103.

> Adult male average length is 3.7 to 5.5 m (12 to 18 ft); average weight is 2,000 to 2,722 kg (4,400 to 6,000 lbs). Adult female average length is 2.7 to 3 m (9 to 10 ft); average weight is 590 to 862 kg (1,300 to 1,900 lbs). Pup average length is 1 to 1.3 m (3.2 to 4 ft); average weight is 27 to 34 kg (60 to 75 lbs).

SCIENTIFIC AND COMMON NAMES: *Mirounga* is derived from an Australian aboriginal word "Miouroung" for its genus, and from the Latin *angustirostris,* or "narrow snout" (fig. 103). *Mirounga angustirostris* occurs in the northern hemisphere, and *M. leonina* in the southern hemisphere; the two species are separated geographically by a mere 2,500 miles (20° S to 20° N). The species was first described by Smithsonian naturalist Theodore Gill in

1866 from a skull of a female collected in Bartholomew's Bay in Baja California, but American whaler Captain Charles Scammon provided the best early descriptions of Northern Elephant Seals. Previous common names were Proboscis Seal and Sea Elephant, but the species is now called the elephant seal in English, Spanish (*foca elephante*), and French (*elephant de mer*).

DESCRIPTION: Northern Elephant Seals are the largest pinniped in the northern hemisphere and are named for the long, inflatable nose of the male. Their body form is the typical spindle shape of earless seals, and the layer of subcutaneous fat is a major part of their body weight. The fur is short, coarse, and uniform in color, ranging from shades of brown and dark gray to tan and silver, with lighter countershading on the belly. When newly molted, their fur color is dark gray to silver and gradually weathers to brown, tan, or blond. Older seals show their age with lighter gray or white fur on the head. There are no spots or patterns to their coloration, but when molting, the fur appears ragged. Newborn pups have longer black fur that molts to silver after 1 month. The foreflippers are short and have long nails, but the rear flippers lack nails and are rigid with cartilage on the outer digits. The small external

Figure 103. Northern Elephant Seal bull, female and pup.

ear opening is nearly invisible above the eye. The few facial vibrissae are short and black, and located on the muzzle, with a few above the eyes.

Elephant seals are an extreme example of sexual dimorphism in mammals, and described by Darwin in the 1870s. In size and shape, the sexes look like two different species. Bulls have a long, pendulous snout, large canines, and a chest shield of horny skin formed by a keratinized and scarred epidermis. The chest shield engorges with blood during the breeding season, becoming pinkish. When calling, the snout inflates with blood and air to form a curved trumpet above the mouth. The skull has a pronounced ridge above the eye for muscle attachment to control the nose. When resting, the long snout is flaccid and the tip hangs below the mouth, nostrils pointed downward, but can be firmly drawn in to form a wedge when in the water. Even immature males have a broader snout than females. Females are much smaller and unremarkable, measuring half the length and a quarter the weight of males, and have the prototypical seal form, lacking any specialized features, except for their big eyes.

Females mature sexually as young as 2 years, but the average is 4 to 5 years. Males reproduce at 3 to 5 years, but bulls are not completely mature, with fully developed chest shield, long notched nose, and bulk, until age 8 to 9. The age of younger bulls is identified by length of snout and size of chest shield. The nose and canine teeth grow with age, but the length of nose also increases during the breeding season, a change unique among pinnipeds. The lifespan of elephant seals is better understood than that of most pinnipeds because of a history of several decades of researchers' marking individuals from birth. The oldest known females were 22 years old, and the oldest males 18, but most seals only live to 12 to 15 years.

FIELD MARKS AND SIMILAR SPECIES: Male elephant seals (both northern and southern species) are the only pinniped with a trunklike nose; females are distinguished from other seals in the region because they are the largest earless seal. Immature elephant seals and adult Harbor Seals are similar in size, but their fur is uniform gray to tan in color, rather than spotted. Also, elephant seals have no claws in their rear flippers. Behaviorally, the two are quite different. Elephant seals resting on land often tolerate humans approaching them, but resting Harbor Seals quickly flee into the water from humans approaching by land or sea.

NATURAL HISTORY: Extreme size, thick blubber, and adaptive physiology enable elephant seals to live almost entirely at sea (80 to 95 percent) and to venture into marine habitats where no other pinnipeds, and few cetaceans, dare go. Until recently, little was known about their distribution or behavior at sea, but advances in microtechnology have yielded astounding facts. Seals of all sex and age groups dive deep, continuously and for a long time, day and night from the continental shelf and beyond. They are the deepest-diving pinniped and can hold their breath longer than any other mammal. The deepest recorded depths exceed 1 mile deep [1,700 m (5,500 ft)], with average depths of around 305 m (1,000 ft). Most amazing is that they dive at these depths continuously but only require a few minutes of breathing at the surface before the next dive. Even on land when resting, they hold their breaths for an average of 7 to 10 minutes, and sometimes for as long as 25 minutes. This adaptation on land might have an added benefit in preventing water loss through respiration during fasting.

Despite their mostly pelagic existence, there are seals year-round at well-established colonies, because attendance fluctuates with sex, age, and season: breeding (December to March), molt (March to August), and juvenile haulout (September to November). At colonies, land-based researchers have made amazing discoveries about the physiology and behavior of elephant seals and, about mammals in general, by understanding the extremes of their physiology in molting, fasting, and diving.

During the breeding season, mostly adults are onshore; they remain on land more or less continuously, fasting and living off their fat reserves. Females give birth and remain on land with pups, fasting for up to 30 days, at which time they mate, wean the pup, and then return to sea only to come onshore again, within 2 months, to molt their fur.

Elephant seals are one of only a few pinniped species that experience a radical molt, which is energetically costly and requires resting onshore. Their ragged appearance comes from the fur's shedding in fist-sized chunks. The accelerated growth of new fur, stimulated by sunlight, rapidly displaces the old, worn coat. The newer fur has a bluish, metallic look. An individual molts over a period of about 2 to 3 weeks, but the molt season is protracted, with seals of different age and sex classes molting in a staggered sequence of young, females, and finally bulls. Numbers are highest in April and May when young and adult females are onshore. Adult males molt in June through August.

In the fall, young seals gather onshore at established colonies, but also in odd places, including marinas and boat ramps. The number of seals onshore in the fall can exceed the maximum number counted during the breeding season, but the reason for this seasonal event remains speculative. Two dominant theories are that young seals (1) require terrestrial gravity for bone development and (2) are cycling in preparation for when they are mature enough to breed. Whatever the reason, by the time that alpha males and pregnant females arrive in late November, juveniles are mostly gone.

When seals congregate onshore during the nonbreeding season, they are very social, constantly vocalizing, touching (making them thigmotactic), piling on top of each other, and play-fighting. The din of grumbling, growling, and warbling is perpetually heard in colonies, day and night. Even during the molt, males reinforce dominance with trumpeting and mock fights between periods of rest.

When onshore, elephant seals regularly flip wet sand onto their backs using their front flippers. This behavioral trait helps cool them on warm days, but they also flip sand when they are agitated, and pups attempt to flip sand within hours of birth. Sand flipping is a giveaway that a pinniped is an elephant seal. On uncommonly hot days, seals will retreat to the water to cool off.

However, seeing an elephant seal at sea is rare, since they spend little time at the surface. When at the surface, they assume the standard "bottle" posture while resting with head tilted up. Most researchers consider elephant seals to be solitary at sea, but recent telemetry studies suggest that many males migrate along similar routes to foraging areas, suggesting that contact may be much greater than previously assumed. The sudden simultaneous appearance of many seals at colonies is also suggestive that they are communicating before arrival.

REPRODUCTION: Elephant seals have a polygynous breeding system. Rather than defending territories on land, though, the large dominant males aggressively defend their position near females. Alpha bulls tolerate less dominant bulls nearby, who often do most of the work of chasing away rival males. Bulls use a sequence of postures to intimidate rivals, beginning with stances that use the least effort. The "face-off" involves aligning the head and body directly at a potential rival; if that is insufficient to intimidate, the bull will rise up on his front flippers, exposing the broad chest shield and opening the mouth to display the large canines; he then "trumpets." Trumpeting is the primary aggressive behavior of males, and is created using the inflated snout, hanging in the open mouth and acting as a resonating chamber that magnifies the bugle. The call is distinctive in the species' creating a popping sound not unlike a single-stroke diesel engine. Males practice trumpeting against hard surfaces that amplify the sound, such as rock faces, cave walls, and the undersides of wharves.

Their strong sexual dimorphism and aggressive behavior are linked to very high levels of testosterone for a mammal. Chasing and attacking a rival is a very energy-expensive effort for a fasting bull, and most fights last less than 5 minutes, unless the option of retreating to the water is available. Small, discrete colonies may have only a few dominant bulls, whereas large, continuous colonies will have an array of bulls and subordinate males at intervals along a beach.

Females give birth within a few days of their arrival at colonies, and the first pup born is usually from late November to mid December. They give birth to a single pup but twins have been recorded several times. Pups cannot swim at birth, and consequently, are vulnerable to washout by storms. Pregnant females continue arriving at the colony into February, and the peak birthing date at colonies is between the last week of January and the first week of February.

While onshore with pup for 33 days, females do not leave on foraging trips. Contrary to the laws of energy efficiency, females fast while nursing, defending their pups, and mating. By the end of the season, they have lost 30 to 40 percent of their body weight, and have little energy to fend off multiple amorous males. They often depart colonies at night to avoid marine predators.

Most females are in their reproductive prime at 5 to 12 years old, but pregnant two-year-olds are not uncommon. They will give birth every year, but will occasionally skip years, especially during El Niño–Southern Oscillation (ENSO) years.

Pups gradually learn to swim at rookeries after weaning, and then depart to sea after some 30 days. They form "weaner pods" away from nursing females and aggressive males. Some weaned pups steal milk from sleeping or inattentive females, gaining added weight of up to 181 kg (400 lbs), and achieving the status of "tic weaner." The added weight is not always beneficial, though, because fat pups cannot evade injury from aggressive, low-status breeding males.

FOOD: Determining what elephant seals eat is difficult because by the time they arrive at colonies, they have digested their food. Researchers, though, can deduce what they are eating from looking at stomach contents of dead seals and by tracking where they feed. The depth of dives is very deep, challenging the deepest divers—beaked whales, and Sperm Whales. By attaching to seals microprocessors and global positioning devices that measure depth, sea surface temperature, time and location, researchers have discovered that seals have distinct dive patterns when they are feeding or traveling. A U-shaped dive suggests that the seal is traveling or searching for food, and a flat-bottom dive shape likely foraging along the sea floor. Another dive pattern indicates that the seal is feeding in the midwater depth of 91 to 213 m (300 to 700 ft) where two water masses of different temperatures converge. Shallower V-shaped dives suggest seals are processing food while traveling to new patches of prey in the vast ocean.

When they discover patches of prey, seals dive continuously, not following a diurnal cycle. Their dive patterns, though, show that the dives are deeper during the day and shallower at night, following the nocturnal migration of prey of the deep-scattering layer. Researchers have also discovered that as seals gain weight, their dives become shallower, presumably because their acquired fat layer makes them more buoyant.

The most common species detected in their diet are squid, octopus, sharks, rays, ratfish, crustaceans, and some boney fish such as hake, salmon, and rockfish. Many of these species are found at depths of more than 91 m (300 ft); however, they are also found in a variety of habitats, including oceanic and benthic waters. Elephant seal eyes have a visual pigment called a "deep-sea rhodopsin" which enables them to detect the bioluminescence of deep-sea species. Females and young males are often pock marked with cookie-cutter shark bites, indicating seals are being bitten at a depth range more than 900 m (3,000 ft).

As in size, the two sexes have very divergent foraging strategies. Adult females tend to forage north-west along the North

Pacific Gyre as far as 173° W longitude, and beyond the Hawaiian Islands. Adult males travel north to the Bering Sea and eastern Aleutians, where they forage along the continental margin. Females mostly forage in the midwater zones, where they likely eat prey such as cephalopods and Pacific Hake. Males mostly forage on the bottom in deeper waters along the continental margin of the Aleutian Islands on sharks, skates, rays and rockfish. Their average daily intake is likely 6 percent of body weight, a measure researchers determined for their southern cousin. An estimated daily diet of squid for a female is 9 to 15 kg (20 to 33 lbs) per day, which calculates to around 800 to 1,200 squid per trip.

MORTALITY: The rapid recovery of elephant seals from near extinction is explained in part by their low rate of mortality. Loss of pups is the primary cause. Pups may be crushed by bulls and washed away by large sea swells, and mortality is aggravated by overcrowding, particularly where seals are packed on cliff backed beaches with no place to retreat. Seals are also affected by major climatic disturbances such as ENSO events when the source and distribution of prey are disrupted, causing seals to skip breeding or die of starvation. During the 1998 ENSO, researchers determined that many adult females spent more time at sea looking for prey, and many females returned to colonies not pregnant.

The popular literature has dramatized shark feeding on elephant seals, and researchers have documented many incidences of shark attacks on seals near colonies; however, the actual population level effects are low since the seal population continues to grow exponentially. Other predators include blue sharks and Killer Whales.

DISTRIBUTION: The range of Northern Elephant Seals covers a vast area of the North Pacific Ocean extending from Central America north to Arctic waters of the Bering Sea and west as far as Japan and the Commander Islands of Russia. The core breeding range, though, extends between Isla Natividad, Mexico, and Cape Arago, in southern Oregon. Three subpopulations are recognized based on geography and colonization; Baja, southern Channel Islands and central California. The southern Channel Islands sub-population was formed by colonizers from Baja and the central California sub-population by seals predominantly from over-crowded rookeries in the southern Channel Islands. Approximately 40 percent of the world total population of

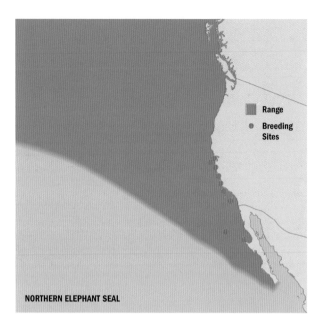

NORTHERN ELEPHANT SEAL

elephant seals occurs in the southern California Bight, and 85 percent of those adults occur on San Miguel Island, the largest rookery. The breeding range continues to expand as new colonies form along the north Pacific.

In Baja, elephant seals breed on offshore islands including Guadalupe, Natividad, Cedros, San Benitos and San Martin. In California, they breed on the southern Channel Islands, Año Nuevo Island, Gorda, Cape San Martin, San Simeon, Southeast Farallon Islands, and Point Reyes. On the California mainland, pups are born at Diablo Cove and San Simeon (San Luis Obispo County), Cape San Martin (Monterey County), Año Nuevo Point (Santa Cruz County), Point Reyes (Marin County), and Point Saint George (Del Norte County). Two small but persistent colonies are near Crescent City, California, and Cape Arago, Oregon. Seals will often return to the rookery where they were born; however, as a colony becomes over-crowded, young breeders will colonize new beaches. New colonies often form on remote beaches where seals are not disturbed when birthing, but once established, seals will overflow onto nearby beaches where they can retreat into back dunes away from high surf and tides.

Unlike any other mammal, Northern Elephant Seals migrate long distances two times each year between foraging areas and colonies. This dual annual migration covers 19,000 to 22,500 km (12,000 to 14,000 mi), the longest annual migration of any mammal. Why they migrate such vast distances is likely driven by an assured supply of abundant, high-quality prey that can sustain rapid weight gain, a pregnancy in females, a long journey back to colonies, and fasting when onshore. The sexes migrate to two very different regions to forage. Females are often associated with the convergence of water masses between the subarctic front and the transition zone in the North Pacific Gyre where prey are concentrated. Males forage more often on the bottom along the continental margin of the Aleutian Island chain. However; this pattern is not inflexible; some females will migrate where males go, particularly during ENSO years, and a few males migrate to the North Pacific Gyre.

Until recently, little was known about elephant seal distribution and activities at sea because researchers rarely saw them. With new gadgetry, though, we now know that this low visibility at sea is because seals spend a mere 3 to 11 minutes at the surface to breathe between continuous dives of up to 90 minutes. From satellite tags and time-depth recorders, researchers have learned that seals disperse rapidly and widely from colonies. One seal tagged at San Miguel Island, for example, was located in the Bering Sea within 2 weeks, and trips of 9 to 11 days from Año Nuevo to foraging grounds in the Bering Sea are common.

HABITAT: At sea the sexes diverge: females forage at midwater depths [305 m (1,000 ft)] in the North Pacific Gyre, and males forage more often on the bottom along the continental margin. On land, the sexes converge and, because of their limited mobility, elephant seals prefer substrates with a gradual slope such as sandy or small cobble beaches or flat rocky reefs. They occupy the beach above the high tide line, but will retreat to the intertidal zone in hot weather and when molting.

VIEWING: Mainland colonies are excellent places to observe seals. The most easily accessed sites in California include San Simeon and Año Nuevo State Parks along Highway One, and Point Reyes Headland in Point Reyes National Seashore. Venturing by boat to San Miguel Island, though, is an extraordinary experience akin to visiting the wild beasts of the Serengeti.

CONSERVATION: Historically, Northern Elephant Seals ranged from Baja California to southern Alaska. Archaeological evidence from Native American middens confirmed that elephant seals were regularly hunted by the Makah in Washington and the Chumash on the Channel Islands. Charles Scammon, a British seal hunter, recorded Northern Elephant Seals from Cabo San Lazaro, Baja, to Point Reyes, California. Their large size, an advantage for foraging, limited their mobility on land and was one factor leading to their rapid decimation by seal hunters. Commercial sealers hunted them for the oil produced from the thick blubber; an average bull was reported to produce 24 gallons of oil, but one bull collected by the whaling brig *Mary Helen* in 1852 supposedly yielded 210 gallons.

Agreement	Conservation Status
U.S. ESA	Not listed
U.S. MMPA	No special status
COSEWIC	Not at risk
IUCN	Least concern

By the later 1880s, the species was generally considered extinct until Charles Townsend from the Smithsonian discovered a small colony on Guadalupe Island. Only around 20 animals were counted onshore by scientists between 1884 and 1892, although numbers were likely higher since the seals spend a majority of their annual cycle at sea. Researchers agree that by the early 20th century, though, less than a few hundred seals survived, and some researchers speculate that fewer than 100 existed at their nadir.

With protection provided first by the Mexican government on Guadalupe Island with designation as a pinniped sanctuary, and later by the United States on the Southern Channel Islands, the seals made an extraordinary recovery, recolonizing first the Southern Channel Islands from the 1930s to the 1950s, Año Nuevo Island in 1961, Southeast Farallon Island in 1972, Point Reyes and Cape San Martin in 1981, San Simeon in 1992, and southern Oregon in the mid 1990s. Annual population growth rates have been estimated at 6 to 8 percent, and new colonies have growth rates of 14 to 17 percent. An extreme example of colonization is the Piedras Blancas colony, which grew to over 1,000 pups within 5 years, with colonizers shifting en mass from the Channel Islands. New colonies have often formed during or just after ENSO years when densely crowded colonies are washed out by large storm events and seals seek new sites.

The narrow genetic variability of Northern Elephant Seals may someday limit their ability to adapt, and thus, have fewer options to compensate for changes in their environment such as climate change. Despite the genetic limitations, individuals display distinct personalities and the population continues to expand. Associated with this rapid increase has been the colonization of many mainland areas along the California coast. Prehistoric colonies on the mainland were rare, likely because of hunting by native peoples and grizzly bears. Researchers generally agree that elephant seals are currently more abundant, and breeding more widely, than prehistorically.

POPULATION: The population estimate for California in 2001 was 101,000 animals, including 21,000 pups, and, with the Baja population, was close to 150,000 to 175,000. In just 4 years, the California population increased to 124,000 in 2005.

COMMENTS: Elephant seals have been more intensively studied than perhaps any other marine mammal. Researchers, taking advantage of the seals' long rest onshore and their habit of often returning to their place of birth, have examined their diving physiology using radio isotopes to measure metabolism and even put seals into Magnetic Resonance Imaging devices (MRIs) to measure the function of the spleen. To truly understand their diving capabilities, though, researchers have attached all types of gizmos with marine epoxy glue to their bodies before their long sojourn at sea. Their breath-holding ability has been the focus of research because it is key to understanding how seals, and people, dive. This ability is one of the greatest of any mammal and often exceeds their aerobic dive limit. But how can they dive more than a mile down and still have energy to pursue prey? To study this, researchers attached a variety of devices to measure heart rate, swim speed and dive depth on seals in the wild. To reduce oxygen consumption, the seals' heart rates dropped to as few as four beats per minute; they were able to adjust heart rate throughout the dive. As they descended to depth, the heart rate continued to decline. This ability to adjust heart rate and restrict blood flow to only those parts of the body that need them on demand means that seals can continuously monitor their own energy needs while diving. Although seals are able to increase swimming speed in deep water, presumably when chasing prey, these changes in speed at depth reduce their overall time spent on the bottom. To reach deep foraging areas, seals swim in a burst and glide pattern to save energy, and rely on negative buoyancy to assist them in their descent. The next question is how do they catch their prey during these cavernous dives? Researchers speculate that seals are diving to the oxygen minimum layer where it is easy to capture prey because prey do not have the oxidative capacity to flee.

REFERENCES: Allen et al. 1989, Andrews et al. 2000, Carretta et al. 2009, Colegrove et al. 2005, Cooper and Stewart 1983, Hassrick et al. 2007, Hindell 2002, Le Boeuf and Crocker 2005, Le Boeuf and Laws 1994, Le Boeuf and Panken 1977, Sanvito et al. 2007, Scammon 1874, Stewart and Huber 1993, Williams et al. 2000.

SEA OTTER   *Enhydra lutris*
Pl. 18; Figs. 94, 104, 105.

> Adult male average length is 1.5 to 1.8 m (5 to 6 ft); average weight is 37 kg (100 lbs). Adult female average length is 1.2 to 1.5 m (4 to 5 ft); average weight is 23 to 28 kg (60 to 75 lbs). Newborn average length is 0.6 m (2 ft); average weight is 1 to 2 kg (3 to 5 lbs), with an average weight of about 9 kg (25 lbs) at 1 year.

SCIENTIFIC AND COMMON NAMES: *Enhydra lutris* is the only fully marine-dwelling member of the otter subfamily Lutrinae (fig. 104). The Greek word *enhydra* means "in water," and *lutris* means "otter" in Latin. Two subspecies occur in the California Current: *Enhydra lutris kenyoni* and *Enhydra lutris nereis*. "Sea sprite," from the Greek subspecies name *nereis*, aptly describes the California Sea Otter (as detailed below); this subspecies is a latitudinal variant of the Alaskan Sea Otter *(Enhydra lutris)*. The eastern Alaskan subspecies (*Enhydra lutris kenyoni*), also called the northern species, is distinguished from the western Alaskan subspecies (*Enhydra lutris lutris*). The former was named after

Figure 104. Sea Otters.

pioneering marine mammalogist Karl Kenyon, who devoted his life and career to the study of otters and other marine mammals. Former scientific names for the species include *Latax lutris,* *Lutra marina,* and *Hydra marina.* Common names include Sea Beaver, Slender Otter, and San Miguel Island Sea Otter. The Spanish name is *nutria marina;* an Oregon Native American name is *elakha.*

DESCRIPTION: Sea Otters are the quintessential charismatic animal, and for decades they have been the poster child for people's concerns about marine mammals. Sea Otters are the largest member of the Mustelidae family but the smallest living marine mammal except for the River Otter, and this position has ensured their success in a restricted marine habitat, the nearshore environment. Sea Otters share the features of mustelids—the weasellike shape and round head common to martins and weasels—but they lack a scent gland, are longer and heavier than the others, and have a shorter, thicker neck. Although otters are the smallest marine mammal, large male otters can exceed the size of pups or juveniles of small pinnipeds in total length. All pinnipeds, however, have tiny tails compared to the long, rudderlike tail of the Sea Otter.

Female and male otters are similar in appearance, but the female tends to be lighter in color, and slightly smaller. In both sexes, the head is rounded, with a short muzzle and blunt nose. The River Otter's teeth are unique among mammals': all their teeth have rounded edges, and the molars are broad and smooth, ready to crush the shells and exoskeletons of invertebrate prey. The two pairs of lower incisors protrude outward, making it easier for the animal to extract soft body parts from shells of its prey. The tiny round ears close up and fold back when the head is submerged. The tail is about one-quarter to one-third the length of the body [25 to 38 cm (10 to 15 in.)], is broad at the base and flattened vertically, and is evolved for swimming in the marine environment. The front and hind limbs also evolved for specialized use. The rear feet are large (equal to one-quarter the body length), webbed, and paddle-shaped (fig. 105). The smaller front paws are padded with retractile claws that are specialized for grooming and holding food. The Sea Otter's many facial whiskers are long and curved and, as with some pinnipeds, are used tactilely along with paws to detect prey in murky waters.

Otter coloration is uniformly dark brown, reddish brown, or black, but with age, the head and neck change in both sexes to blond or white. Males also become grizzled with light-tipped guard hairs all over. Newborn pups, weighing a mere 1 to 2 kg

Figure 105. Sea Otter with large hind feet.

(3 to 5 lbs) and sporting yellowish guard hairs, look like little balls of fluff. The skin of an otter is loose, and under each forelimb the skin forms a fold where the otter holds and carries food while swimming. Most individuals prefer to use the pouch under the left forelimb for this, leaving the other one free to collect prey (suggesting that otters are right-handed).

As the smallest marine mammal, otters face the physiological challenge of maintaining their warm body temperatures. With a larger proportion of surface area, a small mammal loses heat faster than a large one does. Sea Otters lack the thick blubber layer characteristic of most marine mammals; for body heat regulation they rely instead on very fine fur and a high metabolic rate. Like fur seal fur, otter fur consists of outer guard hairs and dense underfur that, when dry, is 3.8 cm (1.5 in.) thick. Indeed, Sea Otters have the densest fur of any mammal, with an estimated 15,500 hairs per square inch, totaling around 800,000 hairs on an adult male. Such dense fur efficiently traps air for buoyancy and insulation; to increase buoyancy, otters actually blow air into their fur while grooming. Their lung capacity is also very high and increases the buoyancy they need to support food, tools, and young as they float on the surface.

FIELD MARKS AND SIMILAR SPECIES: Sea Otters occur in nearshore habitats similar to those of Harbor Seals and California Sea Lions—with which they might be confused. When in the water, their smaller round head, broad triangular nose, and long bushy whiskers are quick distinguishing marks. Both otters and Harbor Seals rest in kelp beds, but otters are easily identified by their ear pinna and their tendency to rest on their backs, tied up in the kelp, whereas Harbor Seals lack external ears and rest vertically in the water, with only their heads exposed. Sometimes, though, either species may float horizontally, with its belly facing down.

The River Otter is the only other species likely to be confused with the Sea Otter, but the River Otter's smaller and flatter head, longer tail (one-third of body length), and smaller overall size quickly separate the two species. Although their habitats may overlap, their behaviors in the water are quite different; River Otters stay close to shore and retreat to land for safety, but Sea Otters are rarely seen onshore and often float on their backs in the water while eating.

NATURAL HISTORY: The Sea Otter has often been called a keystone species or an ecological engineer. Sea Otters inhabit a marine niche in the nearshore where they have a cascading effect on the ecology through their choice of prey, as ecological engineers. Otters are known to reduce the density and average size of some invertebrates, particularly abalone, clams, and sea urchins; this, in turn, can promote the growth of the kelp beds from which otters benefit. This circular link between otters and healthy kelp beds has been studied intensively, and the expansion of kelp beds in California has paralleled the range expansion of the Sea Otter, because otters prey on sea urchins, which suppress kelp. Otters can thereby enrich the rocky nearshore communities that they inhabit. Several studies have discovered that in areas newly colonized by otters, there are dramatic increases, within 2 years, in the abundance, biomass, and diversity of kelp, as well as in the presence and abundance of other species that depend on kelp.

Sea Otters are the only truly marine member of the mustelid family, and their divergence from other otters likely occurred in the late Pliocene or early Pleistocene, perhaps 2 million years ago. Despite their recent evolution, Sea Otters spend almost their entire life at sea, a lifestyle made possible by their luxuriant fur. In contrast to pinnipeds, they infrequently gather in herds onshore and are not driven, physiologically or reproductively, to rest or birth onshore. Rather, they spend much of their life at sea, close to shore, in small groups or alone.

In the course of a day, otters spend their time mostly foraging, grooming, and sleeping. Rarely, otters come onshore; this is mostly in order to feed in the intertidal or to rest, sometimes in the proximity of Harbor Seals. They forage mostly during dawn and dusk, but also at night. They rest nearly half the day, usually during the middle of the day, and in the safety of kelp beds. Because healthy, clean fur is critical to their body temperature regulation, otters frequently groom throughout the day, often before foraging dives and after eating. They also cannot risk molting their fur all at once each year, but instead molt hairs throughout the year, like fur seals do. For the most part males and females remain segregated through the year, except when many females come into estrus in the summer. Otters, however, will form aggregations, called "rafts," of several dozen in suitable habitat; in Alaska, rafts can reach up to several hundred animals.

As highly social animals, otters use complex communication consisting of head jerks and calls. Their vocalizations range from cooing, by mothers with pups, to whistles, grunts, and growls. Researchers can even identify individual otters by their distinct "screams." As yet, biologists have only rudimentary understanding of the meanings of these otter exchanges.

The behavior of Sea Otters reflects their remarkable morphology and physiology. Their movement on land is awkward, not only because of the uneven size and length of their front and back limbs, but also because the longer fifth digit of the rear foot is adapted for swimming, not walking. The Sea Otter is streamlined compared to other otters, and its spinal cord is very flexible, lacking well-developed vertebral spines. When they tuck in their front limbs, their streamlined shape and flexibility allow for undulating body movements through the water. They generate slow movements by the simple sideways sweep of their broad tail. They sustain long-distance travel by an energy-efficient motion consisting of a downward thrust and relaxed recovery of the rear flippers. Swim speeds can thus reach 4.8 kph (3 mph). When diving or traveling short distances, however, Sea Otters generally travel a mere 1.6 to 3.2 kph (1 to 2 mph). All of these modes of locomotion are slow compared to those of most marine mammals, but otters generally feed on slow or sessile organisms and do not chase after fast-moving fish or squid.

Maintenance of body temperature—critical to survival for a mammal submerged in cold seawater—is achieved through several classic otter activities. Grooming fur accounts for 10 percent of their daily activities and includes arranging the fur before foraging, cleaning the fur after eating, and blowing air into the fur while rolling. Another classic posture sees otters resting on the water's surface, belly up and rear feet extended into the air. The hairless rear feet are highly vascularized for rapid absorption of heat from the sun.

When Sea Otters do haul out on shore, they use low-relief rocks with easy access to water; such rocks are often covered with algae and are around 19 m (60 ft) from shore. Groups hauled out together range in size from one to six, though up to 22 have been seen at one time.

Sea Otters feed day and night—indeed, spend an estimated third of their life feeding—but there are differences between sex and age class. Adult males, for instance, are most active in the late afternoon, but juvenile females are active over a broader period, and females with pups feed more often. Seasonally, Sea

Otters also spend more time feeding in winter, likely because of their increased caloric requirements to maintain body temperature in inclement weather and colder water. Unusual for a marine mammal, otters can drink sea water, an activity enabled by their larger kidney.

Otters are one of the rare species that use tools. They collect stones with their front paws and use them as anvils upon which they smash open armored prey such as crabs and sea urchins. Typically, they do this while floating on their backs, pounding the prey with a rock with one paw while holding it on their belly with the other. Otters will apply any handy hard object to dislodge or smash prey. In Monterey Bay, otters have been known to crush shellfish against kayaks, and when in captivity, they use the aquarium wall as the anvil against which to strike their food. From studying individual otters over years, researchers have estimated that their lifespan may extend from 15 to 27 years.

REPRODUCTION: As is the case with many marine mammals, Sea Otters have a polygynous breeding system. Once females reach sexual maturity, at 3 to 5 years of age, and males at 4 to 6, beginning in early summer—males seasonally defend elongated breeding territories, also called "maritories," within female home ranges for 5 to 7 months. The size of male territories average 0.5 km^2 (100 ac) but can be up to 2.2 km^2 (550 acres), forming a band that extends along the shoreline. Maritories overlap, and several researchers have noted that nonterritorial males can feed and travel through a male's territory, but are not allowed to rest within it. Boundary disputes usually involve pushing and shoving, and rarely result in injury. Some males return annually to the same territories for several years, even resting in specific kelp beds each year. Females have overlapping home ranges that also vary in size, to as little as 0.1 km^2 (25 ac). The number of males in overlapping female home ranges is highest when the most females come into estrus; then, a ratio of one male to five females has been noted.

A mating pair will form for a period of 3 days, centering their activity around a rock that they select as a resting place. Their premating play and mating are primarily aquatic, and in some instances, the activities injure a female, leaving it with bloodied nose. The female terminates the relationship by departing.

Mating and pupping may occur year-round, followed by a gestation period of 6 to 7 months. Research indicates that females give birth annually, but not all females give birth during the same time of year. Instead, there are two peaks each year in the number of pups counted: a large one in spring, and a smaller one in fall. In California, most births occur between December and February; in Washington, between March and April.

A pup is completely dependent on its mother at birth, weighing a minute 2.3 kg (5 lbs), and cannot swim until it is 2 months old. Pups can be born on land or in the water, and they have an extradense fur that provides buoyancy as well as insulation. The female–pup bond is strong and enduring—they are in near-constant physical contact, with young pups protected on the bellies of resting mothers. Generally the bond lasts for around 1 year, during which time the pup evolves from a helpless ball of fuzz to an agile juvenile that has learned what, where, and how to eat. They are weaned at 6 months of age, in late summer or early fall, but remain dependent on their mothers for several months longer, until they learn to capture prey and use tools. Consequently, females that successfully raise pups tend to have a long interval of 16 months between births.

FOOD: Researchers have been able to closely examine the prey of otters with binoculars, because otters prepare and consume their food on their bellies while floating on the surface. Their diet primarily consists of macroinvertebrates, including crabs, sea urchins, clams, mussels, spiny lobsters, starfish, and abalone. Otters can gorge on these hard-shelled prey because of their unique crushing teeth, which most other marine mammal species lack. In addition, they consume octopus, squid, turban snails, and a few small, kelp-dependent fish species. More than 150 species of invertebrates and fish, from Alaska to California, have been identified in otters' diet, including more than 20 species of gastropods, more than 30 species of crabs, and more than 40 species of bivalves. Generally, otters eat whatever invertebrates are abundant in the habitat where they reside. In rocky intertidal habitats, they prey mostly on crabs, sea urchins, octopus, and sea cucumbers. In soft bottom habitats of bays and estuaries, they depend almost entirely on bivalves in the mud, such as Butternut Clams. Otters will also respond to seasonal and annual changes in prey abundance, such as spawning squid and pelagic red crabs during El Niño–Southern Oscillation (ENSO) events.

Because of the otter's keystone role as the keystone species in kelp forest ecology, researchers have studied what this species eats as it has expanded its range following near-extinction. When otters first reoccupied former habitat in California, they ate abundant species such as sea urchins, abalone, and clams, but their diet became more diverse over time, extending to fish, chitons, and mussels. In areas where invertebrates are in low abundance, otters' diet can consist of up to 60 percent fish. On rare occasions, otters have been seen eating birds such as gulls, loons, Surf Scoters, phalaropes, and grebes.

Unlike River Otters, Sea Otters are not known to forage cooperatively, except for paired mother and pup. Instead they are solitary foragers, focusing on sessile or very slow-moving prey. On the bottom, otters find prey both visually and by touch, using their forepaws and whiskers. Studies of tagged Sea Otters' diving behavior provide precise estimates of individual diving times and depths: generally, diving abilities vary with age and sex. The maximum dive time of males is around 4 minutes, and depth is 100 m (325 ft) or more, with one documented dive of 300 m (975 ft). Maximum female dive depth was recorded at around 76 m (250 ft). One unfortunate otter was found dead in a crab pot in 91 m (300 ft) of water. Most dives, though, are in waters shallower than 30 m (98 ft); this reflects the otter's preferred habitat and prey, near shore. To meet their energy requirements, otters are estimated to consume 20 to 25 percent of their body weight per day.

MORTALITY: Natural sources for otter mortality include predation by sharks, Coyotes, eagles, and Killer Whales; starvation, especially of young animals; and disease and internal parasites. Estimates of mortality of pups before weaning are as high as 45 percent, and of subadult males, 13 percent. The adult mortality rate is highest among males. Predation by Great White Sharks may depress the population in California, and the population decline in Alaska may be caused by Killer Whale predation. One theory proposes that Killer Whales in Alaska switched from eating pinnipeds to otters coincident with drastic declines in pinniped populations. Diseases have accounted for 35 to 45 percent of otter deaths in California in recent years, and have been linked with domoic acid poisoning, encephalitis, and a suite of diseases involving heart disease, fungi, bacteria, and viruses.

DISTRIBUTION: The current distribution of U.S. west coast Sea Otters is fragmented. The centers of several core areas are around Monterey Bay, California; and, in Washington, from south of Destruction Island at Willoughby Rocks and Kalalouch Point to Cape Flattery and through the Straits of Juan de Fuca, east to Pillar Point. Juveniles and solitary otters, however, disperse widely, from Baja to Washington. The Washington State and British Columbian population represents the southern extension of the eastern Alaskan subspecies, and also includes survivors of an early reintroduction effort conducted in 1969 to 1970. In Oregon, no established population exists, but individual otters are observed intermittently and are likely dispersing from Washington. In California, the small population centered in Monterey Bay (and extending from Point Conception and Santa Barbara north to Half Moon Bay) is the California subspecies. From an earlier reintroduction of the subspecies in the 1980s, a small group of otters remains on San Nicolas Island, in southern California.

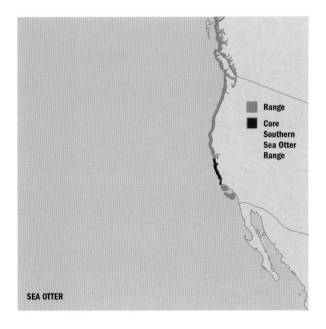

■ Range

■ Core
Southern
Sea Otter
Range

SEA OTTER

As populations recover, otters continue to expand their range. A range extension for this species is recognized when three or more otters become established for 3 or more years in habitat beyond the established range. Deep, open water and the absence of large kelp beds are barriers to Sea Otter range expansion, since Sea Otters spend around 70 percent of their time in kelp beds. Sea ice is believed to be responsible for restriction of the species' northern expansion in Alaska, and warm temperatures for restriction in the south. The southern limit, Point Conception, coincides with the southern limit of cool water upwelling and the distribution of giant kelp, though many Sea Otter sightings have been recorded south of Point Conception. Concentrations of Sea Otters at the ends of the range consist of mostly males.

Sea Otters generally do not follow a distinct annual migratory cycle, and individuals will remain within a home range until food is depleted or the density of otters increases. Nevertheless, Sea Otters do move locally, and some researchers have discovered that tagged males migrate seasonally, between winter areas at the edge of the range and breeding territories in the core areas such as Big Sur. Researchers in Washington also have noted Sea Otter movement between Cape Johnson in the summer and Cape Alava in the winter, and have speculated that such movement was toward more protected habitats in the winter months, to avoid storms. During these transits, males may travel 110 miles or more. Other researchers have determined that some otters of all age and sex classes traveled long distances year-round.

HABITAT: Otters reside in the nearshore rocky, coastal environment and rely on kelp beds (of *Alaria, Macrocystis,* and *Nereocystis*) for protection against weather, ocean swell, and predators. In addition, otters often inhabit soft-bottomed estuaries without kelp, such as Elkhorn Slough in Monterey Bay or San Francisco Bay, where invertebrate prey are abundant. Most of the literature describes Sea Otters as rarely traveling far offshore. Although several researchers have discovered that more than half of juveniles venture a distance from the coast [more than 1,000 m (3,300 ft or 0.6 mi) in California and 4 km (2.5 mi) in Washington], they have also found that the majority of females and adult males are located within 792 m (2,600 ft) of shore, and that regardless of sex, age, or location, most otters reside within 1 mile of shore.

They also rarely venture in water deeper than 61 m (200 ft). Generally, they feed in waters around 40 m (130 ft) deep, but (as detailed above) depths vary with age, sex, diet, and location. Otters in Washington usually feed and rest in waters that are shallower, but further offshore, than those in California; this is likely a function of topography. From tracking activities of radio-tagged otters, researchers discovered that in California the majority of otters were in waters less than 25 m (76 ft) deep, but in Washington average depths ranged from 10 to 16 m (33 to 53 ft). The continental shelf in Washington is wider, with rockier habitat offshore, than around Monterey Bay in California, which likely explains the differences. Regardless of geographic location, kelp beds are otter-preferred habitat, and otters will spend a large proportion of the day (56 percent of a 24-hour period) wrapped in kelp fronds, grooming fur and resting at the surface.

VIEWING: One of the best locations for viewing Sea Otters is Monterey Bay, the epicenter of the Southern Sea Otter. They can also be seen regularly along the remote outer coast of Washington from Destruction Island to Cape Flattery.

CONSERVATION: When Sea Otters were discovered in 1741, they ranged around the Pacific Rim from Viscaino Bay, Baja California, north to the Pribilof Islands, and south to Hokkaido, Japan. In California, they were reportedly very abundant around the Farallon Islands, Monterey Bay, San Francisco Bay, and the Santa Barbara Channel Islands. In Washington, they were abundant from the Columbia River northward, including in bays such as Gray's Harbor. Early estimates had the pre-European population at around 800,000, and in California waters otter numbers may have been 100,000.

Agreement	Conservation Status
U.S. ESA	Threatened
U.S. MMPA	Depleted
COSEWIC	Special concern
IUCN	Endangered

Intensive hunting subsequently fragmented and restricted their distribution, nearly driving the species to extinction. In the 18th and 19th centuries, people hunted the Sea Otter for its luxuriant fur, taking an estimated 1 million pelts over a 170-year period. This began with Spanish explorers to California, who bartered for otter pelts with Native Americans in the 1780s, and continued with the Russian traders, who were more efficient in hunting otters and removed an estimated 50,000 otters in just a few years. By the early 20th century, only isolated groups remained—in the outer Aleutian Islands, along the Alaska Peninsula, and near the Commander and Kurile Islands. Otters were completely extirpated from Oregon and Washington. In 1911, an international treaty between the United States, Russia, Great Britain, and Japan was enacted to protect Sea Otters along with Northern Fur Seals. In 1926, and again in 1938, a remnant population was discovered at Bixby Point, near Point Sur, California, that was estimated at 100 animals.

Under international protection, the Sea Otter in Alaska reoccupied about 30 to 40 percent of its former range, and between 1965 and 1972, 708 otters from Alaska were reintroduced to several locations between Oregon and Alaska, with varying degrees of success. The Sea Otter was one of the first marine mammals protected by California state law: in 1913 it was listed as a "Fully Protected Mammal," and in 1941 the Sea Otter Game Refuge was established at Bixby Point to protect the remnant population.

In Washington the otter population has rapidly expanded, at an average of 8 percent per year since 1989, but in California recovery has been very slow, at a rate of 2 to 5 percent per year and with otters spreading an average of 4 km (2.5 mi) per year along the coastline. Also, there have been periods of decline in the California population, in response to ENSO events, mortality in a gill net fishery, oil spills, and disease. The range has expanded faster to the south than the north, and the current distribution is from Half Moon Bay south to Pismo Beach, including the Channel Islands. Individual otters, including a female and pup, have been reported since the mid-1970s in waters of San Francisco Bay, Point Reyes, and Fort Ross.

Oil spills remain one of the greatest threats to Sea Otter recovery. Because these animals are concentrated within a narrow zone in California, and in fragmented areas in Washington and Oregon, a single oil spill could decimate any of these remnant populations. After the *Exxon Valdez* oil spill in Alaska, much was learned about how oil affects otters: they can die of hypothermia; and even when coated lightly with oil for an 8-day period, their metabolic rates increased 127 percent above normal.

To reduce the risk of a catastrophic loss, a recovery management strategy was initiated to reintroduce otters to new areas within their prehistoric range. The objectives also included attempts to limit abundance of otters in areas where there would be potential conflict with shellfisheries. With these two contradictory objectives, San Nicolas Island was selected as the best site for reintroduction. Officially, San Nicolas Island was preferred because (1) the habitat was highly suitable for otters, (2) it offered the least potential conflict with shellfisheries, (3) it offered the greatest potential for containment, and (4) protection and law enforcement needs were more easily met. Provisions within their recovery also designated an "otter-free" zone where otters would be excluded by nonlethal means, so to protect commercial and sport shellfishery interests south of Point Conception. Accordingly, 200 otters from the California population were reintroduced on San Nicolas Island. Of these, only 20 were counted in later years, and translocation efforts were stopped and officially labeled a failure. Some of the otters traveled back to the mainland; the fate of others was uncertain. Nevertheless, the population currently is increasing there by 9 percent per year.

After Sea Otters were extirpated from much of California in the 1900s, shellfish stocks, including clams and abalone, are believed to have increased in abundance beyond prehistoric times. This has facilitated development of commercial and recreational shellfisheries—and interests that were generally opposed to reintroduction of otters. The primary fishery interests that protested the translocation of Sea Otters were for Dungeness Crabs and Rock Crabs, spiny lobster, abalone, clams, oysters, mussels, and sea urchins. There have been concerns that otters would interfere with commercial oyster operations, but no evidence exists that otters consume oysters near commercial operations. Conclusions from studies of otter effects on fisheries have been contradictory. One concluded that Sea Otter foraging resulted in the loss of the Pismo clam fishery in Central California; another found that otters had no effect on size and distribution of deep-burrowing bivalves.

Long-standing impediments to the otter's recovery are entanglement in fishing gear and effects of pollutants. Otters' exposure to pollutants through consumption of contaminated shellfish likely will increase as this species expands into developed areas such as San Francisco Bay. Past studies have found lead and DDT in otter tissue. An emerging conservation concern is human-caused disease (called pathogen pollution), one of which is toxoplasmosis, a disease that some scientists suspect was transmitted from domestic feral cats or partially treated discharge from treatment plants. Climate change may be especially hard for otters, since they are vulnerable to extreme loss of kelp forest such as took place during the 1998 ENSO event.

POPULATION: Aerial surveys in California recorded more than 3,000 Sea Otters, including 390 dependent pups, in 2007, far lower than the estimated 16,000 otters that marine habitats in California could support. This projection is also far less than the estimated prehistoric population, likely because of the dramatic transformation of much of the California coastline to less habitable waters. The California population is stable or decreasing, according to more recent population estimates. The Washington State population, however, has grown an average of 10 percent per year, and the population count in 2005 was around 800 otters, including 50 pups, with potential habitat that might support 1,300 to 2,700.

COMMENTS: Not only are California Sea Otters genetically distinct from the northern subspecies, but their genetic variability is also extremely low, because the population was hunted to such low levels in the past. One estimate puts otters at retaining only 77 percent of their original genetic diversity. Despite these genetic constraints, otters persist, but mostly in protected areas, such as parks.

REFERENCES: Burns et al. 1985, Estes 1980, Estes and Palmisano 1974, Faurot and Ames 1986, Garshelis et al. 1986, Jessup et al. 2004, Kenyon 1969, Kenyon 1987, Kvitek et al. 1988, Laidre et al. 2001, Laidre et al. 2002, Loughlin 1980, Miller 1980, Morejohn et al. 1975, Siniff et al. 1982, Wendell et al. 1986.

RIVER OTTER   *Lontra canadensis*
Pl. 18; Figs. 106, 107.

Adult average length is 1 to 1.2 m (3 to 4 ft); average weight is 6 to 9 kg (16 to 23 lbs). Newborn average length is 0.3 m (0.9 ft); average weight is 0.1 kg (0.3 lbs).

SCIENTIFIC AND COMMON NAMES: The River Otter, *Lontra canadensis,* is a member of the weasel family (Mustelidae), of the subfamily of otters (Lutrinae), and of the new world genus *Lontra* (fig. 106). *Lontra* and *lutra* are similar terms, from Latin, for "otter," and *canadensis* means Canada (a name derived, in turn, from the Huron–Iroquois word for "village").

The species is widespread across North America, and the subspecies found in the eastern Pacific is *Lontra canadensis pacifica.* A specimen collected in 1905 in the Queen Charlotte Islands, British Columbia, and named *Lutra periclyazomae,* was later recognized as a separate subspecies of River Otter and renamed *Lontra canadensis periclyazomae.* Today, an alternative scientific name for the River Otter is *Lutra canadensis*; in the past, alternative names were *Lutra california* and *Lutra pacifica.* The otter's strong association with rivers accounts for its common name, but people also call this species the California Otter, Canadian Otter, Land Otter, and Fish Otter.

DESCRIPTION: River Otters have features typical of members of the weasel family, with an elongated body, short legs, and a round head. They differ from other members of the family, though, because of their more sinuous appearance, fully webbed feet, short and tapered tail, and dense fur. Female and male otters appear similar in size and weight, but males are slightly heavier. The muscular tail of the River Otter is proportionately

Figure 106. River Otters.

longer than that of the Sea Otter, accounting for one-third (38 to 46 cm; 15 to 18 in.) of the animal's total length; it is broad and round at the base, tapering to the tip. The head is rounded, with a short muzzle and wide, bulbous nose. In profile, the head is flattened, lacking a distinct forehead and presenting an apparent straight line from nose to shoulders. The back arches when the otter is on land, but in the water, when swimming, the otter's streamlined body flexes. The small [< 2.5 cm (1 in.)] round ears are set back on the skull, and the small, beady eyes are closer to the nose than to the ears. The teeth are heavier than those of most members of the family, evolved to crush cartilage of fish and exoskeletons of invertebrate prey, but are not broad and rounded like the Sea Otter's. The facial whiskers, whitish and unremarkable, are a sensory organ—similar to the Harbor Seal's—that can detect movement of prey in turbid water.

River Otter fur is uniformly brown or black, with lighter gray or buff countershading that extends from the lips to the ear, and down across the chest. Young otters typically are darker, and older ones with white-tipped hairs may appear grizzled. As with fur seals and Sea Otters, the dense fur helps maintain body temperature in the water, and consists of outer guard hairs and short dense underfur. The underfur is estimated to have 9,000 hairs per square inch midback; this is about one-third less dense than the Sea Otter's.

FIELD MARKS AND SIMILAR SPECIES: Young Harbor Seals or California Sea Lions might be confused with River Otters if only the animal's head is exposed at the surface, but the otter's distinguishing features are its small round head and pronounced ears. On land, the otter has short legs and a long tail, compared to the flippers and bobbed tail of a pinniped. The species most likely to be confused with the River Otter is the Sea Otter. The River Otter's smaller and flatter head, longer tail (one-third its body length), and smaller overall size are key field marks to separate the species. River Otters also lack thick facial whiskers and the distinctive large, flipperlike hind feet of Sea Otters. Young Sea Otters and River Otters can both have light-colored chests, but the lighter-colored hair of the River Otter extends over a wider area and forms a distinct line extending from the lower jaw back to the ear. Behaviorally, the species can be separated when seen in the water: the Sea Otter's characteristic posture is floating on the surface, often entwined in kelp; the River Otter, in contrast, moves constantly to stay afloat and hugs the shoreline, retreating to land for safety. When on land, River Otters move smoothly compared to Sea Otters, which are hindered by their flipper-shaped rear feet.

NATURAL HISTORY: In form and function, River Otters exemplify the adaptive evolution of sea mammals from both terrestrial and marine origins. Their morphology and physiology tether them to the land to breed and to regulate their body temperature, but they venture into some of the same nearshore marine habitats that Sea Otters and pinnipeds do, particularly in bays and estuaries. They are truly an aquatic mammal, with denser fur than most terrestrial animals and with webbed feet, but they lack many of the features of other marine mammals; the most obvious difference is the lack of modified appendages. Even the Sea Otter has flipper-shaped rear feet that evolved for vigorous swimming. The River Otter's trachea is shorter than most land-based carnivores' but longer than marine mammals', presumably because a short trachea facilitates rapid air exchange. Their nostrils close when these otters submerge, as do the Sea Otter's.

River Otters do not migrate but do move locally, and they will emigrate in response to changes in food availability or to human encroachment. They do not congregate onshore like pinnipeds do, but instead establish overlapping, linear, terrestrial, and aquatic home ranges. A male home range may overlap with that of one or more females, and female home ranges overlap as well. The home ranges are not fixed territories, but River Otters will mark terrestrial rubbing sites and posts along the periphery, and otters of either sex will avoid each other within overlapping home ranges, using latrines as marking sites. Researchers have taken the opportunity provided by these marking posts to collect DNA for identifying individuals and deriving estimates of home range size and population densities.

River Otters use dens for shelter, as many mustelids do, and an individual may have several distributed within its home range. Rather than excavate their own dens, though, River Otters acquire those abandoned by other animals, or use hollow logs or caves, or even abandoned human structures, such as duck blinds. The den consists of an entrance, tunnel, and nest chamber. The entrance can be underwater, which provides greater protection from predators, or above water, in mud banks along estuaries.

The nest chamber is usually lined with soft materials such as moss, leaves, and hair.

Otters are active year-round, day and night, but are most lively at dawn and dusk. They do not hibernate, but one study in Washington State reported that River Otters were seasonally more active in estuaries during spring, and in rivers during the fall. Seasonal activity patterns likely are influenced by food concentration and reproductive status; during the winter, for example, several species of salmon concentrate in estuaries and spawn up rivers, attracting otters.

River Otters are very gregarious, but the primary social unit is the family, composed of a female with pups, also called kits. The mating system is polygynous, but males will rejoin females after pups are 6 months old. Males may be solitary or belong to groups of up to 17 males. The DNA of male group members have shown no familial ties, but rather otters may gather to feed cooperatively on schooling fish such as herring. Males that do not feed mostly on schooling fish do not appear to form groups. Females rarely join these groups, and likely only do so when they forgo breeding.

Aquatic movements of River Otters can range from slow paddling with hind limbs, followed by a glide, to faster undulating motions with simple sideways sweeps of the tail and lower torso. When they emerge from water, otters roll and rub their bodies on sand or grass to clean their fur, and have specific sites where they repeatedly engage in this grooming. Movement on land starts with walking and can advance to running and bounding. With each progressively faster pace, the tail is lifted, aiding in balance. Otters can attain speeds up to 30 km/hr (18 miles/hr) on land, a speed pinnipeds can only achieve in the water. Upon return to the water, otters often toboggan on well-worn mudslides for rapid retreat; several slides may occur along the shoreline of a home range.

River Otters have anal scent glands, as do most mustelids and other carnivores. They use the two glands for marking posts along their home range boundary, near dens, at rolling spots, and around feeding locations. They also mark using urine and feces, but usually fastidiously return to designated toilets. Mariners occasionally grumble about otters' marking lines that tether boats to shore, near where otters forage.

River Otters are intelligent and curious, and their playful antics have been chronicled by naturalists for centuries. Types of play range from group wrestling to solitary tail-chasing and stone-juggling. They have a large repertoire of sounds, including whistles, chuckles, and screams. Socially, they communicate by chortles and growls, and pups chirp for parental attention.

REPRODUCTION: Otters follow a distinct annual breeding season from November through May, and copulation is both aquatic and terrestrial, in contrast to the mostly aquatic copulation of Sea Otters. Both sexes reach sexual maturity at 2 years of age, though yearling females can occasionally produce young, but males are likelier to begin breeding when 3 to 4 years old. Their estimated lifespan is some 14 years in wild populations, and 25 in captivity.

River Otters give birth to litters of one to three pups, and sometimes to as many as six. Pups are usually born around 60 days after mating, but researchers report widely different gestation periods of up to 375 days. This discrepancy is due to their ability to delay implantation of the fertilized egg on the uterine wall. With copulation between December and April, an average delayed implantation of 9 months, and actual gestation of 2 months, total gestation can extend more than 12 months, occurring from November through May. Otters usually breed every year in coastal populations such as in Oregon, but in some areas, they skip years.

Females use dens for birthing and raising pups; these are usually located in the center of the female home range. Usually a chamber within the den is used for birthing; pups venture out after 1 month or so. The pups are entirely dependent, and their eyes do not even open for 2 to 3 weeks. From four teats, pups nurse for up to 12 weeks on milk composed of 24 percent fat, but can eat solid food at 9 weeks. Pups must learn to swim and are taught by the mother at 1.5 to 2 months. Although pups are usually weaned at 3 months, they are dependent for 6 months, and a female will give food to them for up to 9 months. The female–litter relationship endures for at least 1 year, and females will tolerate young of the previous year, even when with pups. The long relationship boosts the survival and socialization of pups, because the mother teaches them to swim and how and where to forage. Occasionally siblings will participate in training and play with a new litter.

FOOD: River Otters eat a variety of fare along the shore, from invertebrates and frogs to fish and birds. On occasion, they also will consume small mammals and plants, such as leafy algae and berries. Their primary marine diet, however, overlaps with that of other marine mammals of the nearshore, where they feed on abundant fish and invertebrates. Because they lack speed and agility in the water, they seek out slow-moving or sedentary prey and feed cooperatively when herding fish. Spawning salmon that are dead or near death as they move from estuaries into rivers and streams are a favored food. Other marine fish in the diet include perch, sculpin, and shiner. Favored invertebrates consist of native and nonnative species of crabs, clams, mussels, snails, and crayfish. Sea ducks and rails are the most common birds eaten, but gull chicks and other nesting bird chicks and eggs have also been documented. Bird watchers at Rodeo Lagoon near San Francisco Bay were stunned one year to observe River Otters killing and eating more than 20 Brown Pelicans; however, biologists speculated that the otters resorted to avian prey when their preferred diet of fish declined in the small water body.

A wide range of food items is typical of species that are mostly resident and feed on whatever is available locally. In one California study, River Otters consumed primarily crustaceans but also birds, insects, and mammals, in order of importance. In another study in Washington, their diet was almost exclusively fish and crustaceans.

When foraging, otters swim at speeds of 11 kph (7 mph) and can cover 400 m (1,300 ft) under water in a single dive while pursuing prey. Diving estimates average less than 1 minute, but males can remain submerged for up to 4 minutes. Depths range up to 20 m (66 ft) but average 3 m (9 ft) in nearshore subtidal waters.

Within a day, River Otters feed mostly in the morning, averaging around 2 to 3 hrs, but will feed cooperatively at any time when fishing on schooling prey. River Otters consume around 0.9 to 1.4 kg (2 to 3 lbs) of live food per day.

MORTALITY: Otters are most vulnerable to predators on land as they exit and enter the waterways. The long list of predators includes Bobcat, Cougar, Dog, Coyote, bear, and large raptors. Once in the water, though, Killer Whales are the only documented marine predator, although likely River Otters are also vulnerable to attack by sharks and Steller Sea Lions.

DISTRIBUTION: Marine River Otters along the Pacific coast of North America have a limited range, in contrast to their inland brethren. Across North America, the range extends throughout the freshwater habitats from latitude 25° to 70° N. Along the eastern Pacific seaboard, though, otters are present from the Aleutian Islands of Alaska south throughout the coastal bays of Canada and the San Juan Archipelago, but south only to central California. They may range far inland via major rivers such as the Columbia, Sacramento, and San Joaquin, but mostly reside within estuaries, bays, and small tributaries of the coast.

River Otters are mostly resident and do not migrate great distances; nevertheless, within a day an otter can move up to 40 km (25 mi), and home range estimates along an Alaskan shoreline were 10 to 40 km (6 to 25 mi). Daily movements of family groups, though, are estimated to average only 1 to 3 miles depending on the season.

HABITAT: Once on land, River Otters usually do not venture more than around 30 m (100 ft) from water, but dens have been reported up to 0.8 km (0.5 mi) inland. Their shore-based habitat is constrained by suitable substrate where dens are

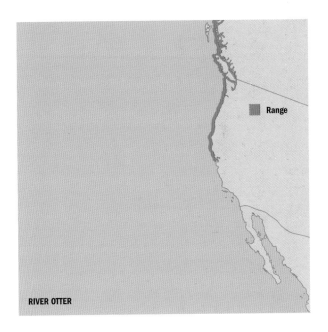

RIVER OTTER

near to water. Forest or dense brush is preferred, including mixed hardwood conifer, conifer, and riparian, but they also reside in fresh and saltwater marshes and the lower reaches of streams. Otters are usually associated with plants such as willows, Douglas fir, spruce, cattails, horsetails, bulrushes, and sedges.

VIEWING: Finding River Otters is difficult even with local knowledge, but they can be irregularly seen anywhere throughout their range.

CONSERVATION: The warm and luxuriant fur of River Otters was sought after for centuries in North America, and otters are still hunted in Canada and several states in the United States, including Oregon and Washington. In 1983, around 33,000 skins were collected in the entire United States. In many states, though, otters were extirpated because of hunting pressure and other human actions. Other major causes for their decline have been development around estuaries, pesticides, and water pollution, particularly due to mercury. Researchers recently discovered that the reproductive organs of young male River Otters along the Columbia River were underdeveloped, and speculated that pollutants disrupted their hormones during growth. Exposure to oil from oil spills, as occurs in San Francisco Bay, can reduce thermal properties of the River Otter fur, much as it does to that of fur seals and Sea Otters. Modern conflicts with humans are associated with otters' consuming raised fish at hatcheries or private ponds and consuming endangered species such as Coho salmon. Otters also often den under houses or in unattended boats, causing a nuisance to owners. Human-related diseases such as toxoplasmosis have been identified in River Otters of the Puget Sound area and may cause mortality (fig. 107).

Agreement	Conservation Status
U.S. ESA	Not listed
State protections	Protected in California; furbearer in Oregon and Washington
COSEWIC	Not listed
IUCN	Least concern

Figure 107. A boatload of River Otters.

POPULATION: River Otters are considered common in western North America, with densities of one per 4 km (2.5 mi) of shoreline. Populations are listed as stable or increasing in Oregon and Washington, and are fully protected in California.

COMMENTS: *Cryptosporidium* sp. and *Giardia* sp. have been isolated in marine-dwelling River Otters in Puget Sound. Ironically, their ability to habituate to humans has promoted the recovery of otters but has also exposed them to disease and pollution, leading to poor health. One study showed that River Otters from areas of higher human density were more likely than otters from areas of low human density to be exposed to diseases that come from domestic and peridomestic animals such as cats and rats.

REFERENCES: Gaydos et al. 2007a, Gaydos et al. 2007b, Jameson and Peeters 1988, Lariviere and Walton 1998, Nolet et al. 1993.

GLOSSARY

Abyssal plain Flat or gently sloping sea bottom of the deep ocean.

Aerobic dive limit (ADL) Time limit of a dive, based on the amount of oxygen stored and the rate of oxygen consumption during a dive.

Alloparenting Raising of young done by individuals other than the parents.

Ambergris A dark waxy material found in the intestines of sperm whales, once highly prized for perfume.

Anthropogenic Indirectly or directly related to a human activity.

Apnea Pause between breaths. In humans, apnea can be lethal, but in marine mammals, it is an adaptation for diving.

Baleen Plates of keratin material that hang from the upper jaw of Mysticeti whales, used during filter feeding to entangle prey. Also called "whalebone."

Bathymetry A measurement of the depth of the ocean floor relative to the water surface.

Beak A part of the cetacean anatomy: the snout with upper and lower jaws.

Benthic Living in, on, or just above the seafloor.

Biodiversity The relative number of species in a taxon or genus or an area.

Blackfish A group of toothed whales that share similar coloration and body and head shape. Blackfish in the California Current include Killer Whales, pilot whales, False Killer Whales, Melon-Headed Whales, and Risso's Dolphins.

Blow The exhalation of cetaceans that creates a cloud of vapor and that can be used to identify species of whale.

Blubber A layer of fat under the skin of marine mammals that, among its other specialized functions, provides insulation.

Bow-riding To ride the pressure waves at the bow or stern of vessels.

Breach To leap out of the water, creating a large splash by landing on the back or side.

Bull Adult male.

Bycatch Animals caught in fishery operations that are not the target of the fishery but that usually die accidentally from drowning or entanglement.

Calf Offspring of a whale, less than one year of age, usually attended by an adult.

Callosity Patch of hardened skin on the head of right whales that becomes a surface to which ectoparasites attach.

Caudal Associated with the tail or rear of an animal.

Cephalopod A group of invertebrates that includes cuttlefish, octopuses, and squid.

Chin slap To create a splash by slapping the chin on the surface of the water.

Clan Maternally related.

Click A very brief sound emitted in echolocation or communication by marine mammals (frequency range of 110–150 kHz).

Continental shelf The edge of the continental plate, adjacent to the landmass, submerged in relatively shallow water. The continental shelf break is the edge of the continental shelf, where the bathymetry drops to the abyssal plain.

Convergence A front where water masses of different temperatures or salinities meet.

Copepod Small marine or freshwater crustacean, also called flea of the sea; eaten by some baleen whales.

Cow Adult female.

Crepuscular Associated with dawn or dusk; usually used to describe feeding activity.

Deep-scattering layer A group of organisms that migrate up and down the water column within a day, migrating to the surface at night to feed and down to below the light level during the day.

Delayed implantation Implantation of a fertilized egg to the uterine wall after a delay; occurs in many mammals, including pinnipeds and bats.

Depleted species A species with population below a sustainable level.

Diatoms A single-celled algae with cell walls of silica.

Dimorphism The presence of two morphs, such as difference in sex or color, in a group or species. *See also* Sexual dimorphism.

Dinoflagellates Single-celled algae having whiplike tails.

Dispersal Movement of individuals away from birthplace within the first year of life; also, any movement away from a point of origin.

Dorsal Pertaining to the backside of an animal, such as can be said of the back, head, or tail.

Dorsal cape A dark color on the backs of some dolphins that gives the appearance of a cape draped across the back from just behind the head to just past the dorsal fin.

Dorsal fin The fin on the back of most species of cetacean and fish, positioned approximately at the middle of the back.

Eared seal Seals with external ear pinnae, including sea lions and fur seals.

Eastern Pacific The waters of the Pacific Ocean that extend from latitude 180° E to the American continents.

Echolocation The production and reception of sound for locating objects or scanning the environment.

Ecoregion A relatively large area of land or water that contains geographically distinct communities as perceived through abiotic factors, such as climate, landform, soil, and hydrology.

Ecosystem A system that is stable in terms of the interactions of its physical and biological elements.

Ecotype A variant in the ecological form of a species that is not distinct enough to be a subspecies but that indicates regional or behavioral differences between members of the same species.

El Niño-Southern Oscillation (ENSO) A climate pattern related to major temperature fluctuations in the surface waters of the eastern Pacific Ocean.

Epipelagic Waters located from the surface to a depth of 200 meters.

Euphausiids Shrimplike crustaceans also called *krill.*

Falcate Curved toward the back; usually used to describe the shape of the dorsal fin.

Flipper Front or rear appendages of marine mammals used for moving through the water.

Flipper wave To slap a flipper on the surface of the water, or to wave the pectoral flipper in the air.

Fluke prints A series of smooth circles on the water surface created by the motion of flukes in shallow water, somewhat like footprints on sand.

Flukes The tail of a cetacean.

Front A zone where rapid change occurs in the chemical and physical properties of seawater.

Gastroliths Stones held in the stomachs of animals, thought to aid in digestion.

Habitat From the Latin for "it inhabits": a place where a species lives.

Haul out To come on shore; also, a terrestrial site where pinnipeds or otters gather on land. The term was used as early as 1880 by Joel Allen.

Lanugo Soft, downy hair that covers the fetus of a pinniped and that is usually shed in the uterus or shortly after birth.

Lateral Away from the middle of the body, toward the sides.

Latitude The location of a place on Earth as north or south of the equator; measurements range from 0° at the equator to 90° at the poles (90° N or 90° S).

Lek A place where reproductive males gather and display to attract females.

Lobtailing To slap the tail or lower portion of body on the surface of the water, creating a large, loud splash.

Maritory Underwater display territories of marine organisms, including humpback whales and harbor seals.

Mass stranding The stranding onshore of two or more animals (not including a female with calf) at the same time.

Matrilineal Said of a group of females descended from a single female.

Mesopelagic The middle of the water column, from depths of 200 m to 1000 m.

Migration Seasonal or annual movements of animals, usually between feeding and breeding areas.

Molt Shedding of old fur or skin.

Morphology The physical appearance (e.g., shape, size, features) of an animal, species, or genus.

Nares Nasal openings.

Nitrogen narcosis Elevated concentrations of nitrogen gas in the brain, which reduce mental ability and can lead to death.

Oceanic Deep ocean waters off the continental shelf.

Optimum sustainable yield A measure of young produced, used for management of a population to ensure continued and sustained optimum production.

Otolith Ear bone of bony fish, used in species and age identification.

Pantropical Occurring throughout the tropical regions of the world.

Pectoral flippers Erstwhile arms, evolved into flippers, that aid in movement through the water or that provide hydrodynamic form for swimming.

Pelagic Waters of the open sea beyond the continental shelf and away from the bottom; also used to describe things located in those waters.

Peduncle Tail stock of whales; looks like bumps as a whale arches to dive.

Pelage Hair of mammals.

Pelagic Occurring in, or living in, the open ocean.

Philopatric Tending to return to its place of birth to give birth.

Pinna Ear flap supported by cartilage that extends from the head where the ear canal is exposed. Pinnae is plural.

Piscivorous Eating fish.

Plankton Plant and animal life that passively floats or weakly swims in water.

Pod Loosely refers to a group of whales; in resident killer whales, refers specifically to a group of related matrilines.

Polygynous A breeding system wherein dominant males usually mate with multiple females.

Population A group of animals breeding freely with each other but not with other groups.

Porpoising To leap out of the water, as some cetaceans and pinnipeds do.

Precautionary principle A conservative conservation strategy recommending that when an activity or product could threaten the environment, cautious study should first be undertaken to understand its possible consequences.

Pteropods Small mollusks that swim through the water using "wings."

Pup A young pinniped or otter that still stays with its mother.

Purse seine A type of fishing gear that encloses fish within a circle of netting that forms a vertical curtain trapping fish, as well as non-target animals such as dolphins.

Range The geographic extent of a species' occurrence.

Reproductive rate The number of offspring (calves or pups in the case of marine mammals) in a population.

Rookery A terrestrial area used by pinnipeds for breeding, whether seasonally or year-round.

Rorqual A member of the Balaenopteridae family of baleen whales that includes one of six species (Blue Whale, Fin Whale, Sei Whale, Humpback Whale, Bryde's Whale, Common Minke Whale).

Rostrum The nose or projected beak of a cetacean.

ROV Acronym for "remote operating vehicle," an underwater exploratory probe controlled by computer.

Sexual dimorphism Physical differences between sexes of the same species, such as of size, teeth, head shape, or other. Elephant seals exhibit extreme dimorphism.

Song In whales, a series of sounds repeated in a pattern.

Shelf Part of the ocean adjacent to the continent, usually with a depth of 100–200 m.

Spout The exhalation of the whale that forms a cloud of mist formed of combined exhaled air and seawater, also called a *blow*.

Spyhop To raise the head vertically out of the water, as whales do.

Stocks A particular group of animals that breed freely within their own group and that can also breed with members of other groups. There can be several stocks within a population having no genetic differences between stocks; the stocks may simply be geographically separated.

Sympatric Coexisting in the same geographic place.

Tag A marker or tracking device attached to an animal, such as a satellite tag.

Tail slap To slap the tail against the surface of the water, as dolphins do.

Territory A region (usually on land) occupied and defended by individuals or groups.

Teutophagous Eating squid.

Thermocline A zone of rapid vertical change in water temperature. In waters off California, thermocline depth varies seasonally from 10 m to 100 m.

Thermoregulation The ability of an animal to control its body temperature.

Trematode A parasitic worm, also called a *fluke*.

Trophic Related to feeding; trophic level pertains to where an organism is in the food chain.

Upwelling Movements of waters from deep in the water column toward the surface, generated by winds and bathymetric features.

Ventral Toward the belly or front.

Vibrissae Tactile hairs on the upper lip of mammals.

Weaning Period when offspring stop nursing.

Whalebone Baleen.

Whale lice A family of crustaceans called Cyamidae that live on the skin of whales.

Zoonosis An animal disease that can be transmitted to humans. Leptospirosis (swamp fever) is an example.

REFERENCES AND FURTHER READING

Aguilar, A. 2002. Fin whale, *Balaenoptera physalus*. In *Encyclopedia of Marine Mammals,* ed. W.F. Perrin, B. Wursig, and J.G.M. Thewissen, 435–438. San Diego: Academic Press.

Ainley, D.G., H.R. Huber, and K.M. Bailey. 1982. Population fluctuations of California sea lions and the Pacific whiting fishery off central California. *Fishery Bulletin* 80:253–258.

Allen, J.A. 1880. History of the North American pinnipeds. U.S. Geological and Geographical Survey of the Territories, Misc. Public. No. 12, Washington Government Printing Office.

Allen, S.G., D.G. Ainley, G.W. Page, and C.A. Ribic. 1985. The effect of disturbance on harbor seal haul out patterns at Bolinas Lagoon, California. *Fishery Bulletin* 82:493–500.

Allen, S.G., S. Peaslee, and H.R. Huber. 1989. Colonization by northern elephant seals of the Point Reyes Peninsula, California. *Marine Mammal Science* 5:298–302.

Allen, S.G., R.W. Risebrough, L. Fancher, M. Stephenson, and D. Smith. 1993. Red harbor seals of San Francisco Bay. *Journal of Mammalogy* 74:588–593.

Alonso, M.K., S. Pedraza, A.C.M. Schiavini, R.N. Goodhall, and E.A. Crespo. 1999. Stomach contents of false killer whales (*Pseudorca crassidens*) stranded on the coasts of the Strait of Magellan, Tierra del Fuego. *Marine Mammal Science* 15:712–724.

Alter, S.E., E. Rynes, and S.R. Palumbi. 2007. DNA evidence for historic population size and past ecosystem impacts of gray whales. *Proceedings of the National Academy of Sciences* 104 (38):15162–15167.

Amano, M., and N. Miyazaki. 2004. Composition of a school of Risso's dolphin, *Grampus griseus*. *Marine Mammal Science* 20:152–160.

Anderson, P. K. 2001. Marine mammals in the next one hundred years: Twilight for a Pleistocene megafauna? *Journal of Mammalogy* 82 (3):623–629.

Andrews, R. D., D. R. Jones, J. D. Williams, P. H. Thorson, G. W. Oliver, D. P. Costa, and B. J. Le Boeuf. 2000. Heart rates of northern elephant seals diving at sea and resting on the beach. *Journal of Experimental Biology* 200:2083–2095.

Angliss, R. P., and B. M. Allen. 2007. Steller sea lion (*Eumetopias jubatus*): Eastern U.S. Stock. In *Alaska Marine Mammal Stock Assessments,* U.S. Dep. Commer., NOAA Tech. Memo. NMFS-AFSC-193.

Angliss, R. P., and B. M. Allen. 2008. Gray whale (*Eschrichtius robustus*): Eastern North Pacific Stock. In *Alaska Marine Mammal Stock Assessments.* U.S. Dep. Commer., NOAA Tech. Memo. NMFS-AFSC-193.

Antonelis, G. A. Jr., and C. Fiscus. 1980. The pinnipeds of the California current. *CalCOFI Report* 21:68–78.

Antonelis, G. A. Jr., C. H. Fiscus, and R. L. DeLong. 1983. Spring and summer prey of California sea lions, *Zalophus californianus*, at San Miguel Island, California, 1978–79. *Fishery Bulletin* 82:67–76.

Antonelis, G. A. Jr., and M. A. Perez. 1984. Estimated annual food consumption by northern fur seals in the California current. *CalCOFI Report* 25:135–145.

Arai, K., T. K. Yamada, and Y. Takano. 2004. Age estimation of male Stejneger's beaked whales (*Mesoplodon stejnegeri*) based on counting of growth layers in tooth cementum. *Mammal Study* 29:125–136.

Archer, F., T. Gerrodette, S. Chivers, and A. Jackson. 2004. Annual estimates of the unobserved incidental kill of pantropical spotted dolphin (*Stenella attenuata*) calves in the tuna purseseine fishery in the eastern tropical Pacific. *Fishery Bulletin* 102:233–244.

Archer, F. I., and W. F. Perrin. 1999. *Stenella coeruleoalba. Mammalian Species* 603:1–9.

Arnbom, T., V. Papastavrou, L. W. Weilgart, and H. Whitehead. 1987. Sperm whales react to an attack by killer whales. *Journal of Mammalogy* 68 (2):450–453.

Arnold, J. 1992. Complex hunter-gatherer-fishers of prehistoric California: Chiefs, specialists and maritime adaptations of the Channel Islands. *American Antiquity* 57:60–84.

Au, W. W. L. 2002. Echolocation. In *Encyclopedia of Marine Mammals,* ed W. F. Perrin, B. Wursig, and J. G. M. Thewissen, 358–367. San Diego: Academic Press.

Au, W.W.L., J.K.B. Ford, J.K. Horne, and K.A. Newman Allman. 2004. Echolocation signals of free-ranging killer whales (*Orcinus orca*) and modeling of foraging for Chinook salmon (*Oncorhynchus tshawytscha*). *Journal of Acoustical Society of America* 115:901–909.

Baird, R.W. 2002. *Killer Whales of the World: Natural History and Conservation*. Stillwater, MN: Voyageur Press.

Baird, R.W., M.B. Hanson, and L.M. Dill. 2005. Factors influencing the diving behavior of fish-eating killer whales: Sex differences and diel and interannual variation in diving rates. *Canadian Journal of Zoology* 83 (2):257–267.

Baird, R.W., P.R. Stacey, D.A. Duffus, and K.M. Langelier. 2002. An evaluation of gray whale (*Eschrichtius robustus*) mortality incidental to fishing operations in British Columbia, Canada. *Journal of Cetacean Research and Management* 4:289–296.

Ballance, L.T., and R.L. Pitman. 1998. Cetaceans of the western tropical Indian Ocean: Distribution, relative abundance, and comparisons with cetacean communities of two other tropical ecosystems. *Marine Mammal Science* 14 (3):429–459.

Barlow, J. 1988a. Harbor porpoise, *Phocoena phocoena*, abundance estimation for California, Oregon and Washington: I. Ship surveys. *Fishery Bulletin* 86:417–432.

Barlow, J. 1988b. Harbor porpoise, *Phocoena phocoena*, abundance estimation for California, Oregon and Washington: II. Aerial surveys. *Fishery Bulletin* 86:433–444.

Barlow, J., and G.A. Cameron. 2003. Field experiments show that acoustic pingers reduce marine mammal bycatch in the California drift gillnet fishery. *Marine Mammal Science* 19 (2):265–283.

Baumgartner, M.F., and B.R. Mate. 2003. Summertime foraging ecology of North Atlantic right whales. *Marine Ecology Progress Series* 264:123–135.

Baumgartner, M.F., C.A. Mayo, and R.D. Kenny. 2006. Enormous carnivores, microscopic food, and a restaurant that's hard to find. In *The Urban Whale: North Atlantic Right Whales at the Crossroads*, ed. S.D. Kraus and R.M. Rolland, 138–171. President and Fellows of Harvard College.

Bearzi, M. 2005. Habitat partitioning by three species of dolphins in Santa Monica Bay, California. *Bulletin of the Southern California Academy of Sciences* 104 (3):113–124.

Beatson, E. 2007. The diet of pygmy sperm whales, Kogia breviceps, stranded in New Zealand: Implications for conservation. *Reviews in Fish Biology and Fisheries* 17:295–303.

Belcher, R.L., and T.E. Lee. 2002. Arctocephalus townsendi. *Mammalian Species* 700:1–5.

Benoit-Bird, K.J., and W.W.L. Au. 2009. Phonation behavior of cooperatively foraging spinner dolphins. *Journal of Acoustical Society of America* 125:539–546.

Berger, W.H. 2009. *Ocean: Reflections on a Century of Exploration.* Berkeley: University of California Press.

Bernard, H.J., and S.B. Reilly. 1999. Pilot whales, Globicephala (Lesson, 1828). In *Handbook of Marine Mammals,* Vol. 6, ed. S.H. Ridgway and R. Harrison, 245–279. New York: Academic Press.

Berta, A., C.E. Ray, and A.R. Wyss. 1989. Skeleton of the oldest known pinniped, *Enaliarctos mealsi. Science* 244:60–62.

Berta, A., J.L. Sumich, and K.M. Kovacs. 2006. *Marine Mammals: Evolutionary Biology,* 2nd ed. San Diego: Academic Press.

Bertao, D.E. 2006. *The Portuguese Shore Whalers of California, 1854–1904.* San Jose, CA: Portuguese Heritage Publications of California, Inc.

Best, P.B. 1982. Seasonal abundance, feeding, reproduction, age and growth in minke whales off Durban (with incidental observations from the Antarctic). *Report of International Whaling Commission.* 32:759–786.

Bigg, M.A. 1981. Harbor seal. In *Handbook of Marine Mammals,* Vol II, ed. S.H. Ridgway and R.J. Harrison. New York: Academic Press.

Bigg, M.A. 1985. Status of the Steller sea lion (*Eumetopias jubatus*) and California sea lion (*Zalophus californianus*) in British Columbia. *Canadian Special Publication for Fisheries and Aquatic Sciences* (77).

Bloodworth, B.E., and D.K. Odell. 2008. *Kogia breviceps* (Cetacea: Kogiidae). *Mammalian Species* (819):1–12.

Boehlert, G.W., D.P. Costa, D.E. Crocker, P. Green, T. O'Brien, S. Levitus, and B.J. Le Boeuf. 2001. Autonomous pinniped environmental samplers: Using instrumented animals as oceanographic data collectors. *Journal of Atmospheric and Oceanic Technology* 18:1182.

Bonnell, M.L., M.O. Pierson, and G.D. Farrens. 1983. Pinnipeds and sea otters of central and northern California, 1980–1983: Status, abundance and distribution. Final Report prepared by Center for Marine Studies, University of California, Santa Cruz, for the Minerals Management Service, Contract 14-12-0001-29090. OCS Study MMS 84-0044.

Bortolotti, D. 2008. *Wild Blue: A Natural History of the World's Largest Animal.* New York: St. Martin's Press.

Boulva, J., and I.A. McLaren. 1979. Biology of the harbor seal, *Phoca vitulina*, in eastern Canada. *Bulletin of the Fisheries Research Board of Canada* 200:1–24.

Brown, R.F., B.E. Wright, S.D. Riemer, and J. Laake. 2005. Trends in abundance and current status of harbor seals in Oregon: 1977–2003. *Marine Mammal Science* 21:657–670.

Brownell, R.L. Jr., P.J. Clapham, T. Miyashita, and T. Kasuya. 2001. Conservation status of North Pacific right whales. *Journal of Cetacean Research and Management* 2 (special issue):269–286.

Brownell, R.L. Jr., K. Ralls, S. Baumann-Pickering, and M.M. Poole. 2009. Behavior of melon-headed whales, *Peponocephala electra*, near oceanic islands. *Marine Mammal Science* 25 doi: 10.1111/j.1748–7692.2009.00281.x.

Brownell, R.L. Jr., W.A. Walker, and K.A. Forney. 1999. Pacific white-sided dolphin *Lagenorhynchus obliquidens* (Gill, 1865). In *Handbook of marine mammals,* Vol. 6, ed. S.H. Ridgway and R. Harrison, 57–84. New York: Academic Press.

Bryden, M.M., R.J. Harrison, and R.J. Lear. 1977. Some aspects of the biology of *Peponocephala electra* (Cetacea: Delphinidae): I. General and reproductive biology. *Australian Journal of Marine and Freshwater Research* 28:703–715.

Buckland, S.T., and J.M. Breiwick. 2002. Estimated trends in abundance of eastern Pacific gray whales from shore counts (1967/68 to 1995/96). *Journal of Cetacean Research* 4 (1):41–48.

Burns, J. 2002. Harbor seal and spotted seal. In *Encyclopedia of Marine Mammals,* ed. W.F. Perrin, B. Wursig, and J.G.M. Thewissen, 552–560. San Diego: Academic Press.

Burns, J.J., K.J. Frost, and L.F. Lowry, ed. 1985. Marine mammal species accounts. *Alaska Department of Fish and Game, Game Technical Bulletin* (7).

Calambokidis, J., J.D. Darling, V. Deecke, P. Gearin, M. Gosho, W. Megill, C.M. Tombach, D. Goley, C. Toropova, and B. Gisborne. 2002. Abundance, range and movements of a feeding aggregation of gray whales (*Eschrichtius robustus*) from California to southern Alaska in 1998. *Journal of Cetacean Research and Management* 4 (3):267–276.

Calambokidis, J., G.S. Schorr, G.H. Steiger, J. Francis, M. Bakhtiari, G. Marshall, E. Oleson, D. Gendron, and K. Robertson. 2008. Insights into the underwater diving, feeding and calling behavior of blue whales from a suction-cup attached video-imaging tag (CRITTERCAM). *Marine Technology Society Journal* 41 (4):19029.

Calambokidis, J., G.H. Steiger, and D.K. Ellfrit. 2004. Distribution and abundance of humpback whales (Megaperta novaeangliae) and other marine mammals off the northern Washington coast. *Fishery Bulletin* 102:563–580.

Carretta, J.V., K.A. Forney, M.S. Lowry, J. Barlow, J. Baker, B. Hanson, and M. M. Muto. 2007. Minke whale (*Balaenoptera acutorostrata*): California/Oregon/Washington stock. In *Marine Mammal Stock Assessments: 2007,* U.S. Dep. Commer., NOAA Tech. Memo., NMFS-SWFSC-414, 197–201.

Carretta, J.V., K. A. Forney, M.S. Lowry, J. Barlow, J. Baker, D. Johnston, B. Hanson, M.M. Muto, D. Lynch, and L. Carswell. 2009. U.S. Marine Mammal Stock Assessments: 2008. U.S. Dep. Commer. NOAA Tech. Memo. NMFS-SWFC 434.

Carretta, J.V., B.L. Taylor, and S.J. Shivers. 2001. Abundance and depth distribution of harbor porpoise, *Phocoena phocoena,* in northern California determined from a 1995 ship survey. *Fishery Bulletin* 99:29–39.

Carson, R.L. 1950. *The Sea Around Us.* New York: Mentor Books.

Carson, R.L. 1962. *Silent Spring.* Boston: Houghton Mifflin Co.

Cass, V. 1985. Exploitation of California sea lions, *Zalophus californianus,* prior to 1972. *Marine Fisheries Review* 47:36–38.

Castro, P., and M.E. Huber. 2007. *Marine Biology.* New Jersey: McGraw Hill Companies.

Chivers, S.J., R.W. Baird, D.J. McSweeney, D.L. Webster, N.M. Hedrick, and J.C. Salinas. 2007. Genetic variation and evidence for population structure in eastern North Pacific false killer whales (*Pseudorca crassidens*). *Canadian Journal of Zoology* 85:783–794.

Chivers, S.J., R.G. Leduc, K.M. Robertson, N.B. Barros, and A.E. Dizon. 2005. Genetic variation of *Kogia* spp. with preliminary evidence for two species of *Kogia sima. Marine Mammal Science* 21:619–634.

Chivers, S.J., and A.C. Myrick. 1993. Comparison of age at sexual maturity and other reproductive parameters for two stocks of spotted dolphin, *Stenella attenuata. Fishery Bulletin* 91:611–618.

Clapham, P.J. 2002. Humpback whale *Megaptera novaeangliae.* In *Encyclopedia of Marine Mammals,* ed. W.F. Perrin, B. Wursig, and J.G.M. Thewissen, 589–592. San Diego: Academic Press.

Clapham, P.J., C. Good, S.E. Quinn, R.R. Reeves, J.E. Scarf, and R.L. Brownell Jr. 2004. Distribution of North Pacific right whale (*Eubalaena japonica*) as shown by 19th and 20th century whaling catch and sighting records. *Journal of Cetacean Research and Manage*ment 61 (1):1–6.

Clapham, P.J., and J.G. Mead. 1999. Megaperta novaeangliae. *Mammalian Species* (604):1–9.

Clarke, M.R. 2003. Production and control of sound by the small sperm whales *Kogia breviceps* and *K. sima* and their implications for other Cetacea. *Journal of the Marine Biological Association of the United Kingdom* 83:241–263.

Colegrove, K.M., D.J. Greig, and F.M.D. Gulland. 2005. Causes of live strandings of northern elephant seals (*Mirounga angustirostris*) and Pacific harbor seals (*Phoca vitulina*) along the central California coast, 1992–2001. *Aquatic Mammals* 31 (1):1–10.

Coll, M., S. Libralato, S. Tudela, I. Palomera, and F. Pranovi. 2008. Ecosystem overfishing in the ocean. *PLoSOne* 3 (12):e3881.

Committee on Taxonomy. 2009. List of marine mammal species and subspecies. Society of Marine Mammalogy, *www.marinemam malscience.org*.

Connor, R., R. Wells, J. Mann, and R. Read. 2000. The bottlenose dolphin: Social relationships in a fission-fusion society. In *Cetacean Societies: Field Studies of Dolphins and Whales,* ed. J. Mann, R.C. Connor, P.L. Tyack, and H. Whitehead, 91–127. Chicago: University of Chicago Press.

Cooper, C.F., and B.S. Stewart. 1983. Demography of northern elephant seals, 1911–1982. *Science* 219:969–971.

Cooper, L.N., M.T. Clementz, S. Bajpai, and B.N. Tiwari. 2007. Whales originated from aquatic artiodactyls in the Eocene epoch of India. *Nature* 450:1190–1194.

COSEWIC. 2003. COSEWIC assessment and status report on the sei whale *Balaenoptera borealis* in Canada. Ottawa: Committee on the Status of Endangered Wildlife in Canada.

Croll, D., C.W. Clark, A. Acevedo, B. Teshy, S. Flores, J. Gedamke, and J. Urgan. 2002. Only male fin whales sing loud songs. *Nature* 417:809.

Curtis, E.S. 1924. *The North American Indian,* Vols. 9, 11, 12. Seattle: E.S. Curtis.

Dahlheim, M.E., and J.E. Heyning. 1999. Killer whale, *Orcinus orca* (Linnaeus, 1758). In *Handbook of Marine Mammals,* Vol. 6, ed. S.H. Ridgway and R. Harrison, 281–322. New York: Academic Press.

Daily, M., F. M. D. Gulland, L. Lowenstine, P. Silvagni, and D. Howard. 2000. Prey, parasites and pathology associated with the mortality of a juvenile gray whale (*Eschrichtius robustus*) stranded along the northern California coast. *Diseases of Aquatic Organisms* 42:111–117.

Dalebout, M. L., C. S. Baker, V. G. Cockcroft, J. G. Mead, T. K. Yamada. 2004. A comprehensive and validated molecular taxonomy of beaked whales, family Ziphiidae. *Journal of Heredity* 95:459–473.

Dalebout, M. L., J. G. Mead, C. Scott Baker, A. N. Baker, and A. L. van Helden. 2002. A new species of beaked whale *Mesoplodon perrini* sp. n. (Cetacea: Ziphiidae) discovered through phylogenetic analyses of mitochondrial DNA sequences. *Marine Mammal Science* 18:577–608.

Dalebout, M. L., G. J. B. Ross, C. S. Baker, R. C. Anderson, P. B. Best, V. G. Cockcroft, H. L. Hinsz, V. Peddemors, and R. L. Pitman. 2003. Appearance, distribution and genetic distinctiveness of Longman's beaked whale, *Indopacetus pacificus. Marine Mammal Science* 19:421–461.

Dalton, R. 2003. Archaeology: The coast road. *Nature* 422:10–12.

Danil, K., and S. J. Chivers. 2007. Growth and reproduction of female short-beaked common dolphins, *Delphinus delphis*, in the eastern tropical Pacific. *Canadian Journal of Zoology* 85:108–121.

Dehnhardt, G., B. Mauck, and H. Bleckmann. 1998. Seal whiskers detect water movements. *Nature* 394:235–236.

DeMaster, D. P. D. Miller, J. R. Henderson, and J. M. Coe. 1985. Conflicts between marine mammals and fisheries off the coast of California. In *Marine Mammals and Fisheries,* ed. J. R. Beddington, R. J. Beverton, and D. M. Lavigne. Boston: George Allen and Unwin.

Deméré, T. A., A. Berta, and P. J. Adams. 2003. Pinnipedomorph evolutionary biogeography. *Bulletin of the American Museum of Natural History* 279:32–76.

Dolar, M. L. L. 2002. Fraser's dolphin *Lagenodelphis hosei.* In *Encyclopedia of Marine Mammals,* ed. W. F. Perrin, B. Wursig, and J. G. M. Thewissen, 485–487. San Diego: Academic Press.

Dorsey, E. M. 1983. Exclusive adjoining ranges in individually identified minke whale (*Balaenoptera acuturostrata*) in Washington State. *Canadian Journal of Zoology* 61:174–181.

Dudzinski, K. M., and T. Frohoff. 2008. *Dolphin Mysteries: Unlocking the Secrets of Communication.* New Haven, CT: Yale University Press.

Erlandson, J.M. 2002. Anatomically modern humans, maritime voyaging, and the Pleistocene colonization of the Americas. In *The First Americans: The Pleistocene Colonization of the New World*, ed. N.G. Jablonski. *California Academy of Sciences* (27):59–92.

Estes, J.A. 1980. Enhydra lutris. *Mammalian Species* (133):1–8.

Estes, J.A., D.P. DeMaster, D.F. Doak, T.M. Williams, and R.L. Brownell. 2007. *Whales, Whaling, and Ocean Ecosystems.* Berkeley: University of California Press.

Estes, J.A., and J. Palmisano. 1974. Sea otters: Their role in structuring nearshore communities. *Science* 185:1058–1160.

Etnier, M.A. 2002. Occurrences of Guadalupe fur seals on the Washington coast over the past 500 years. *Marine Mammal Science* 18:551–557.

Evans, W.E. 1994. Common dolphin, white-bellied porpoise, *Delphinus delphis* (Linnaeus, 1758). In *Handbook of Marine Mammals*, Vol. 5, ed. S.H. Ridgway and R. Harrison, 191–224. New York: Academic Press.

Fadley, B.S., B.W. Robson, J.T. Sterling, A. Greig, and L.A. Call. 2005. Immature Steller sea lion (*Eumetopias jubatus*) dive activity in relation to habitat features of the eastern Aleutian Islands. *Fisheries Oceanography* 14:243–258.

Faurot, E., and J. Ames. 1986. Analysis of sea otter, *Enhydra lutris*, scats from a California haulout. *Marine Mammal Science* 2:223–227.

Feldkamp, S., R.L. DeLong, and G.A. Antonelis. 1989. Diving patterns of California sea lions, *Zalophus californianus. Canadian Journal of Zoology* 67:872–883.

Ferrero, R.C., and W.A. Walker. 1993. Growth and reproduction of the northern right whale dolphin, *Lissodelphis borealis*, in the offshore waters of the North Pacific Ocean. *Canadian Journal of Zoology* 71:2335–2344.

Fiscus, C.H., D.W. Rice, and A.A. Wolman. 1989. Cephalopods from the stomachs of sperm whales taken off California. U.S. Dept. Commerce. NOAA Tech. Rpt. NMFS-83.

Fish, F.E., A.J. Nicastro, and D. Weihs. 2006. Dynamics of the aerial maneuvers of spinner dolphins. *Journal of Experimental Biology* 209:590–598.

Flinn, R.D., A.W. Trites, E.J. Gregr. 2002. Diets of fin, sei, and sperm whales in British Columbia: An analysis of commercial whaling records, 1963–1967. *Marine Mammal Science* 18:663–679.

Ford, J.K.B. 2002. Killer whale *Orcinus orca*. In *Encyclopedia of Marine Mammals,* ed. W.F. Perrin, B. Wursig, and J.G.M. Thewissen, 669–676. San Diego: Academic Press.

Ford, J.K.B., G.M. Ellis, D.R. Matkin, K.C. Balcomb, D. Biggs, and A.B. Morton. 2006. Killer whale attacks on minke whales: Prey capture and antipredator tactics. *Marine Mammal Science* 21:603–618.

Garshelis, D.L., J.A. Garshelis, and A. Kimker. 1986. Sea otter time budgets and prey relationships in Alaska. *Journal of Wildlife Management* 50:637–647.

Gaskin, D.E., P.W. Arnold, and B.A. Blair. 1974. Phocoena phocoena. *Mammalian Species* (42):1–8.

Gaspar, C., R. Lenzi, M.L. Reddy, and J. Sweeney. 2000. Spontaneous lactation by an adult *Tursiops truncatus* in response to a stranded *Steno bredanensis* calf. *Marine Mammal Science* 16 (3):653–658.

Gaydos, J.K., P.A. Conrad, K.V.K. Gilardi, G.M. Blundell, and M. Ben-David. 2007a. Does human proximity affect antibody prevalence in marine-foraging river otters (*Lontra canadensis*)? *J. Wildlife Diseases* 43:116–123.

Gaydos, J.K., W.A. Miller, K.V.K. Gilardi, A. Melli, H. Schwantje, C. Engelstoft, H. Fritz, and P.A. Conrad. 2007b. *Cryptospiridium* and *Giardia* in marine-foraging river otters (*Lontra canadensis*) from the Puget Sound Georgia Basin Ecosystem. *Journal of Parasitology* 93:198–202.

Gendron, D., S. Lanham, and M. Carwardine. 1999. North Pacific right whale (*Eubalaena glacialis*) sighting south of Baja California. *Aquatic Mammals* 25:31–34.

Gentry, R.L. 2002. Northern fur seal (*Callorhinus ursinus*). In *Encyclopedia of Marine Mammals,* ed. W.F. Perrin, B. Wursig, and J.G.M. Thewissen, 813–817. San Diego: Academic Press.

Gentry, R.L., and G. Kooyman, eds. 1986. *Fur Seals: Maternal Strategies on Land and at Sea.* Princeton, NJ: Princeton University Press.

George, J.C., J. Bada, J. Zeh, L. Scott, S.E. Brown, T. O'Hara, and R. Suydam. 1999. Age and growth estimates of bowhead whales (*Balaena mysticetus*) via aspartic acid racemization. *Canadian Journal of Zoology* 77:571–580.

Gerrodette, T., and J. Forcada. 2005. Non-recovery of two spotted and spinner dolphin populations in the eastern tropical Pacific Ocean. *Marine Ecology Progress Series* 291:1–21.

Gilmartin, W. G., R. L. DeLong, R. L. Smith, and A. W. Sweeney. 1976. Premature parturition in the California sea lion. *Journal of Wildlife Diseases* 12:104–115.

Glanz, M. H. 2003. *Climate Affairs: A Primer.* Washington, DC: Island Press.

Goldbogen, J. A., N. D. Pyenson, and R. E. Shadwick. 2007. Big gulps require high drag for fin whale lunge-feeding. *Marine Ecology Progress Series* 349:281–301.

Goley, P. D. 1999. Behavioral aspects of sleep in Pacific white-sided dolphins (*Lagenorhynchus obliquidens,* Gill 1865). *Marine Mammal Science* 15:1054–1064.

Goley, P. D., and J. M. Straley. 1994. Attack on gray whales (*Eschrictius robustus*) in Monterey Bay, California, by killer whales (*Orcinus orca*) previously identified in Glacier Bay, Alaska. *Canadian Journal of Zoology* 72:1528–1530.

Gosho, M. E., D. W. Rice, and J. M. Breiwick. 1984. The sperm whale, *Physeter macrocephalus. Marine Fisheries Review* 46 (4):54–64.

Götz, T., U. K. Verfuss, H. U. Schnitzler. 2005. "Eavesdropping" in wild rough-toothed dolphins (*Steno bredanensis*)? *Biology Letters* 2:1–3.

Götz, T., U. K. Verfuss, H. U. Schnitzler. 2006. "Eavesdropping" in wild rough-toothed dolphins (*Steno bredanensis*)? *Biology Letters* 2:5–7.

Gray, J. E. 1864. Notes on the seals (Phocidae), including the description of a new seal (*Halcyon richardii*), from the west coast of North America. *Proceedings of the Zoological Society of London* 1864:27–34.

Gregr, E. J., L. Nichol, J. K. B. Ford, G. Ellis, and A. W. Trites. 2000. Migration and population structure of northeastern Pacific whales off coastal British Columbia: An analysis of commercial whaling records from 1908–67. *Marine Mammal Science* 16:699–727.

Grigg, E. K., S. G. Allen, D. E. Green, and H. Markowitz. 2002. Diurnal and nocturnal haul out patterns of harbor seals (*Phoca vitulina richardii*) at Castro Rocks, San Francisco Bay, California. *California Fish and Game* 88:15–27.

Gulland, F. M. D., and A. J. Hall. 2005. The role of infectious diseases in influencing status and trends. In *Marine Mammal Research: Conservation beyond Crisis,* ed. J. Reynolds III, W. F. Perrin, R. R. Reeves, S. Montgomery, and T. J. Ragen. Baltimore: Johns Hopkins University Press.

Gulland, F. M. D., M. Haulena, D. Fauquier, G. Langlois, M. Lander, and T. Zabka. 2002. Domoic acid toxicity in California sea lions

(*Zalophus californianus*): Clinical signs, treatment and survival. *Veterinary Record* 150:475–480.

Haley, D. 1986. *Marine Mammals of the Eastern North Pacific and Arctic Waters.* Seattle: Pacific Search Press.

Halpern, B.S., S. Walbridge, and K.A. Selkoe. 2008. A global map of human impact on marine ecosystems. *Science* 319:948–952.

Hanni, K.D., D.J. Long, R.E. Jones, P. Pyle, and L.E. Morgan. 2007. Sightings and strandings of Guadalupe fur seals, central and northern California, 1988–1995. *Journal of Mammalogy* 78 (2):684–687.

Harington, C.R. 2008. Arctic marine mammals and climate change. *Ecological Applications* 18:S23–S40.

Hartman, K.L., F. Visser, and A.J.E. Hendricks. 2008. Social structure of Risso's dolphins (*Grampus griseus*) at the Azores: A stratified community based on highly organised social units. *Canadian Journal of Zoology* 86:294–306.

Hassrick, J.L., D.E. Crocker, R.L. Zeno, S.B. Blackwell, D.P. Costa, and B.J. Le Boeuf. 2007. Swimming speed and foraging strategies of northern elephant seals. *Deep-Sea Research II* 54:369–383.

Hayes, S.A., D.P. Costa, J.T. Harvey, and B.J. Le Boeuf. 2004. Aquatic mating strategies of the male Pacific harbor seal (*Phoca vitulina richardii*): Are males defending the hotspot? *Marine Mammal Science* 20 (3):639–656.

Heath, C. 2002. California, Galapagos, and Japanese sea lions. In *Encyclopedia of Marine Mammals,* ed. W.F. Perrin, B. Wursig, and J.G.M. Thewissen, 180–186. San Diego: Academic Press.

Heckel, G., S.B. Reily, J.L. Sumich, and I. Espejel. 2001. The influence of whale watching on the behavior of migrating gray whales (*Eschrichtius robustus*) in Todos Santos Bay and surrounding waters, Baja California, Mexico. *Journal of Cetacean Research and Management* 3 (3):227–237.

Heyning, J.E. 1984. Functional morphology involved in interspecific fighting of the beaked whale, *Mesoplodon carlhubbsi. Canadian Journal of Zoology* 62:1645–1654.

Heyning, J.E. 2002. Cuvier's beaked whale. In *Encyclopedia of Marine Mammals,* ed. W.F. Perrin, B. Wursig, and J.G.M. Thewissen, 305–307. San Diego: Academic Press.

Heyning, J.E., and M.E. Dahlheim. 1988. *Orcinus orca. Mammalian Species* (304):1–9.

Heyning, J.E., and J.G. Mead. 1996. Suction feeding in beaked whales: Morphological and observational evidence. *Contributions in Science*, Natural History Museum of Los Angeles County (464):1–12.

Heyning, J.E., and W.F. Perrin. 1994. Evidence for two species of common dolphins (genus Delphinus) from the eastern North Pacific. *Contributions in Science*, Natural History Museum of Los Angeles County (442).

Hickey, B.M. 1979. The California Current System. *Progress in Oceanography*. 8:191–279.

Higgins, L.V., D. Costa, A. Huntley, and B.J. Le Boeuf. 1988. Behavioral and physiological measurements of maternal investment in the Steller sea lion, *Eumetopias jubatus. Marine Mammal Science* 4:44–58.

Hindell, M.A. 2002. Elephant seals *Mirounga angustirostris* and *M. leonina.* In *Encyclopedia of Marine Mammals,* ed. W.F. Perrin, B. Wursig, and J.G.M. Thewissen, 370–373. San Diego: Academic Press.

Hoelzel, A.R., and S.J. Stern. 2000. *Minke Whales.* Stillwater, MN: Voyageur Press.

Horwood, J. 1987. *The Sei Whale: Population Biology, Ecology and Management.* London: Croom Helm.

Horwood, J. 2002. Sei whale (*Balaenoptera borealis*). In *Encyclopedia of Marine Mammals,* ed. W.F. Perrin, B. Wursig, and J.G.M. Thewissen, 1069–1071. San Diego: Academic Press.

Houck, W.J., and T.A. Jefferson. 1999. Dall's porpoise, *Phocoenoides dalli* (True, 1885). In *Handbook of Marine Mammals,* Vol. 6, ed. S.H. Ridgway and R. Harrison, 443–472. New York: Academic Press.

Hubbs, C.L. 1946. First records of two beaked whales, *Mesoplodon bowdoini* and *Ziphius cavirostris*, from the Pacific Coast. *Journal of Mammalogy* 27 (3):242–255.

Hubbs, C.L. 1956. Back from oblivion: Guadalupe fur seal still a living species. *Pacific Discovery*, California Academy of Sciences 9:14–21.

Hui, C.A. 1979. Undersea topography and distribution of dolphins of the genus Delphinus in the Southern California Bight. *Journal of Mammalogy* 60 (3):521–527.

Irving, L., L.J. Peyton, C.H. Bahn, and R.S. Peterson. 1962. Regulation of temperature in fur seals. *Physiological Zoology* 35:275–284.

Jackson, J.B.C. 2008. Ecological extinction and evolution in the brave new ocean. *Proceedings of the National Academy of Sciences* 105:11458–11465.

Jameson, E.W., and H.J. Peeters. 1988. *California Mammals.* Berkeley: University of California Press.

Jameson, R. J., and K. W. Kenyon. 1977. Prey of sea lions in the Rogue River, Oregon. *Journal of Mammalogy* 58:672.

Jefferson, T. 1988. Phocoenoides dalli. *Mammalian Species* (319):1–7.

Jefferson, T. A. 2002a. Rough-toothed dolphin, *Steno bredanensis.* In *Encyclopedia of Marine Mammals,* ed. W. F. Perrin, B. Wursig, and J. G. M. Thewissen, 1055–1059. San Diego: Academic Press.

Jefferson, T. A. 2002b. Dall's porpoise, *Phocoenoides dalli.* In *Encyclopedia of Marine Mammals,* ed. W. F. Perrin, B. Wursig, and J. G. M. Thewissen, 308–310. San Diego: Academic Press.

Jefferson, T. A., and N. B. Barros. 1997. *Peponocephala electra. Mammalian Species* (553):1–6.

Jefferson, T. A., and S. Leatherwood. 1994. *Lagenodelphis hosei. Mammalian Species* (470):1–5.

Jefferson, T. A., S. Leatherwood, and M. A. Webber. 1993. Marine Mammals of the World. FAO Species Identification Guide. UNEP and FAO.

Jefferson, T. A., and M. W. Newcomer. 1993. *Lissodelphis borealis. Mammalian Species* (425):1–6.

Jefferson, T. A., M. A. Webber, and R. L. Pitman. 2008. *Marine Mammals of the World.* San Diego: Academic Press.

Jeffries, S., H. Huber, J. Calambokidis, and J. Laake. 2005. Trends and status of harbor seals in Washington State: 1978–1999. *Journal of Wildlife Management* 67:207–218.

Jessup, D. A., M. Miller, J. Ames, M. Harris, C. Kreuder, P. A. Conrad, and J. A. K. Mazet. 2004. Southern sea otter as a sentinel of marine ecosystem health. *EcoHealth* 1 (3):239–245.

Johnson, A., and A. Acevedo-Gutierrez. 2007. Regulation compliance by vessels and disturbance of harbor seals (*Phoca vitulina*). *Canadian Journal of Zoology* 85:290–294.

Johnson, M., P. T. Madsen, W. M. X. Zimmer, N. Aguilar Soto, and P. L. Tyack. 2006. Foraging Blainville's beaked whales (*Mesoplodon densirostris*) produce distinct click types matched to different phases of echolocation. *Journal of Experimental Biology* 209:5038–5050.

Jones, M. L., and S. L. Swartz. 2002. Gray whale *Eschrichtius robustus.* In *Encyclopedia of Marine Mammals,* ed. W. F. Perrin, B. Wursig, and J. G. M. Thewissen, 524–536. San Diego: Academic Press.

Jones, M. L., S. L. Swartz, and S. Leatherwood, ed. 1984. *The Gray Whale (Eschrichtius robustus).* New York: Academic Press.

Kasuya, T. 2002. Giant beaked whales *Berardius bairdii* and *B. arnuxii*. In *Encyclopedia of Marine Mammals,* ed. W.F. Perrin, B. Wursig, and J.G.M. Thewissen, 519–524. San Diego: Academic Press.

Kato, H. 2002. Bryde's whale (*Balaenoptera edeni* and *B. brydei*). In *Encyclopedia of Marine Mammals,* ed. W.F. Perrin, B. Wursig, and J.G.M. Thewissen, 171–177. San Diego: Academic Press.

Kenny, R.D. 2004. North Atlantic, North Pacific, and southern right whale, *Eubalaena gracialis, E. japonica,* and *E. australis.* In *Encyclopedia of Marine Mammals,* ed. W.F. Perrin, B. Wursig, and J.G.M. Thewissen, 806–813. San Diego: Academic Press.

Kenny, R.D., C.A. Mayo, and H.E. Winn. 2001. Migration and foraging strategies at varying spatial scales in western North Atlantic right whales: A review of hypotheses. *Journal of Cetacean Research and Manage*ment Special Issue 2:251–260.

Kenyon, K.W. 1969. The sea otter in the eastern Pacific Ocean. *North American Fauna* (68):352.

Kenyon, K.W. 1987. Sea Otter. In *Wild mammals of North America,* ed. J.A. Chapman and G.A. Feldhamer, 769–827. Baltimore: Johns Hopkins University Press.

Kenyon, K.W., and D.W. Rice. 1961. Abundance and distribution of the Steller sea lion. *Journal of Mammalogy* 42:223–234.

Kjeld, M. 2006. Salt and water balance of modern baleen whales: Rate of urine production and food intake. *Canadian Journal of Zoology* 81:606–616.

Knowlton, A.R., S.D. Kraus, and R.D. Kenny. 1994. Reproduction in North Atlantic right whales (*Eubalaena gracialis*). *Canadian Journal of Zoology* 72:1297–1305.

Kruse, S., D.K. Caldwell, and M.C. Caldwell. 1999. Risso's dolphin, *Grampus griseus* (G. Cuvier, 1812). In *Handbook of Marine Mammals,* Vol. 6, ed. S.H. Ridgway and R. Harrison, 183–212. New York: Academic Press.

Kuczaj, S.A., and D.B. Yeater. 2007. Observations of rough-toothed dolphins (*Steno bredanensis*) off the coast of Utila, Honduras. *Journal of the Marine Biological Association of the United Kingdom.* 87:141–148.

Kvitek, R.G., A.K. Fukayama, B.S. Anderson, and B.K. Grimm. 1988. Sea otter foraging on deep-burrowing bivalves in a California coastal lagoon. *Marine Biology* 98:157–167.

Lagerquist, B.A., K.M. Stafford, and B.R. Mate. 2000. Dive characteristics of satellite-monitored blue whales (*Balaenoptera musculus*) off the central California coast. *Marine Mammal Science* 16:375–391.

Laidre, K. L., R. J. Jameson, and D. P. DeMaster. 2001. An estimation of carrying capacity for sea otters along the California Coast. *Marine Mammal Science* 17:294–309.

Laidre, K. L., R. J. Jameson, S. J. Jeffries, R. C. Hobbs, C. E. Bowlby, and G. R. VanBlaricom. 2002. Estimates of carrying capacity for sea otters in Washington State. *Wildlife Society Bulletin* 30:1172–1181.

Laist, D. W., A. M. Knowlton, J. G. Mead, A. S. Collet and M. Podesta. 2001. Collisions between ships and whales. *Marine Mammal Science* 17:35–75.

Lariviere, S., and L. R. Walton. 1998. *Lontra canadensis. Mammalian Species* (587):1–8.

Leatherwood, S., W. E. Evans, and D.W. Rice. 1972. *The Whales, Dolphins, and Porpoises of the Eastern North Pacific: A Guide to Their Identification in the Water.* San Diego: Naval Undersea Center.

Leatherwood, S., and W. A. Walker. 1979. The northern right whale dolphin *Lissodelphis borealis* Peale in the Eastern Pacific. In *Behavior of Marine Animals,* Vol. 3, ed. H. E. Winn and B. L. Olla, 85–141. New York: Plenum Publishing.

Le Boeuf, B. J., and D. E. Crocker. 2005. Ocean climate and seal condition. *BMC Biology* 3.

Le Boeuf, B. J., and R. M. Laws. 1994. *Elephant Seals: Population Ecology, Behavior and Physiology.* Berkeley: University of California Press.

Le Boeuf, B. J., and K. Panken. 1977. Elephant seals breeding on the mainland in California. *Proceedings of the California Academy of Sciences* 31:601–612.

Le Boeuf, B. J., M. H. Perez-Cortes, R. J. Urban, B. R. Mate, and U. F. Ollervides. 2000. High gray whale mortality and low recruitment in 1999: Potential causes and implications. *Journal of Cetacean Research Management* 2 (2):85–99.

Lightfoot, K. G., and O. Parrish. 2009. *California Indians and Their Environment.* Berkeley: University of California Press.

Lilly, J. 1967. *The Mind of the Dolphin.* New York: Doubleday.

Lilly, J.C. 1962. *Transactions of the American Philosophical Society* 106:520–529.

Lipsky, J.D. 2002. Right whale dolphins *Lissodelphis borealis* and *L. peronii.* In *Encyclopedia of Marine Mammals,* ed. W. F. Perrin, B. Wursig, and J.G.M. Thewissen, 1030–1033. San Diego: Academic Press.

Locker, C., and T. Waters. 1986. Weights and anatomical measurements of northeastern Atlantic fin and sei whales. *Marine Mammal Science* 2:169–185.

Longhurst, A. R. 1998. *Ecological Geography of the Sea.* San Diego: Academic Press.

Loughlin, T. R. 1980. Home range and territoriality of sea otters near Monterey, California. *Journal of Wildlife Management* 44:576–582.

Loughlin, T. R., J. L. Bengtson, and R. L. Merrick. 1986. Characteristics of feeding trips of female northern fur seals. *Canadian Journal of Zoology* 65:2079–2084.

Loughlin, T. R., M. A. Perez, and R. L. Merrick. 1987. *Eumetopias jubatus. Mammalian Species* (283):1–7.

Lowry, M. S., and K. A. Forney. 2005. Abundance and distribution of California sea lions (*Zalophus californianus*) in central California during 1998 and summer 1999. *Fishery Bulletin* 103:331–343.

Lui, H., M. Dagg, and S. Strom. 2005. Grazing by the calanoid copepod *Neocalanus cristatus* on the microbial food web in the coastal Gulf of Alaska. *Journal of Plankton Research* 27:647–662.

Lyman, R. L. 1988. Zoogeography of Oregon coast marine mammals: The last 3,000 years. *Marine Mammal Science* 4:247–264.

MacLeod, C. D. 2003. Intraspecific scarring in odontocete cetaceans: An indicator of male "quality" in aggressive social interactions? *Journal of Zoology* [London] 244:71–77.

Macleod, K., R. Fairbairns, A. Gill, B. Fairbairns, J. Gordon, C. Blair-Myers, and E. C. M. Parsons. 2004. Seasonal distribution of minke whales *Balaenoptera acutorostrata* in relation to physiography and prey off the Isle of Mull, Scotland. *Marine Ecology Progress Series* 277:263–274.

Mann, J., R. C. Connor, P. L. Tyack, and H. Whitehead. 2000. *Cetacean Societies: Field Studies of Dolphins and Whales.* Chicago: University of Chicago Press.

Mate, B. R., and J. Urban-Ramirez. 2003. A note on the route and speed of a gray whale on its northern migration from Mexico to central California, tracked by satellite-monitored radio tag. *Journal of Cetacean Resource Management* 5:155–157.

Mayo, C. A., B. H. Letcher, and S. Scott. 2001. Zooplankton filtering efficiency of the baleen of a North Atlantic right whale, *Eubalaena gracialis. Journal of Cetacean Resource Management.* 2(special issue):225–229.

McAlpine, D. F. 2002. Pygmy and dwarf sperm whales *Kogia breviceps* and *K. sima.* In *Encyclopedia of Marine Mammals,* ed. W. F. Perrin, B. Wursig, and J. G. M. Thewissen, 1007–1009. San Diego: Academic Press.

McAuliffe, K., and H. Whitehead. 2005. Eusociality, menopause and information in matrilineal whales. *Trends in Ecology and Evolution* 20:650–650.

Mead, J.G., and R.L. Brownell. 2005. Order Cetacea. In *Mammal species of the world,* ed. D.E. Wilson, and D.M. Reeder, 723–736. Baltimore: Johns Hopkins University Press.

Miller, D.J. 1980. The sea otter in California. *CalCOFI* 21:79–81.

Minamikawa, S., T. Iwasaki, and T. Kishiro. 2007. Diving behavior of a Baird's beaked whale, *Berardius bairdii,* in the slope water region of the western North Pacific: First dive records using a data logger. *Fisheries Oceanography* 16 (6):573–577.

Miyazaki, N., and W.F. Perrin. 1994. Rough-toothed dolphin, *Steno bredanensis* (Lesson, 1828). In *Handbook of Marine Mammals,* Vol. 5, ed. S.H. Ridgway and R. Harrison, 1–21. New York: Academic Press.

Montagu, A., and J.C. Lilly. 1963. *The Dolphin in History.* Los Angeles: William Andrews Clark Memorial Library.

Moore, M.J., and G.A. Early. 2004. Cumulative sperm whale bone damage and the bends. *Science* 306 (5705):2215.

Moore, S.E., and J.T. Clarke. 2002. Potential impact of offshore human activities on gray whales (*Eschrichtius robustus*). *Journal of Cetacean Resource Management.* 4 (1):19–25.

Morejohn, G.V. 1979. The natural history of Dall's porpoise in the North Pacific Ocean. In *Behavior of Marine Animals,* Vol. 3, ed. H. Winn and B.L. Olla, 45–83. New York: Plenum Press.

Morejohn, G.V., J. Ames, and D. Lewis. 1975. Post mortem studies of sea otters in California. Calif. Depart. Fish and Game. Marine Resources Technical Rept. No. 30.

Morgan, L., S. Maxwell, F. Tsao, T.A.C. Wilkinson, and P. Etnoyer. 2005. *Marine Priority Conservation Areas: Baja California to the Bering Sea.* Montreal: Commission for Environmental Cooperation of North America and the Marine Conservation Biology Institute.

Morton, A.B. 2000. Occurrence, photo-identification and prey of Pacific white-sided dolphins (*Lagenorhyncus obliquidens*) in the Broughton Archipelago, Canada 1984–1998. *Marine Mammal Science* 16:80–93.

Motani, R. 2000. Rulers of the Jurassic seas. *Scientific American* 283 (6):52–59.

Nachtigall, P.E., and A.Y. Supin. 2008. A false killer whale adjusts its hearing when it echolocates. *Journal of Experimental Biology* 211:1714–1718.

Nagorsen, D. 1985. *Kogia simus. Mammalian Species* (239):1–6.

Newell, C. L., and T. J. Cowles. 2006. Unusual gray whale *Eschrichtius robustus* feeding in the summer of 2005 off the central Oregon coast. *Geophysical Research Letters* 33:L22S11.

Newsome, S. D., M. A. Etnier, D. Gifford-Gonzalez, D. L. Phillips, M. van Tuinen, E. A. Hadly, D. P. Costa, D. J. Kennett, T. P. Guilderson, and P. L. Koch. 2007. The shifting baseline of northern fur seal ecology in the northeast Pacific Ocean. *Proceedings National Academy of Sciences* 104 (104):9709–9714.

Nishiwaki, M., and K. S. Norris. 1966. A new genus, Peponocephala, for the odontocete cetacean species *Electra electra. Scientific Report of the Whale Research Institute* 20:95–100.

Nolet, B. A., D. E. H. Wansink, and H. Kruck. 1993. Diving of otter (*Lutra lutra*) in a marine habitat use of depth by a single prey loader. *Journal of Animal Ecology* 62:22–32.

Noren, S. R., G. Biedenbach, J. V. Redfern, and E. F. Edwards. 2008. Hitching a ride: The formation locomotion strategy of dolphin calves. *Functional Ecology* 2:278–283.

Noren, S. R., and E. F. Edwards. 2007. Physiological and behavioral development in dolphin calves: Implications for calf separation and mortality due to tuna purse-seine sets. *Marine Mammal Science* 23 (1):15–29.

Norris, K. D., and B. Mohl. 1983. Can odontocetes debilitate prey with sound? *American Naturalist* 112:85–103.

Nowak, R. M. 2003. *Walker's Marine Mammals of the World.* Baltimore: Johns Hopkins University Press.

Odell, D. K. 1981. California sea lion, *Zalophus californianus* (Lesson, 1828). In *Handbook of Marine Mammals,* Vol. 1, ed. S. H. Ridgway and R. J. Richards, 67–97. New York: Academic Press.

Odell, D. K., and K. M. McClure. 1999. False killer whale, *Pseudorca crassidens.* In *Handbook of Marine Mammals,* Vol. 6, ed. S. Ridgway and R. Harrison, 213–244. San Diego: Academic Press.

Oleson, E. M., J. Barlow, J. Gordon, S. Rankin, and J. Hildebrand. 2003. Low frequency calls of Bryde's whales. *Marine Mammal Science* 19:407–419.

Oleson, E. M., J. Calambokidis, W. C. Burgess, M. A. McDonald, C. A. LeDuc, and J. A. Hildebrand. 2007. Behavioral context of call production by eastern North Pacific blue whales. *Marine Ecology Progress Series* 330:269–284.

Oliver, J. S., and P. N. Slattery. 1985. Destruction and opportunity on the sea floor: Effects of gray whale feeding. *Ecology* 66 (6):1965–1975.

Olsen, P.A., and S.B. Reilly. 2002. Pilot whales *Globicephala melas* and *G. macrorhynchus*. In *Encyclopedia of Marine Mammals,* ed. W.F. Perrin, B. Wursig, and J.G.M. Thewissen, 898–903. San Diego: Academic Press.

Orr, R.T. 1963. A northern record for the Pacific bottlenose dolphin. *Journal of Mammalogy* 44:424.

Orr, R.T., and R.C. Helm. 1989. *Marine Mammals of California.* University of California Press.

Oswald, J.N., J. Barlow, and T.F. Norris. 2003. Acoustic identification of nine delphinid species in the eastern tropical Pacific Ocean. *Marine Mammal Science* 19:20–37.

Palacios, D.M., S.J. Bograd, R. Mendelssohn, and F.B. Schwing. 2004. Long-term and seasonal trends in stratification in the California Current, 1950–1993. *Journal of Geophysical Research* 109. C10016, doi:10.1029/2004JC002380.

Pastene, L.A., M. Goto, N. Kanda, A.N. Zerbini, D. Kerem, K. Watanabe, Y. Bessho, M. Hasegawa, R. Nielsen, F. Larsen, and P.J. Palsboll. 2007. Radiation and speciation of pelagic organisms during periods of global warming: The case of the common minke whale, *Balaenoptera acuturostrata. Molecular Ecology* 16:1481–1500.

Pauly, D., and J. Maclean. 2003. *In a Perfect Ocean.* Washington, DC: Island Press.

Payne, R., and S. McVay. 1971. Songs of humpback whales. *Science* 173 (3997):585–597.

Peacock, T., and E. Bradley. 2008. Going with (or against) the flow. *Science* 320:1302–1303.

Perrin, W.F. 1994. Common dolphin, white-bellied porpoise, *Delphinus delphis* (Linnaeus, 1758). In *Handbook of Marine Mammals,* Vol. 5, ed. S.H. Ridgway and R. Harrison, 191–224. New York: Academic Press.

Perrin, W.F. 1998. *Stenella longirostris. Mammalian Species* (599):1–7.

Perrin, W.F. 2002a. Spinner dolphin (*Stenella longirostris*). In *Encyclopedia of Marine Mammals,* ed. W.F. Perrin, B. Wursig, and J.G.M. Thewissen, 1174–1178. San Diego: Academic Press.

Perrin, W.F. 2002b. Common dolphin (*Delphinus delphis, D. capensis* and *D. tropicalis*). In *Encyclopedia of Marine Mammals,* ed. W.F. Perrin, B. Wursig, and J.G.M. Thewissen, 245–248. San Diego: Academic Press.

Perrin, W.F. 2009. World Cetacea Database. www.marinespecies .org/cetacea.

Perrin, W.F., and J.W. Gilpatrick. 1994. Spinner dolphin, *Stenella longirostris* (Gray, 1828). In *Handbook of Marine Mammals,* Vol. 5, ed. S.H. Ridgway and R. Harrison, 99–128. New York: Academic Press.

Perrin, W.F., and A.A. Hohn. 1994. Pantropical spotted dolphin, *Stenella attenuata.* In *Handbook of Marine Mammals,* Vol. 5, ed. S.H. Ridgway and R. Harrison, 71–98. New York: Academic Press.

Perrin, W.F., S. Leatherwood, and A. Collet. 1994a. Fraser's dolphin, *Lagenodelphis hosei* (Fraser, 1956). In *Handbook of Marine Mammals,* Vol. 5, ed. S.H. Ridgway, and R. Harrison, 225–240. New York: Academic Press.

Perrin, W.F., C.E. Wilson, and F.I. Archer II. 1994b. Striped dolphin, *Stenella coeruleoalba* (Meyen, 1833). In *Handbook of Marine Mammals,* Vol. 5, ed. S.H. Ridgway and R. Harrison, 129–160. New York: Academic Press.

Perry, S.L., D.P. DeMaster, and G.K. Silber. 1999. The great whales: History and status of six species listed as endangered under the U.S. Endangered Species Act of 1973: A special issue of the *Marine Fisheries Review. Marine Fisheries Review* 61 (1):1–74.

Perryman, W.L. 2002. Melon-headed whale. In *Encyclopedia of Marine Mammals,* ed. W.F. Perrin, B. Wursig, and J.G.M. Thewissen, 733–735. San Diego: Academic Press.

Perryman, W.L., and M.S. Lynn. 1993. Identification of geographic forms of common dolphin (*Delphinus delphis*) from aerial photography. *Marine Mammal Science* 9:119–137.

Perryman, W.L., and M.S. Lynn. 1994. Examination of stock and school structure of striped dolphin (*Stenella coeruleoalba*) in the eastern Pacific from aerial photogrammetry. *Fishery Bulletin* 92:122–131.

Peterson, R.S., and G.A. Bartholomew. 1967. The natural history and behavior of the California sea lion. Special Public. No. 1 of the Am. Soc. of Mamm.

Philbrick, N. 2000. *In the Heart of the Sea: The Tragedy of the Whaleship Essex.* New York: Penguin Books.

Philips, J.D., P.E. Nachtigall, W.W.L. Au, J.L. Pawloski, and H.L. Roitblat. 2003. Echolocation in the Risso's dolphin, *Grampus griseus. Journal of the Acoustical Society of America* 113:605–616.

Pitcher, K.W., and D.G. Calkins. 1981. Reproductive biology of Steller sea lions in the Gulf of Alaska. *Journal of Mammalogy* 62:599–605.

Pitcher, K.W., P.F. Olesiuk, R.F. Brown, M.S. Lowry, S.J. Jeffries, J.L. Sease, W.L. Perryman, C.E. Stinchcomb, and L.F. Lowry.

2007. Abundance and distribution of the eastern North Pacific Steller sea lion (*Eumetopias jubatus*) population. *Fishery Bulletin* 107:102–115.

Pitman, R.L. 2002. Mesoplodont whales (Mesoplodon ssp.). In *Encyclopedia of Marine Mammals,* ed. W.F. Perrin, B. Wursig, and J.G.M. Thewissen, 738–742. San Diego: Academic Press.

Pitman, R.L. 2009. Killer snowballs. *Natural History.* December 2008/January 2009. Endpaper.

Pitman, R.L., L.A. Aguayo, and R. Urban. 1987. Observations of an unidentified beaked whale (Mesoplodon sp) in the eastern tropical Pacific. *Marine Mammal Science* 3:345–352.

Pitman, R.L., and S.J. Chivers. 1999. Terror in black and white. *Natural History* 107:26–29.

Pitman, R.L. and J.W. Durban. 2009. Save the seal! *Natural History* (November):48.

Pitman, R.L., and M.S. Lynn. 2001. Biological observations of an unidentified mesoplodont beaked whale in the eastern tropical Pacific and probable identity: *Mesoplodon peruvianus. Marine Mammal Science* 17 (3):648–657.

Pitman, R.L., D.M. Palacios, P.L. Brennan, K.C. Balcomb, and T. Miyashita. 1999. Sightings and possible identity of a bottlenose whale in the tropical Indo-Pacific: *Indopacetus pacificus?* *Marine Mammal Science* 15:531–549.

Pitman, R., and C. Stinchcomb. 2002. Rough-toothed dolphins (*Steno bredanensis*) as predators of Mahi mahi (*Coryphaena hippurus*). *Pacific Science* 56:447–450.

Ragen, T., H. Huntington, and G. Hovelsrud. 2008. Conservation of Arctic marine mammals faced with climate change. *Ecological Applications* 18(suppl):S166–S174.

Rankin, S., and J. Barlow. 2005. Source of the North Pacific "boing" sound attributed to minke whales. *Journal of the Acoustical Society America* 118:3346–3351.

Rasmussen, K., D.M. Palacios, J. Calambokidas, M.T. Saborio, L.D. Rosa, E.R. Secchi, G.H. Steiger, J.M. Allen, and G.S. Stone. 2007. Southern Hemisphere humpback whales off Central America: Insights into the longest mammalian migration. *Biology Letters* doi:10.1098/rsbl.2007.0067.

Read, A.J. 1999. Harbour porpoise, *Phocoena phocoena* (Linnaeus, 1758). In *Handbook of Marine Mammals,* Vol. 6, ed. S.H. Ridgway and R. Harrison, 323–356. New York: Academic Press.

Read, A. J., P. N. Halpin, L. B. Crowder, B. D. Best, and E. Fujioka, eds. 2009. OBIS-SEAMAP: Mapping marine mammals, birds and turtles. http://seamap.env.duke.edu/species/.

Read, A. J., and A. J. Westgate. 1997. Monitoring the movements of harbour porpoises, *Phocoena phocoena*, with satellite telemetry. *Marine Biology* 130:315–322.

Ream, R. R., J. T. Sterling, and T. R. Loughlin. 2005. Oceanographic features related to northern fur seal migratory movements. *Deep-Sea Research II* 52:823–843.

Reilly, S. B., J. L. Bannister, P. B. Best, M. Brown, R. L. Brownell Jr., D. S. Butterworth, P. J. Clapham, J. Cooke, G. P. Donovan, J. Urbán, and A. N. Zerbini. 2008. *Eschrichtius robustus.* In IUCN 2009. IUCN Red List of Threatened Species. Version 2009.1.

Reyes, J. C., J. G. Mead, and K. Van Waerebeek. 1991. A new species of beaked whale *Mesoplodon peruvianus* sp. n. (Cetacea: Ziphiidae) from Peru. *Marine Mammal Science* 7 (1):1–24.

Reyes, J. C., K. Van Waerebeek, J. C. Cárdenas, and J. L. Yañez. 1996. *Mesoplodon bahamondi* sp. n. (Cetacea, Ziphiidae), a living beaked whale from the Juan Fernandez Archipelago, Chile. *Bol. Mus. Nac. Hist. Nat. Chile.* 45:31–44.

Reynolds, J. E., and S. A. Rommel. 1999. *Biology of Marine Mammals.* Washington, DC: Smithsonian Institution Press.

Reynolds, J. E., R. S. Wells, and S. D. Eide. 2000. *The Bottlenose Dolphin: Biology and Conservation.* Gainesville: University Press of Florida.

Rice, D. W. 1998. *Marine Mammals of the World: Systematics and Distribution.* The Society of Marine Mammalogy. Special Publication No. 4.

Rice, D. W., and A. A. Wolman. 1971. The life history and ecology of the gray whale (*Eschrichtius robustus*). *American Society of Mammalogists* Special Publication No. 3. Stillwater, OK.

Rice, D. W., A. A. Wolman, B. Mate, and J. T. Harvey. 1986. A mass stranding of sperm whales in Oregon: Sex and age composition of the school. *Marine Mammal Science* 21:64–69.

Rick, T. C., J. M. Erlandson, T. J. Braje, J. A. Estes, M. H. Graham, and R. L. Vellanoweth. 2008. Historical ecology and human impacts on coastal ecosystems of the Santa Barbara Channel, California. In *Human Impacts on Ancient Marine Ecosystems: A Global Perspective,* ed. T. C. Rick and J. M. Erlandson, 77–101. Berkeley: University of California Press.

Rick, T. C., J. M. Erlandson, R. L. Vellanoweth, and T. J. Braje. 2005. From Pleistocene mariners to complex hunter-gatherers: The archaeology of the California Channel Islands. *Journal of World Prehistory* 19:169–228.

Ripley, W. E., K. Cox, and J. L. Baxter. 1962. California sea lion census for 1958, 1960 and 1961. *California Fish and Game* 48:228–231.

Sanvito, S., F. Galimberti, and E. H. Miller. 2007. Having a big nose: Structure, ontogeny, and function of the elephant seal proboscis. *Canadian Journal of Zoology* 85:207–220.

Scammon, C. M. 1874. *The Marine Mammals of the Northwest Coast of North America.* Reprint of edition in 1968. Putnam, NY: Dover.

Scarff, J. E. 1986. Historic and present distribution of the right whale (Eubalaena glacialis) in the Eastern North Pacific South of 50° North and East of 180° West. *Report of the International Whaling Commission* (Special Issue 10):43–63.

Scarff, J. E. 2001. Preliminary estimates of whaling-induced mortality in the 19th century North Pacific right whale (*Eubalaena japonicus*) fishery, adjusting for struck-but-lost whales and non-American whaling. *Journal of Cetacean Resource Management.* 2(special issue):261–268.

Scattergood, L. 1949. Notes on the little piked whale. *Murrelet* 30:3–13.

Scheffer, V. B. 1958. *Seals, Sea Lions, and Walruses.* Standford: Stanford University Press.

Sears, R. 2002. Blue whale *Balaenoptera musculus.* In *Encyclopedia of Marine Mammals,* ed. W. F. Perrin, B. Wursig, and J. G. M. Thewissen, 112–116. San Diego: Academic Press.

Sears, R., and J. Calambokidis. 2002. Update COSEWIC status report on the Blue Whale *Balaenoptera musculus* in Canada. In *COSEWIC Assessment and Update Status Report on the Blue Whale* Balaenoptera musculus *in Canada,* 1–32. Ottawa: Committee on the Status of Endangered Wildlife in Canada.

Shelden, K. E., S. E. Moore, J. M. Waite, P. R. Waite, and D. J. Rugh. 2005. Historic and current habitat use by North Pacific right whales *Eubalaena japonica* in the Bering Sea and Gulf of Alaska. *Mammal Review* 35 (2):129–155.

Shelden, K. E., D. J. Rugh, and A. Schulman-Janiger. 2004. Gray whales born north of Mexico: Indicator of recovery or consequence of regime shift? *Ecological Applications* 14 (6):1789–1805.

Shilling, M. R., I. Seipt, M. T. Weinrich, S. E. Frohock, A. E. Kuhlbrg, and P. J. Clapham. 1992. Behavior of individually-identified sei whales *Balaenoptera borealis* during an episodic influx into the southern Gulf of Maine in 1986. *Fishery Bulletin* 90:749–755.

Shirihai, H., and B. Jarrett. 2006. *Whales, Dolphins and Other Marine Mammals of the World.* Princeton, NJ, and Oxford: Princeton Field Guides.

Simmonds, M. P., and S. J. Isaac. 2007. The impacts of climate change on marine mammals: Early signs of significant problems. *Orynx* 41:19–26.

Siniff, D. B., T. D. Williams, A. Johnson, and D. L. Garshelis. 1982. Experiments on the response of sea otters *Enhydra lutris* to oil contamination. *Biological Conservation* 23:261–272.

Smith, M. H. 2005. First census of northbound migration of gray whales (*Eschrichtius robustus*) from Goleta, California. American Cetacean Society. 16th Biennial Conf. on the Biology of Marine Mammals.

Soldevilla, M. S., E. E. Henderson, G. S. Campbell, S. M. Wiggins, and J. A. Hildebrand 2008. Classification of Risso's and Pacific white-sided dolphins using spectral properties of echolocation clicks. *Journal of the Acoustical Society of America* 124 (1):609–624.

Soto, N. A., M. P. Johnson, P. T. Madsen, F. Díaz, I. Domínguez, A. Brito, and P. Tyack. 2008. Cheetahs of the deep sea: Deep foraging sprints in short-finned pilot whales off Tenerife (Canary Islands). *Journal of Animal Ecology* 77 (5):936–947.

Springer, A. M., J. A. Estes, G. B. van Vliet, T. M. Williams, D. F. Doak, E. M. Danner, K. A. Forney, and B. Pfister. 2003. Sequential mega-faunal collapse in the North Pacific Ocean: An ongoing legacy of industrial whaling. *Proceedings of the National Academy of Sciences* 100:12223–12228.

Stacey, P. J., S. Leatherwood, and R. W. Baird. 1994. *Pseudorca crassidens. Mammalian Species* (456):1–6.

Steiger, G. H., and J. Calambokidas. 2000. Reproductive rates of humpback whales off California. *Marine Mammal Science* 16:220–239.

Stewart, B. S., G. A. Antonelis Jr., R. L. DeLong, and P. K. Yochem. 1988. Abundance of harbor seals on San Miguel Island, California, 1927 through 1986. *Bulletin of the Southern California Academy of Sciences* 87:39–43.

Stewart, B. S., and H. R. Huber. 1993. *Mirounga angustirostris. Mammalian Species* (449):1–10.

Stewart, B. S., and P. Yochem. 1987. Entanglement of pinnipeds in synthetic debris and fishing net and line fragments at San

Nicolas and San Miguel Islands, California, 1978–1986. *Marine Pollution Bulletin* 18:336–339.

Stewart, B.S., and P.K. Yochem. 1984. Seasonal abundance of pinnipeds at San Nicolas Island, California, 1980–1982. *Bulletin of the Southern California Academy of Sciences* 83:121–132.

Sullivan, R.M., J.D. Stack, and W.J. Houck. 1983. Observations of gray whales (*Eschrichtius robustus*) along northern California. *Journal of Mammalogy* 64 (4):689–692.

Suryan, R.M., and J.T. Harvey. 1999. Variability in reactions of Pacific harbor seals, *Phoca vitulina richardii*, to disturbance. *Fishery Bulletin* 97:332–339.

Sydeman, W.J., and S.G. Allen. 1999. Pinnipeds in the Gulf of the Farallones: 25 years of monitoring. *Marine Mammal Science* 15:446–461.

Taylor, B.L., R. Baird, J. Barlow, S.M. Dawson, J. Ford, J.G. Mead, G. Notarbartolo di Sciara, P. Wade, and R.L. Pitman. 2008. *Mesoplodon ginkgodens*. In IUCN 2009. IUCN Red List of Threatened Species. Version 2009.1.

Thewissen, J.G.M., L.N. Cooper, M.T. Clementz, S. Bajpai, and B.N. Tiwari. 2007. Whales originated from aquatic artiodactyls in the Eocene epoch of India. *Nature* 450:1190–1195.

Thompson, P.M., S. Van Parijs, and K.T. Kovacs. 2001. Local declines in the abundance of harbour seals: Implications for the designation and monitoring of protected areas. *Journal of Applied Ecology* 38:117–125.

Tinker, S.W. 1988. *Whales of the World*. New York: E.J. Brill Publishing Co.

Towell, R.G., R.R. Ream, and A.E. York. 2006. Decline in northern fur seal (*Callorhinus ursinus*) pup production on the Pribilof Islands. *Marine Mammal Science* 22 (2):486–491.

Trenberth, K.E. 1997. The definition of El Niño. *Bulletin of the American Meteorological Society* 78 (12):2771–2777.

Trillmich, F., and K. Ono. 1991. Pinnipeds and El Niño: Responses to environmental stress. *Ecological Studies* 88. Springer-Verlag, Berlin.

Tsuneo, K., and Y. Tadasu. 2003. Genetic variability of Stejneger's beaked whale (*Mesoplodon stejnegeri*) stranded on the shore of Sea of Japan based on mitochondrial DNA sequences. *Mammalian Science* 85:93–96.

Twiss, J.R., and R.R. Reeves. 1999. *Conservation and Management of Marine Mammals*. Washington, DC, and London: Smithsonian Institution Press.

Tyack, P. 1999. Communication and cognition. In *Biology of Marine Mammals,* ed. J.E. Reynolds and S.A. Rommel, 287–323. Smithsonian Institution Press.

Tyack, P.L., M. Johnson, N.A. Soto, A. Sturlese, and P.T. Madsen. 2006. Extreme diving of beaked whales. *Journal of Experimental Biology* 209:4238–4253.

Uhen, M.D. 2007. Evolution of marine mammals: Back to the sea after 300 million years. *Anatomical Record* 290:514–522.

Van Dolah, F.M. 2005. Effects of harmful algal blooms. In *Marine mammal research: Conservation beyond crisis,* ed. J. Reynolds III, W.F. Perrin, R.R. Reeves, S. Montgomery, and T.J. Ragen. Baltimore: Johns Hopkins University Press.

Van Waerebeek, K., and B. Wursig. 2002. Pacific white-sided dolphin and dusky dolphin *Lagenorhyncus obliquidens* and *L. obscurus.* In *Encyclopedia of Marine Mammals,* ed. W.F. Perrin, B. Wursig, and J.G.M. Thewissen, 859–861. San Diego: Academic Press.

Wade, P.R., M.P. Heide-Jorgensen, K. Shelden, J. Barlow, J. Carretta, J. Durban, R. LeDuc, L. Munger, S. Rankin, A. Sauter, and C. Stinchcomb. 2006. Acoustic detection and satellite tracking leads to discovery of rare concentration of endangered North Pacific right whales. *Biology Letters* 2 (3) [10.1098/rsbl.2006.0460 (published online)].

Wade, P.R., G.M. Watters, T. Gerrodette, and S.B. Reilly. 2007. Depletion of spotted and spinner dolphins in the eastern tropical Pacific: Modeling hypotheses for their lack of recovery. *Marine Ecology Progress Series* 343:1–14.

Walker, M.M. 2002. Biomagnetism. In *Encyclopedia of Marine Mammals,* ed. W.F. Perrin, B. Wursig, and J.G.M. Thewissen, 104–105. San Diego: Academic Press.

Walker, M.M., J.L. Kirschvink, G. Ahmed, and A. Diction. 1992. Evidence that fin whales respond to geomagnetic field during migration. *Journal of Experimental Biology* 171:67–78.

Walker, W.A., and M.B. Hanson. 1999. Biological observations on Stejneger's beaked whale, *Mesoplodon stejnegeri,* from strandings on Adak Island, Alaska. *Marine Mammal Science* 15 (4):1314–1329.

Walker, W.A., J.G. Mead, and R.L. Brownell Jr. 2002. Diets of Baird's beaked whales, *Berardius bairdii,* in the southern Sea of Okhotsk and off the Pacific Coast of Honshu, Japan. *Marine Mammal Science* 18 (4):902–919.

Watkins, W.A., M.A. Daher, N.A. Dimarzio, A. Samuels, D. Wartzok, K.M. Fristrup, P.W. Howey, and R.R. Maiefski. 2002. Sperm whale dives tracked by radio tag telemetry. *Marine Mammal Science* 18:55–68.

Watkins, W. A., M. A. Daher, K. M. Fristrup, and T. J. Howald, 1993. Sperm whales tagged with transponders and tracked underwater by sonar. *Marine Mammal Science* 9, 55–67.

Watkins, W. A., M. A. Daher, A. Samuels, and D. P. Gannon. 1997. Observations of *Peponocephala electra*, the melon-headed whale, in the southeastern Caribbean. *Caribbean Journal of Science* 33:34–40.

Watkins, W. A., P. Tyack, K. E. Moore, and J. E. Bird. 1987. The 20-Hz signals of finback whales (*Balaenoptera physalus*). *Journal of the Acoustical Society of America* 82:1901–1912.

Watwood, S. L., P. J. O. Miller, M. Johnson, P. Madsen, and P. L. Tyack. 2006. Deep diving foraging behavior of sperm whales (*Physeter macrocephalus*). *Journal of Animal Ecology* 75 (3):814–825.

Weber, D. S., B. S. Stewart, and N. Lehman. 2004. Genetic consequences of a severe population bottleneck in the Guadalupe fur seal (*Arctocephalus townsendi*). *Journal of Heredity* 95 (2):144–153.

Weihs, D., F. E. Fish, and A. J. Nicastro. 2007. Mechanics of remora removal by dolphin spinning. *Marine Mammal Science* 23:707–714.

Weilgart, L. S., and H. Whitehead. 1986. Observations of a sperm whale (*Physeter catadon*) birth. *Journal of Mammalogy* 67 (2):399–401.

Weise, M. J., D. P. Costa, and R. M. Kudela. 2007. Movement and diving behavior of male California sea lion (*Zalophus californianus*) during anomalous oceanographic conditions of 2005 compared to those of 2004. *Geophysical Research Letters* 33. LS22S10.

Wells, R. S., and M. D. Scott. 1999. Bottlenose dolphin, *Tursiops truncatus* (Montagu, 1821). In *Handbook of Marine Mammals*, Vol. 6, ed. S. H. Ridgway and R. Harrison, 137–182. New York: Academic Press.

Wendell, F. E., J. Ames, and R. T. Burge. 1986. Temporal and spatial patterns in sea otter, *Enhydra lutris*, range expansion and in the loss of Pismo clam fisheries. *California Fish and Game* 72:197–212.

Whitehead, H. 1998. Cultural selection and genetic diversity in matrilineal whales. *Science* 282:1708–1711.

Whitehead, H. 2002a. *Sperm Whales: Social Evolution in the Ocean.* London: University of Chicago Press.

Whitehead, H. 2002b. Estimates of the current global population size and historical trajectory for sperm whales. *Marine Ecology Progress Series* 242:295–304.

Whitehead, H. 2002c. Breaching. In *Encyclopedia of Marine Mammals,* ed. W. F. Perrin, B. Wursig, and J. G. M. Thewissen, 162–164. San Diego: Academic Press.

Whitehead, H. 2003. *Sperm Whales: Social Evolution in the Ocean.* Chicago: University of Chicago Press.

Whitehead, H., and J. E. Carlson. 1985. Predicting inshore whale abundance—whale and capelin off the Newfoundland coast. *Canadian Journal of Fisheries and Aquatic Science* 42:976–981.

Whitfield, J. 2008. Does "junk food" threaten marine predators in northern seas? *Science* 322:1786–1787.

Williams, T. M., R. W. Davis, L. A. Fuiman, J. Francis, B. J. Le Boeuf, M. Horning, J. Calambokidis, and D. A. Croll. 2000. Sink or swim: Strategies for cost efficient diving by marine mammals. *Science* 288:133–136.

Willis, P. M., and W. M. Dill. 2006. Mate guarding in male Dall's porpoises. *Ethology* 113:587–597.

Willis, P. M, B. J. Crespi, L. M. Dill, R. W. Baird, and M. B. Hanson. 2005. Natural hybridization between Dall's porpoises (*Phocoenoides dalli*) and harbor porpoises (*Phocoena phocoena*). *Canadian Journal of Zoology* 82:828–834.

Wursig, B. 2002. Bow-riding. In *Encyclopedia of Marine Mammals,* ed. W. F. Perrin, B. Wursig, and J. G. M. Thewissen, 131–133. San Diego: Academic Press.

Yochem, P. K., B. S. Stewart, R. L. DeLong, and D. P. DeMaster. 1987. Diel haul-out patterns and site fidelity of harbor seals (*Phoca vitulina richardsi*) on San Miguel Island, California in Autumn. *Marine Mammal Science* 3:323–332.

Yoshida, H., and H. Kato. 1999. Phylogenetic relationships of Bryde's whales in the western North Pacific and adjacent waters inferred from mitochondrial DNA sequences. *Marine Mammal Science* 15:1269–1286.

Young, N. M., G. M. Hope, W. W. Dawson, and R. L. Jenkins. 1998. The tapetum fibrosum in the eyes of two small whales. *Marine Mammal Science* 4:281–290.

ART CREDITS

All artwork is courtesy of Sophie Webb unless otherwise noted. All photos are by Sophie Webb, courtesy of the Protected Resources Division, NOAA Southwest Fisheries Science Center, unless otherwise noted.

RICHARD D. ALLEN   Fig. 90

SCOT ANDERSON/PO Box 390, Inverness CA 94937
Display pp. ii–iii; figs. 12, 14, 36, 60, 101

EILEEN AVERY/Channel Islands Naturalist Corps   Fig. 58

ROBIN BAIRD/www.cascadiaresearch.org   Figs. 15, 67, 71, 77, 89

BERGER 2009   Fig. 9

CARL BUELL   Figs. 4, 5

JOHN CALAMBOKIDIS/www.cascadiaresearch.org   Fig. 13

STEVE CHOY/SIMoN NOAA   Fig. 102

SARAH CODDE/Point Reyes National Seashore   Fig. 35

W. H. DAWBIN/Sea Watch Foundation   Fig. 62

CHARLES DEERING MCCORMICK LIBRARY of Special Collections, Northwestern University Library   Fig. 27

ANNIE DOUGLAS/www.cascadiaresearch.org   Fig. 68

CLAIRE FACKLER/Photograph provided by NOAA National Marine Sanctuaries Program   Display p. 488

DAVID FIERSTEIN/Copyright © 2001, MBARI: Monterey Bay Aquarium Research Institute   Fig. 24

HOWARD FOSTER/University of California Digital Library, Berkeley, California   Display p. xiv

HOWARD GOLDSTEIN/Photograph provided by Southwest Fisheries Science Center, NOAA Fisheries Service   Fig. 82

FRANCES GULLAND/The Marine Mammal Center   Fig. 17

JAMES T. HARVEY   Fig. 41

THOMAS A. JEFFERSON   Figs. 56, 66b, 79a

HEATHER JENSEN/Point Reyes National Seashore   Figs. 34, 103

CHRIS JOHNSON/Director, EarthOCEAN Media   Fig. 59

R. E. JONES/Personal collection, Napa, California   Fig. 29

CHAD KING/Photograph provided by Monterey Bay National Marine Sanctuary   Fig. 42

BARBARA LACORTE/ Photograph provided by Channel Islands Naturalist   Fig. 57

GALEN LEEDS   Fig. 106

STEVE LONHART/ Photograph provided by Monterey Bay National Marine Sanctuary   Fig. 94

FORREST LYMAN   Fig. 45

THE MARINE MAMMAL CENTER   Figs. 17, 104

DAN MCSWEENEY/Wild Whales Research Foundation Figs. 66a, 69

SHARON R. MELIN/NOAA Alaska Fisheries Science Center Fig. 91

MOJOCOAST   Fig. 16

PATRICIA C. MORRIS/University of California Reserve System   Display pp. i, 386–387; figs. 25, 32, 95a, 95b, 96, 97b

PHIL MYERS/Animal Diversity Web, http://animaldiversity .org   Fig. 10

NASA/GODDARD SPACE FLIGHT CENTER, The SeaWiFS Project and Geoeye, Scientific Visualization Studio   Fig. 21

CORNELIA OEDEKOVEN/NOAA Fisheries Service, Southwest Fisheries Science Center   Fig. 1, 7, 47

CORNELIA OEDEKOVEN/U.S. Navy   Fig. 54

DANIEL PAULY AND ARTIST RACHEL "AQUE" ATANACIO/Pauly and Maclean 2003   Fig. 33

JOSH PEDERSON/Photograph provided by NOAA Monterey Bay NMS   Fig. 105

PHOEBE A. HEARST MUSEUM of Anthropology and the Regents of the University of California, Catalogue No. 2-5852   Fig. 26

ROBERT L. PITMAN   Display pp. 94–95; figs. 49, 51, 53, 64, 65, 70, 79b, 85b, 86, 100

PROTECTED RESOURCES DIVISION, Southwest Fisheries Science Center, NOAA Fisheries Service   Figs. 52, 55, 81, 98, 99

CHARLES SCAMMON, *The Marine Mammals of the North-western Coast of North America,* 1874/Bancroft Library, University of California, Berkeley   Fig. 28

BARBARA TAYLOR/NOAA Fisheries, Southwest Fisheries Science Center   Fig. 63

MARC A. WEBBER   Fig. 97a

JIM WESTER   Fig. 107

Redrawn Figures

Figure 3, redrawn from Ronald Blakey/Northern Arizona University

Figures 9, 18, 19, redrawn from Berger 2009

Figure 20, redrawn from NOAA, Cordell Bank National Marine Sanctuary

Figure 22, redrawn from Berta et al. 2006

Figure 23, redrawn from Morgan et al. 2005, Commission for Environmental Cooperation

Figure 30, redrawn from A.M. Springer et al. 2003, copyright © 2003 National Academy of Sciences, U.S.A.

Figure 31, redrawn from figure courtesy of Timothy Gerrodette, NOAA Fisheries

INDEX

Bolded page numbers indicate main articles.

ADDITIONAL CAPTIONS

PAGE I   Steller Sea Lion bull.

PAGES II–III   Common Bottlenose Dolphin making eye contact with a researcher.

PAGE VI   Feeding Humpback Whale with an enormous gulp of fish and water, with pleats fully distended after lunging toward the surface.

PAGE XIV   The California Current Province from British Columbia to Baja, California, out to 500 miles. A digital map with U.S. Geological Survey digital data.

PAGES 8–9   Baird's Beaked Whales traveling in tight formation.

PAGES 94–95   Male Killer Whale, the Einstein of the sea.

PAGES 386–387   Pinniped colony at Año Nuevo Island with Elephant Seals, Steller Sea Lions, and California Sea Lions.

PAGE 478   Gray Whale, spyhopping.

PAGE 488   California Sea Lion in underwater kelp forest.

PAGE 518   Immature Northern Fur Seal.

PAGE 524   A school of Common Bottlenose Dolphins.

ABOUT THE AUTHORS
AND ILLUSTRATOR

Sarah Allen   As a child, Sarah Allen was inspired by the rich marine life of California. Since then, she has spent a lifetime in the study of marine birds and mammals. Along the way, she acquired undergraduate and graduate degrees from the University of California at Berkeley and was fortunate to be guided by insightful mentors in both school and work. Although pinnipeds are her passion, she has studied wildlife on land and sea from California to Antarctica, and in a few places in between. For many years, she worked with the Point Reyes Bird Observatory, and for the past 17 years, she has worked as a scientist with a federal environmental agency, continuing to explore, restore, and preserve native ecosystems of the Pacific. In her spare time, she found that her long enchantment with marine mammals inspired the evolution of this guide.

Joe Mortenson   In graduate school at the University of Michigan, Joe Mortenson began watching nocturnal South American knifefish with an infrared monitoring system, listening to their weakly electric discharges. His slide show "Joe Mortenson and His Electric Fish" earned him a position at Dalhousie University in Nova Scotia, where he plotted the slow DC discharges of little and winter skates from the nearby Atlantic and tried to answer Darwin's question about the evolution of these signals. Subsequently, he strayed from his laboratory to remote Sable Island and began to observe seals, something that has become a lifelong passion. In his book *Whale Songs and Wasp Maps: The Mystery of Animal Thinking,* he reflects on seals, electric fishes, and various other creatures. Joe, an electronic explorer of an uncharted ocean, like Drake and Magellan of old, has also become an urgent advocate for the conservation of our water planet.

Sophie Webb   Since her childhood, Sophie Webb has drawn and painted wildlife. As a biologist, she has studied and painted birds from the Amazon to the Arctic and Antarctic. Almost all her work is based in observation and field sketching, in combination with the occasional use of museum specimens. Her knowledge of how animals move in nature helps her capture an animal's essence in a two-dimensional medium. In 1995 she coauthored and illustrated *A Guide to the Birds of Mexico and Northern Central America,* and in 2000 she published an award-winning children's book, *My Season with Penguins: An Antarctic Journal.* She followed this first children's book with a second, *Looking for Seabirds: Journal from an Alaskan Voyage,* and a third children's book, about dolphins in the Eastern Tropical Pacific, is currently in press. Sophie has worked on numerous cruises, both as a researcher and as a naturalist, and is a director of Oikonos: Ecosystem Knowledge. To view her art and photographs, visit www.sophiewebb.com.

Series Design:	Barbara Haines
Indexer:	Publication Services, Inc.
Composition:	Publication Services, Inc.
Text:	9/10.5 Minion